A CENTURY OF **INNOVATION**

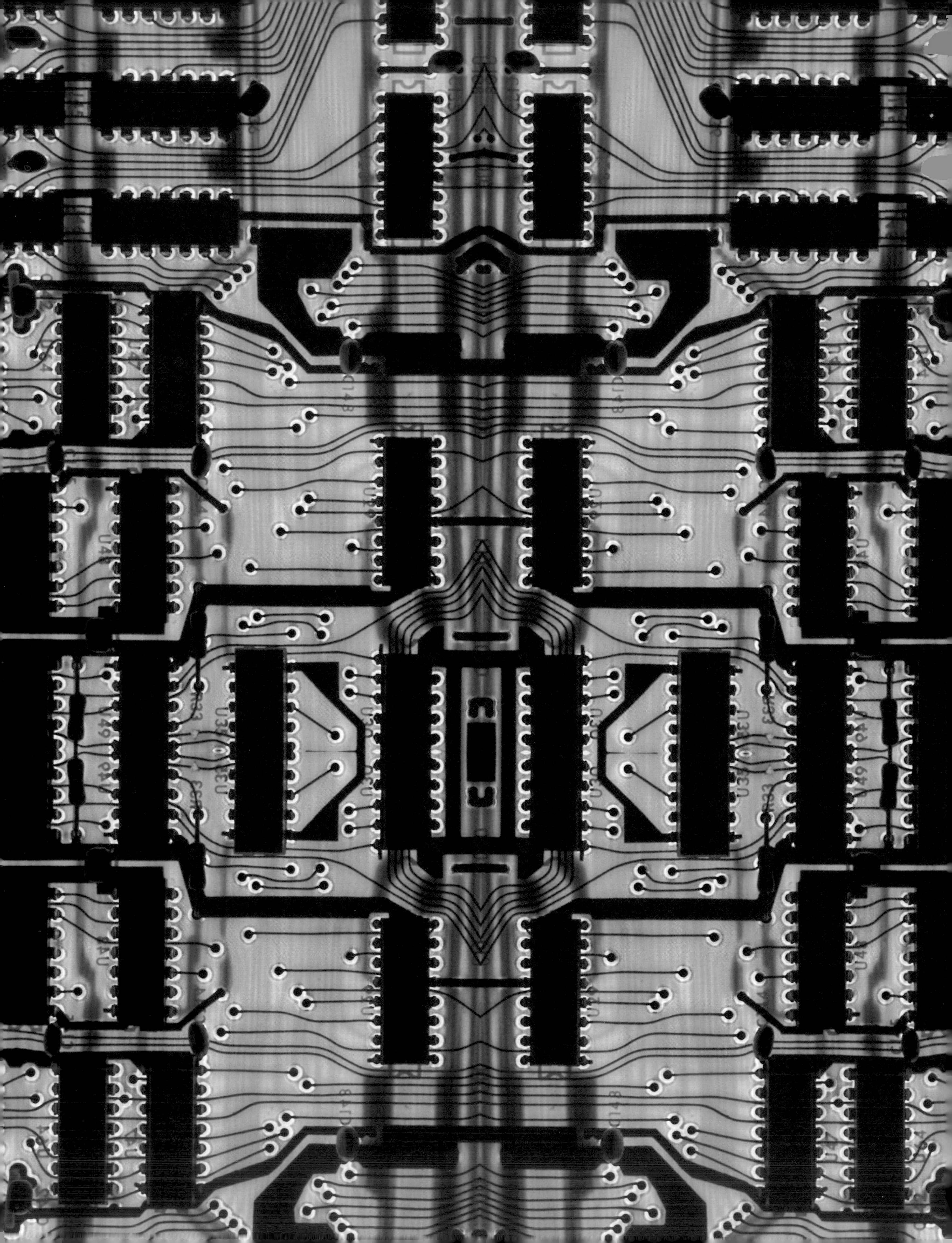

George Constable and Bob Somerville

A CENTURY OF INNOVATION

Twenty Engineering Achievements That Transformed Our Lives

Foreword by

NEIL ARMSTRONG

Afterword by

ARTHUR C. CLARKE

JOSEPH HENRY PRESS
WASHINGTON, D.C.

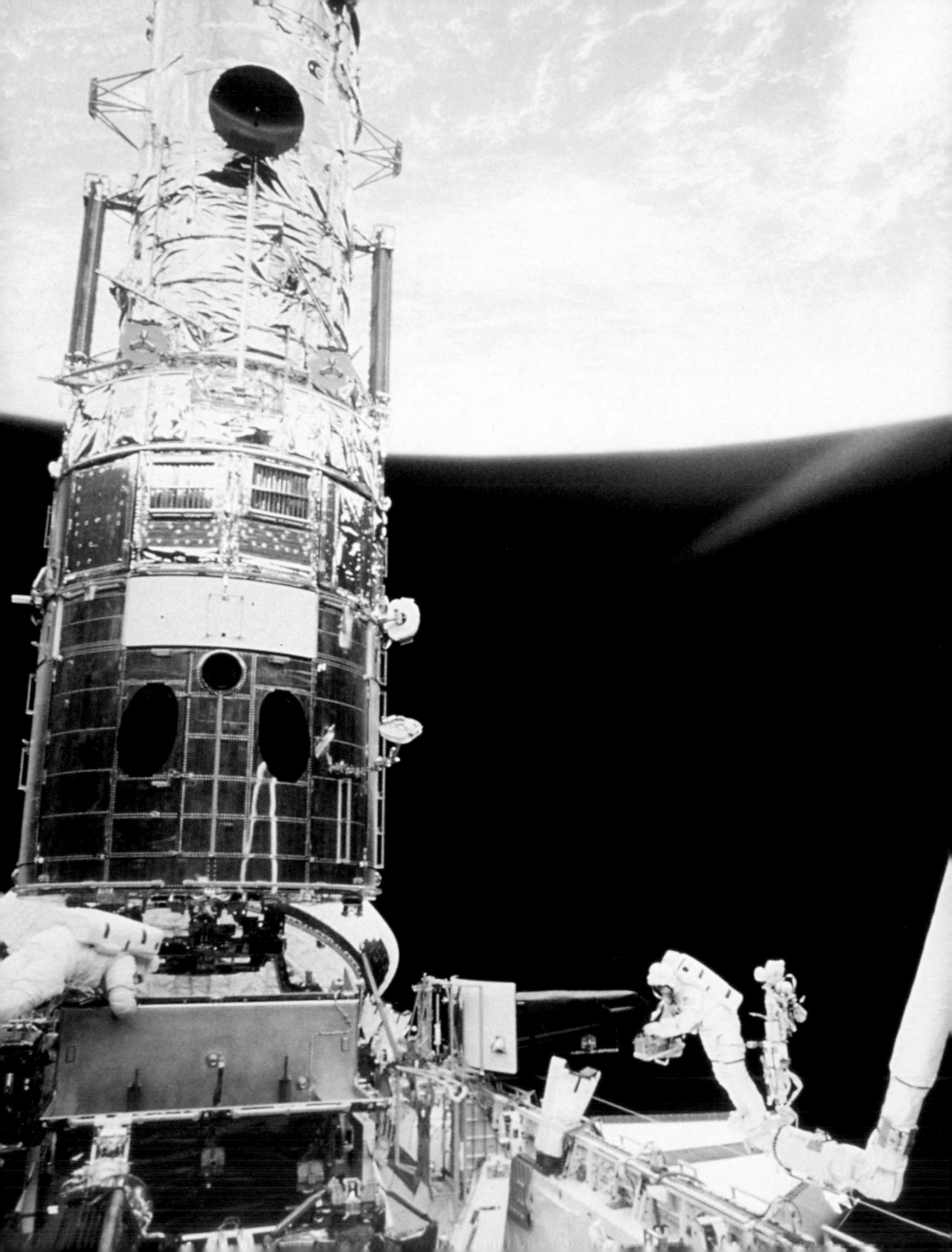

CONTENTS

FOREWORD

Engineering is often associated with science and understandably so. Both make extensive use of mathematics, and engineering requires a solid scientific basis. Yet as any scientist or engineer will tell you, they are quite different. Science is a quest for "truth for its own sake," for an ever more exact understanding of the natural world. It explains the change in the viscosity of a liquid as its temperature is varied, the release of heat when water vapor condenses, and the reproductive process of plants. It determines the speed of light. Engineering turns those explanations and understandings into new or improved machines, technologies, and processes—to bring reality to ideas and to provide solutions to societal needs.

A Century of Innovation is a book about engineering, a profession whose roots lie in the building of the bridges and cathedrals of the Middle Ages and whose growth expanded rapidly with the powered machines and the increased productivity of the Industrial Revolution. But this book is not about those beginnings or even about that rapid growth. Rather it is about the remarkable achievements of engineering in the 20th century.

During that century, the engineering disciplines, which had been principally civil and mechanical, broadened with the addition of a number of new sectors such as metallurgical, automotive, electrical, and aeronautical and deepened with the development of new methods, powerful computational tools, and dependable testing techniques. This evolution has left its imprint on our society in countless ways. Certainly the ability to cross continents and oceans in a matter of hours has changed our sense of the size of the world—it is a much smaller place than it was in our great-grandparents' day or even our grandparents' day. The world is also smaller by virtue of our being able to communicate instantly with people at a distance, whether across town, across the state, or on the other side of the globe by telephone or electronic mail.

There are myriad other ways in which engineering has affected society and our quality of life, ways we so take for granted they are virtually invisible to us. In part that may be because we each have our own definition of the term "quality of life." One person might think having no need to work would be ideal, while another person would think having a great deal of work to do would be ideal. But most of us would probably acknowledge that certain living conditions are essential to a preferred quality in our own lives. Think for a moment about what your day would be like if you turned on the tap in the morning and there was no water. Of if you knew that the water coming from the tap needed to be boiled before it was safe to drink or to use to brush your teeth. A century ago, typhoid fever, cholera, and other waterborne diseases could strike anyone at any time. Today, clean, safe water is something everyone can expect in most of the developed world.

True, the 20th was a century punctuated by two world wars and an ongoing epidemic of smaller hostilities that took countless lives. And it was marked by societal struggles to overcome injustice. But it was also the first century in which technology brought those traumas to the awareness of distant viewers and listeners, to touch people in previously unimagined ways. John Pierce, the engineer who fathered Telstar, the first satellite to relay television signals across the Atlantic, said that engineering helped create a world in which no injustice could be hidden.

In the closing year of the 20th century, a rather impressive consortium of 27 professional engineering societies, representing nearly every engineering discipline, gave its time, resources, and attention to a nationwide effort to identify and communicate the ways that engineering has affected our lives. Each organization independently polled its membership to learn what individual engineers believed to be the greatest achievements in their respective fields. Because these professional societies were unrelated to each other, the American Association of Engineering Societies and the National Academy of Engineering (NAE) helped to coordinate the effort. The NAE, in particular, took a leadership role in this effort because of its unique ability to convene the world's greatest engineering minds under the congressional charter that it shares with the National Academy of Sciences.

The NAE issued the call for nominations to the societies, convened a selection committee of leading engineers from all fields, and set about the laborious job of qualifying and quantifying the information in the nominations. After several rounds of narrowing the nominations, the committee met for two full days to determine which engineering achievements of the 20th century had the greatest positive effect on mankind. While intercontinental ballistic missiles and laser-guided bombs were undoubtedly technological marvels with important justifiable reasons for their existence, projects of this type were somewhat disadvantaged on the "positive effect on mankind" basis. Other engineering marvels (e.g., spectacular bridges, dams, skyscrapers, tunnels, and canals) did not qualify because their impact was largely local.

As you read this book, you'll find that the choices the committee made have one thing in common —that is, if any of them were removed, our world would be a very different and less hospitable place. Each of these achievements has been important to the transformation of society in the past hundred years. These are technologies that have become inextricable parts of the fabric of our lives—some spectacular, some nearly invisible, but all critically important.

If you've leafed through the book or looked at the table of contents, you already know that electrification was the top-rated engineering improvement to the life of earthlings in the past century. The majority of the top 20 achievements would not have been possible without widespread distribution of electricity. Electrification changed this country's economic development and gave rural populations the same opportunities and amenities as people in the cities. It provides the power for small appliances in the home, for computers in control rooms that route power and telecommunications, and for the machinery that produces capital goods and consumer products. Clearly, the ready availability of the electric power we use in our homes and businesses exemplifies how engineering changed the world during the 20th century.

I hope this book will remind you of the breadth, the depth, and the importance of engineering to human existence and progress. The likelihood that the era of creative engineering is past is nil. It is not unreasonable to suggest that, with the help of engineering, society in the 21st century will enjoy a rate of progress equal to or greater than that of the 20th. It is a worthy goal.

Neil Armstrong
Chairman Emeritus
EDO Corporation

ELECTRIFICATION

Scores of times each day, with the merest flick of a finger, each one of us taps into vast sources of energy—deep veins of coal and great reservoirs of oil, sweeping winds and rushing waters, the hidden power of the atom and the radiance of the Sun itself—all transformed into electricity, the workhorse of the modern world.

We mostly take the ready availability of electricity for granted, turning off the radio alarm in the morning, switching on the bedside lamp, pouring a cup of coffee from the machine that automatically started brewing it a few minutes before the alarm went off, tuning in to the morning news show on TV. Yet without the building of myriad power plants and the stringing of countless miles of wire, without the constant monitoring of the electric power grid and the juggling of supply and demand, that ready surge of electrons would not exist—nor would the modern world as we know it. Without a reliable supply of electricity, we couldn't use the lightweight, powerful electric motors that make elevators possible; without elevators, skyscrapers wouldn't exist—and the dramatic skylines of the world's major metropolises would be considerably more modest. Without a reliable supply of electricity, kidney dialysis machines and other life support equipment would be useless to the many patients who depend on them. Without electricity to power traffic lights, the commute to and from work would be mayhem—or maybe not. Without electricity to power automobile factories, we wouldn't have streets and highways full of automobiles either. Indeed, more than half the engineering achievements celebrated in this book would not have been possible without the widespread electrification that occurred in the 20th century, not only in the United States but also in other industrialized nations around the world.

The story of electrification in the United States is a story of public and private investment and of the engineers whose innovations moved the industry forward.

Early in the century the distribution of electric power was largely concentrated in cities served by privately owned utility companies (known today as investor-owned utilities, or IOUs). After Thomas Edison's work led to the first commercial power plant in 1882, these companies played a dominant role in constructing an advanced and complex electrical system that would become a model for the world. As the 20th century dawned, capacity expanded rapidly while continuous innovation improved the system. In 1903, for example, the first steam turbine generator, pioneered by Charles Curtis, was put into operation at the Newport Electric Corporation in Newport, Rhode Island. In 1917 an IOU known as American Gas & Electric (AG&E) established the first long-distance high-voltage transmission line—and the plant from which the line ran was the first major steam plant to be built at the mouth of the coal mine that supplied its fuel, virtually eliminating transportation costs. A year later pulverized coal was used as fuel for the first time at the Oneida Street Plant in Milwaukee. All of these innovations, and more, emerged from engineers working in the private sector. By the end of the century, IOUs would still account for almost 75 percent of electric utility generating capacity in the United States, even as they came to be outnumbered by other types of utilities and nonutility power producers. In 1998, for example, the country's 3,170 electric utilities produced 728 gigawatts of power—530 gigawatts of which were produced by 239 IOUs. (The approximately 2,110 nonutilities generated another 98 gigawatts.)

Developed by GE engineer Charles Curtis, this Volt Curtis steam turbine generated power for the Twin City Rapid Transit Company in Minneapolis in the early 1900s.

Before electrification, rural America cooked meals over open fires in the fireplace, while city folks had electric power galore (opposite). By the end of the century, a satellite's view reveals an electrified Earth at night (previous pages).

Successful at building plants to service large concentrated markets, IOUs in the first third of the century made relatively limited forays into rural America, where scattered farm families were isolated by distance from urban generating plants. As the inhabitants of New York, Chicago, and other cities across the country enjoyed the gleaming lights and the new labor-saving devices powered by electricity, life in rural America remained difficult. On 90 percent of American farms the only artificial light came from smoky, fumy lamps. Water had to be pumped by hand and heated over wood-burning stoves. Virtually every chore required manual labor; for many farm wives the most tiresome of all was the seemingly endless backbreaking drudgery of washing and ironing the family's clothes and linens.

In the 1930s President Franklin Delano Roosevelt saw the solution of this hardship as an opportunity to create new jobs, stimulate manufacturing, and begin to pull the nation out of the despair and hopelessness of the Great Depression. On May 11, 1935, he signed an executive order establishing the Rural Electrification Administration (REA). One of the key pieces of Roosevelt's New Deal initiatives, the REA would provide loans and other assistance so that rural cooperatives—basically, groups of farmers—could build and run their own electrical distribution systems.

The model for the system came from an engineer. In 1935, Morris Llewellyn Cooke, a mechanical engineer who had devised efficient rural distribution systems for power companies in New York and Pennsylvania, had written a report that detailed a plan for electrifying the nation's rural regions. Appointed by Roosevelt as the REA's first administrator, Cooke applied an engineer's

Thomas Edison (left) and Charles Steinmetz, shown here in 1922, were on opposite sides of the DC vs. AC debate.

approach to the problem, instituting what was known at the time as "scientific management"—essentially systems engineering. Rural electrification became one of the most successful government programs ever enacted. Within 2 years it helped bring electricity to some 1.5 million farms through 350 rural cooperatives in 45 of the 48 states. By 1939 the cost of a mile of rural line had dropped from $2,000 to $600. Almost half of all farms were wired by 1942 and virtually all of them by the 1950s.

Getting electric power from where it is generated to customers who need it remains a critical factor in the electrification of the world in general. The basic system on which the electrical supply depends—the power grid—hasn't changed much since its earliest days, except in scale. Power plants equipped with generators convert a source of energy—fossil fuel, falling water, wind, the sun, a nuclear reactor—into electricity. That electrical power is then transmitted through the distribution system to individual buildings, factories, homes, and farms (see pages 6 and 7).

Generation, transmission, and distribution—the same then as now. But back at the very beginning, transmission was a matter of intense debate. On one side were proponents of direct current (DC), in which electrons flow in only one direction. On the other were those who favored alternating current (AC), in which electrons oscillate back and forth. The most prominent advocate of direct current was none other than Thomas Edison. If Benjamin Franklin was the father of electricity, Edison was widely held to be his worthy heir. Edison's inventions, from the lightbulb to the electric fan, were almost single-handedly driving the country's—and the world's—hunger for electricity.

However, Edison's devices ran on DC, and as it happened, research into AC had shown that it was much better for transmitting electricity over long distances. Championed in the last 2 decades of the 19th century by inventors and theoreticians such as Nikola Tesla and Charles Steinmetz and the entrepreneur George Westinghouse, AC won out as the dominant power supply medium. Although Edison's DC devices weren't made obsolete—AC power could be readily converted to run DC appliances—the advantages AC power offered made the outcome virtually inevitable.

With the theoretical debate settled, 20th-century engineers got to work making things better—inventing and improving devices and systems to bring more and more power to more and more people. Most of the initial improvements involved the generation of power. An early breakthrough was the transition from reciprocating engines to turbines, which took one-tenth the space and weighed as little as one-eighth an engine of comparable output. Typically under the pressure of steam or flowing water, a turbine's great fan blades spin, and this spinning action generates electric current.

Steam turbines—powered first by coal, then later by oil, natural gas, and eventually nuclear reactors—took a major leap forward in the first years of the 20th century. Key improvements in design increased generator efficiency many times over. By the 1920s high-pressure steam generators were the state of the art. In the mid-1920s the investor-owned utility Boston Edison began using a high-pressure steam power plant at its Edgar Station. At a time when the common rate of power generation by steam pressure was 1 kilowatt-hour per 5 to 10 pounds of coal, the Edgar Station—operating a boiler and turbine unit at 1,200 pounds of steam pressure—generated electricity at the rate of 1 kilowatt-hour per 1 pound of coal. And the improvements just kept coming. AG&E introduced a key

A massive turbine rotor, being hoisted into place at a power plant in 1952, weighed 42,200 pounds but would drive the turbine generator with the precision of a fine watch.

enhancement with its Philo plant in southeastern Ohio, the first power plant to reheat steam, which markedly increased the amount of electricity generated from a given amount of raw material. Soon new, more heat-resistant steel alloys were enabling turbines to generate even more power. Each step along the way the energy output was increasing. The biggest steam turbine in 1903 generated 5,000 kilowatts; in the 1960s steam turbines were generating 200 times that.

Partnering the steam turbines were the hydroelectric turbogenerators in dams. The physical principle of the turbine was first applied by the ancient Greeks, who used water wheels to convert spinning to power—in this case a millstone turning to grind grain. Hydroelectric projects apply the same principle to create electric power on a grand scale. In the most basic terms, the power in a given amount of water comes from its flowing action as it falls; the more water falling or the greater the fall, the more power. As demonstrated by the many hydroelectric projects that began in the 1930s, that meant building taller dams or building dams across larger rivers—or both.

The two biggest dams—and by far the largest construction projects of their day—were the Grand Coulee Dam on the Columbia River in Washington and the Hoover Dam on the Nevada-Arizona border on the upper reaches of the Colorado River. Like the dams cropping up on other rivers, these would harness more power for the nation while also providing irrigation and flood control. The Grand Coulee, more than four times wider than the Hoover to span the mighty Columbia River, far outstrips the Hoover as a hydroelectric plant. It generates more than 6.5 million kilowatts with its 24 turbines, compared to just over 2 million kilowatts from the Hoover's 17 turbines. Completed in 1942, the Grand Coulee still ranks as the third-largest producer of hydroelectric power in the world.

In some regions, engineers have combined multiple sources of power to fuel growing needs. The Tennessee Valley Authority, another New Deal project, today manages numerous small dams, 11 steam turbine power plants, and two nuclear power plants, altogether producing 125 billion kilowatt-hours of electricity a year.

New engineering challenges have continued to arise, among them how to transmit electricity at higher and higher voltages for maximum efficiency. Improvements in both materials and systems have brought transmission voltages up from the 220 volts of the 1880s to the 765,000 volts of today. And still the search goes on for new and better ways to harness energy from sources that now include everything from nuclear reactors to the wind, the Sun, and even the geothermal energy of Earth itself. Wind farms, with scores of sleek, narrow-bladed, computer-controlled wind turbines, have become increasingly productive; improvements in efficiency have brought the cost of wind-produced electricity down significantly in the past 15 years. IOUs are also devoting more research dollars to improving solar power. Photovoltaic cells that generate electricity directly are becoming more efficient, but engineers are also working on other innovative approaches, including a technique known as solar thermal, in which arrays of mirrored parabola-shaped collectors focus sunbeams to heat oil to as high as 750°F to drive steam turbines.

Stretching 1,244 feet across the Black Canyon, Hoover Dam's arch-gravity concrete structure rises 726 feet above bedrock. The first of the dam's 17 generators began operation in October 1937; the last one came on line in 1961. Together, they have a combined capacity of 2 million kilowatts.

New distribution schemes have followed apace, linking the world into flexible grids that can deliver electricity across thousands of miles. These grids, as well as computer-controlled routing and switching systems, are intended to reduce the possibility of the kinds of

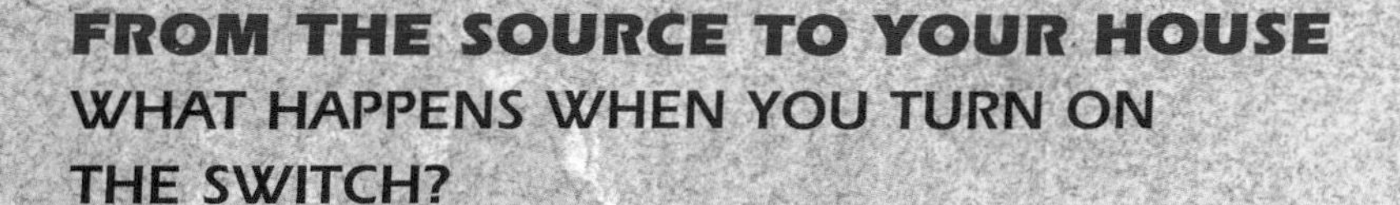

FROM THE SOURCE TO YOUR HOUSE

WHAT HAPPENS WHEN YOU TURN ON THE SWITCH?

The magic of readily available electricity depends on the power grid—an intricate orchestration of equipment and processes that connects power plants generating the electrical supply with the individual houses, businesses, and infrastructure systems (such as street lamps and traffic lights) that use it. The grid includes power plants, transformers, high-voltage transmission lines, substations, and local distribution networks.

Shown below is a steam-turbine power plant, which typically burns coal, oil, or natural gas to heat water and create steam; cooling towers help cool the hot water for recycling. The steam causes turbines to spin, which in turn drives generators that create electricity (below right). Transformers then increase the voltage of the electric current for more efficient transmission via high-voltage transmission lines. The voltage is reduced by other transformers at substations (opposite center) and then distributed through local networks of power lines to individual users. A flick of the switch completes the process.

HOW TRANSFORMERS WORK. Transformers increase or decrease the voltage of an electric current, known respectively as stepping it up or stepping it down. Here, a higher voltage, entering the six windings on the left, induces an alternating magnetic field in the transformer's iron core, which in turn induces an electric field in the three windings on the right, stepping down the voltage of the current that flows out.

Transformers

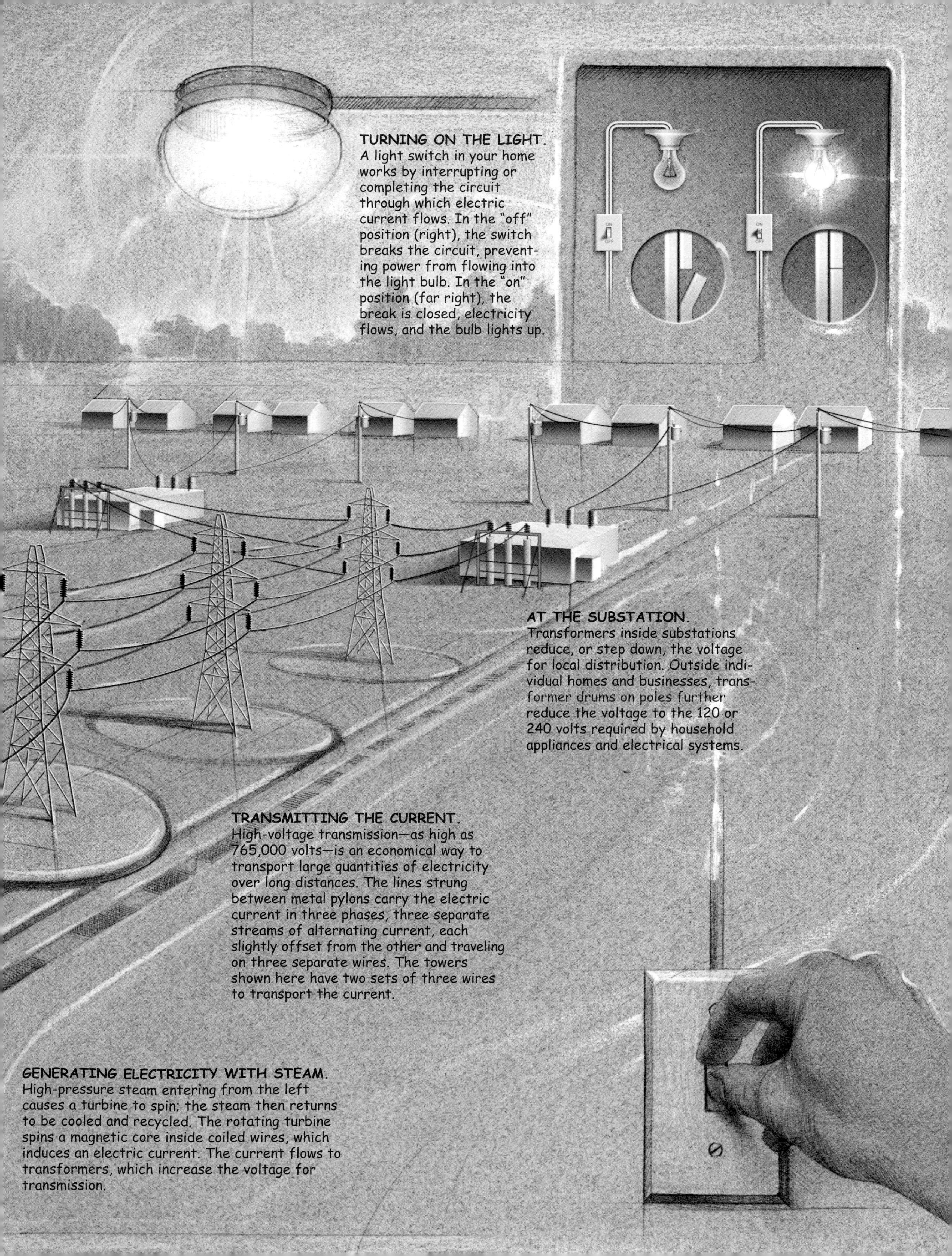

TURNING ON THE LIGHT. A light switch in your home works by interrupting or completing the circuit through which electric current flows. In the "off" position (right), the switch breaks the circuit, preventing power from flowing into the light bulb. In the "on" position (far right), the break is closed, electricity flows, and the bulb lights up.

AT THE SUBSTATION. Transformers inside substations reduce, or step down, the voltage for local distribution. Outside individual homes and businesses, transformer drums on poles further reduce the voltage to the 120 or 240 volts required by household appliances and electrical systems.

TRANSMITTING THE CURRENT. High-voltage transmission—as high as 765,000 volts—is an economical way to transport large quantities of electricity over long distances. The lines strung between metal pylons carry the electric current in three phases, three separate streams of alternating current, each slightly offset from the other and traveling on three separate wires. The towers shown here have two sets of three wires to transport the current.

GENERATING ELECTRICITY WITH STEAM. High-pressure steam entering from the left causes a turbine to spin; the steam then returns to be cooled and recycled. The rotating turbine spins a magnetic core inside coiled wires, which induces an electric current. The current flows to transformers, which increase the voltage for transmission.

Interest is growing in methods of generating electricity without burning fossil fuels, such as geothermal plants (below) and wind farms (right), which are booming international industries.

blackouts that struck the densely populated northeastern United States at wide intervals in the latter part of the 20th century and early in the 21st.

Instrumental in a whole host of improvements has been the Electric Power Research Institute (EPRI), established by public- and investor-owned energy producers in the wake of the 1965 blackout and now including member organizations from some 40 countries. EPRI investigates and fosters ways to enhance power production, distribution, and reliability, as well as the energy efficiency of devices at the power-consuming end of the equation. Reliability has become more significant than ever. In an increasingly digital, networked world, power outages as short as 1/60th of a second can wreak havoc on a wide variety of microprocessor-based devices, from computer servers running the Internet to life support equipment. EPRI's goal for the future is to improve the current level of reliability of the electrical supply from 99.99 percent (equivalent to an average of one hour of power outage a year) to a standard known as the 9-nines, or 99.9999999 percent reliability.

As the demand for the benefits of electrification continues to grow around the globe, resourcefulness remains a prime virtue. In some places the large-scale power grids that served the 20th century so well are being supplemented by decentralized systems in which energy consumers—households and businesses—produce at least some of their own power, employing such renewable resources as solar *(left)* and wind power. Where they are available, schemes such as net metering, in which customers actually sell back to utility companies extra power they have generated, are gaining in popularity. Between 1980 and 1995, 10 states passed legislation establishing net metering procedures and another 26 states have done so since 1995. Citizens of the 21st-century world, certainly no less hungry for electrification than their predecessors, eagerly await the next steps.

Perspective

E. Linn Draper, Jr.

Chairman, President and CEO
American Electric Power Company, Inc.

When I embarked on a career in engineering, I hoped and believed that it would be a fascinating and rewarding occupation. It certainly has lived up to expectations. Playing a role in electrification's transformation of America and the world has been tremendously exciting. But I had no idea that my engineering background would ultimately lead to the chairmanship of a multibillion-dollar corporation with a long tradition of pioneering achievements in the generation and transmission of electricity.

Early in my career I spent 10 years as a member of the nuclear engineering faculty at the University of Texas. Those years involved immersion in highly technical subjects such as nuclear reactor design and nuclear waste management, an intense training that has been invaluable in the senior management roles I've held since then, whether supervising engineering and construction or managing a fleet of fossil-fueled power plants. I now head a company whose generating capacity is two-thirds coal fired. My nuclear engineering background notwithstanding, it's difficult for me to envision another fuel in America's energy mix supplanting coal as our nation's workhorse fuel.

Today more than half of America's electric energy comes from coal. The good news is that coal is being burned today more cleanly than ever before, thanks to a series of remarkable engineering advancements. Electrostatic precipitators remove more than 99 percent of the particulates from a plant's emissions, while scrubbers reduce emissions of sulfur dioxide by up to 98 percent. Low-NOx (nitrogen oxide) burners and selective catalytic reduction technology reduce emissions of nitrogen oxide, a precursor of smog, by between 25 and 90 percent. Since 1970 emissions from coal-fired power plants have fallen by nearly 50 percent, even while the amount of electricity generated has tripled.

I believe that further advances in engineering will enable the next generation of coal-fueled power plants to use coal with even greater efficiency and fewer environmental impacts, by converting it into a stream of gas. At American Electric Power Company (AEP) we're already taking voluntary measures to reduce, offset, or sequester emissions of greenhouse gases from coal-fired plants. We're participating in a project to study the injection of carbon dioxide emissions under the ground, and we're involved in carbon sequestration projects with forests in Bolivia, Brazil, and Louisiana.

Because we take electricity for granted, it's sometimes hard to imagine that two billion people on our planet live without access to electric power. Yet history has shown that the supply of reliable, affordable electricity is an essential prerequisite to economic and social progress. Recently, I've had the privilege of working with electrification projects in remote parts of the world as chairman of the E7 organization. A group that includes leading electric utilities from the G7 nations, the E7 is committed to promoting sustainable development through electrification and through projects to build human capacity in developing countries.

Looking ahead, our nation's energy future depends on the engineering students studying at colleges and universities today. There are many talented people in these schools but not enough, in my view, to meet our future needs. At AEP we're partners in education with scores of local elementary and secondary schools, helping to boost interest in science and math. We also have co-op and internship programs for promising engineering students. If electrification's next century is to be as successful as its last, we need to do everything we can to encourage our best and brightest to pursue careers in engineering and science.

ELECTRIFICATION TIMELINE

At the beginning of the 20th century, following a struggle between the direct-current systems favored by Thomas Edison and the alternating-current systems championed by Nikola Tesla and George Westinghouse, electric power was poised to become the muscle of the modern world. Today it keeps our factories running—as well as the telecommunications industry, the appliances in our homes, and the lifesaving equipment in our hospitals. In myriad other ways the ready access to electricity helps maintain the well-being of billions of people around the globe.

1903 The steam turbine generator invented by Charles G. Curtis and developed into a practical steam turbine by William Le Roy Emmet is a significant advance in the capacity of steam turbines. Requiring one-tenth the space and weighing one-eighth as much as reciprocating engines of comparable output, it generates 5,000 kilowatts and is the most powerful plant in the world.

1908 William J. Bailley of the Carnegie Steel Company invents a solar collector with copper coils and an insulated box.

1910s Irving Langmuir of General Electric experiments with gas-filled lamps, using nitrogen to reduce evaporation of the tungsten filament, thus raising the temperature of the filament and producing more light. To reduce conduction of heat by the gas, he makes the filament smaller by coiling the tungsten.

1913 Southern California Edison puts into service a 150,000-volt line to bring electricity to Los Angeles. Hydroelectric power is generated along the 233-mile-long aqueduct that brings water from Owens Valley in the eastern Sierra.

1917 The first long-distance high-voltage transmission line is established by American Gas & Electric (AG&E), an investor-owned utility. The line originates from the first major steam plant to be built at the mouth of a coal mine, virtually eliminating fuel transportation costs.

1920s Boston Edison's Edgar Station becomes a model for high-pressure steam power plants worldwide by producing electricity at the rate of 1 kilowatt-hour per pound of coal at a time when generators commonly use 5 to 10 pounds of coal to produce 1 kilowatt-hour. The key was operating a boiler and turbine unit at 1,200 pounds of steam pressure, a unique design developed under the supervision of Irving Moultrop.

In Philo, Ohio, AG&E introduces the first plant to reheat steam, thereby increasing the amount of electricity generated from a given amount of raw material. Soon new, more heat-resistant steel alloys are enabling turbines to generate even more power.

Windmills with modified airplane propellers marketed by Parris-Dunn and Jacobs Wind are used to drive 1- to 3-kilowatt DC generators on farms in the U.S. Plains states. At first these provide power for electric lights and power to charge batteries for crystal radio sets, but later they supply electricity for motor-driven washing machines, refrigerators, freezers, and power tools.

1931 The 100-kilowatt Balaclava wind generator on the shores of the Caspian Sea in Russia marks the introduction of bulk-power, utility-scale wind energy conversion systems. This machine operates for about 2 years, generating 200,000 kilowatt-hours of electricity. A few years later, other countries, including Great Britain, the United States, Denmark, Germany, and France, begin experimental large-scale wind plants.

1933 Congress passes legislation establishing the Tennessee Valley Authority (TVA). Today the TVA manages numerous dams, 11 steam turbine power plants, and two nuclear power plants. Altogether these produce 125 billion kilowatt-hours of electricity a year.

1935 The first generator at Hoover Dam along the Nevada-Arizona border begins commercial operation. More generators are added through the years, the 17th and last one in 1961.

President Roosevelt issues an executive order to create the Rural Electrification Administration (REA), which forms cooperatives that bring electricity to millions of rural Americans. Within 6 years the REA has aided the formation of 800 rural electric cooperatives with 350,000 miles of power lines.

1942 Grand Coulee Dam on the Columbia River in Washington State is completed. With 24 turbines (*above*), the dam eventually brings electricity to 11 western states and irrigation to more than 500,000 acres of farmland in the Columbia Basin.

1953 The American Electric Power Company (AEP) commissions a 345,000-volt system that interconnects the grids of seven states. The system reduces the cost of transmission by sending power where and when it is needed rather than allowing all plants to work at less than full capacity.

1955 On July 17, Arco, Idaho, becomes the first town to have all its electrical needs generated by a nuclear power plant. Arco is 20 miles from the Atomic Energy Commission's National Reactor Testing Station, where Argonne National Laboratory operates BORAX (Boiling Reactor Experiment) III, an experimental nuclear reactor.

The same year the Niagara-Mohawk Power Corporation grid in New York draws electricity from a nuclear generation plant, and 3 years later the first large-scale nuclear power plant in the United States comes on line in Shippingport, Pennsylvania. The work of Duquesne Light Company and the Westinghouse Bettis Atomic Power Laboratory, this pressurized-water reactor supplies power to Pittsburgh and much of western Pennsylvania.

1959 New Zealand opens the first large geothermal electricity-generating plant driven by steam heated by nonvolcanic hot rocks. The following year electricity is produced from a geothermal source in the United States at the Geysers, near San Francisco, California.

1961 France and England connect their electrical grids with a cable submerged in the English Channel. It carries up to 160 megawatts of DC current, allowing the two countries to share power or support each other's system.

1964 The Soviet Union completes the first large-scale magnetohydrodynamics plant. Based on pioneering efforts in Britain, the plant produces electricity by shooting hot gases through a strong magnetic field.

1967 The highest voltage transmission line to date (750,000 volts) is developed by AEP. The same year the Soviet Union completes the Krasnoyansk Dam power station in Siberia, which generates three times more electric power than the Grand Coulee Dam.

1978 Congress passes the Public Utility Regulatory Policies Act (PURPA), which spurs the growth of nonutility unregulated power generation. PURPA mandates that utilities buy power from qualified unregulated generators at the "avoided cost"—the cost the utility would pay to generate the power itself. Qualifying facilities must meet technical standards regarding energy source and efficiency but are exempt from state and federal regulation under the Federal Power Act and the Public Utility Holding Company Act. In addition, the federal government allows a 15 percent energy tax credit while continuing an existing 10 percent investment tax credit.

1980s In California more than 17,000 wind machines, ranging in output from 20 to 350 kilowatts, are installed on wind farms. At the height of development, these turbines have a collected rating of more than 1,700 megawatts and produce more than 3 million megawatt-hours of electricity, enough at peak output to power a city of 300,000.

1983 Solar Electric Generating Stations (SEGs) producing as much as 13.8 megawatts are developed in California and sell electricity to the Southern California Edison Company.

1990s The bulk power system in the United States evolves into three major power grids, or interconnections (*below*), coordinated by the North American Electric Reliability Council (NERC), a voluntary organization formed in 1968. The ERCOT (Electric Reliability Council of Texas) interconnection is linked to the other two only by certain DC lines.

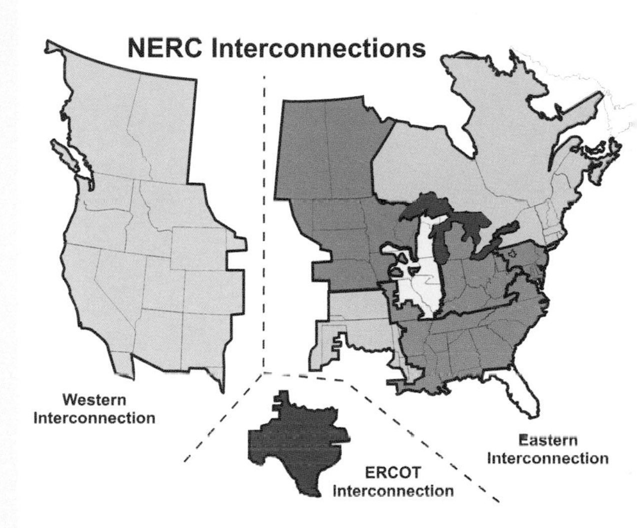

1992 The Energy Policy Act establishes a permanent 10 percent investment tax credit for solar and geothermal power-generating equipment as well as production tax credits for both independent and investor-owned wind projects and biomass plants using dedicated crops.

A joint venture of Sandia National Laboratories and Cummins Power Generation develops an operational 7.5-kilowatt solar dish prototype system using an advanced stretched-membrane concentrator.

2000 By the end of the century, semiconductor switches are enabling the use of long-range DC transmission.

TAPE

AUTOMOBILE

When Thomas Edison did some future gazing about transportation during a newspaper interview in 1895, he didn't hedge his bets. "The horseless carriage is the coming wonder," said America's reigning inventor. "It is only a question of a short time when the carriages and trucks in every large city will be run with motors." Just what kind of motors would remain unclear for a few more years.

Of the 10,000 or so cars that were on the road by the start of the 20th century, three-quarters were electric or had external combustion steam engines, but the versatile and efficient gas-burning internal combustion power plant was destined for dominance. Partnered with ever-improving transmissions, tires, brakes, lights, and other such essentials of vehicular travel, it redefined the meaning of mobility, an urge as old as the human species. The United States alone—where 25 million horses supplied most local transportation in 1900—had about the same number of cars just three decades later. The country also had giant industries to manufacture them and keep them running and a vast network of hard-surfaced roads, tunnels, and bridges to support their conquest of time and distance. By century's end, the average American adult would travel more than 10,000 miles a year by car.

Other countries did much of the technological pioneering of automobiles. A French military engineer, Nicholas-Joseph Cugnot, lit the fuse in 1771 by assembling a three-wheeled, steam-powered tractor to haul artillery. Although hopelessly slow, his creation managed to run into a stone wall during field trials—history's first auto accident. About a century later, a German traveling salesman named Nicholaus Otto constructed the first practical internal combustion engine; it used a four-stroke cycle of a piston to draw a fuel-air mixture into a cylinder, compress it, mechanically capture energy after ignition, and expel the exhaust before beginning the cycle anew. Shortly thereafter, two other German engineers,

Gottlieb Daimler and Karl Benz, improved the design and attached their motors to various vehicles.

These ideas leaped the Atlantic in the early 1890s, and within a decade all manner of primitive cars—open-topped, bone-jarring contraptions often steered by tillers—were chugging along the streets and byways of the land. They were so alarming to livestock that Vermont passed a state law requiring a person to walk in front of a car carrying a red warning flag, and some rural counties banned them altogether. But even cautious farmers couldn't resist their appeal, memorably expressed by a future titan named Henry Ford: "Everybody wants to be somewhere he ain't. As soon as he gets there he wants to go right back."

Behind Ford's homespun ways lay mechanical gifts of a rare order. He grew up on a farm in Dearborn, Michigan, and worked the land himself for a number of years before moving to Detroit, where he was employed as a machinist and then as chief engineer of an electric light company. All the while he tinkered with cars, displaying such obvious talents that he readily found backers when he formed the Ford Motor Company in 1903 at the age of 40.

The business prospered from the start, and after the introduction of the Model T in 1908, it left all rivals in the dust. The Tin Lizzie, as the Model T was affectionately called, reflected Ford's rural roots. Standing seven feet high, with a four-cylinder, 20-horsepower engine that produced a top speed of 45 miles per hour, it was unpretentious, reliable, and remarkably sturdy. Most important from a marketing point of view, it was cheap—an affordable $850 that first year—and became astonishingly cheaper as the years passed, eventually dropping to the almost irresistible level of $290. "Every time I lower the price a dollar, we gain a thousand new buyers," boasted Ford. As for the cost of upkeep, the Tin Lizzie was a marvel. A replacement muffler cost 25 cents, a new fender $2.50.

What made such bargain prices possible was mass production, a competitive weapon that Henry Ford honed with obsessive genius. Its basis, the use of standardized, precision-made parts, had spun fortunes for a number of earlier American industrialists—armaments maker Samuel Colt and harvester king Cyrus McCormick among them. But that was only the starting point for Ford and his engineers. In search of efficiencies they created superb machine tools, among them a device that could simultaneously drill 45 holes in an engine block. They mechanized steps that were done by hand in other factories, such as the painting of wheels. Ford's painting machine could handle 2,000 wheels an hour. In 1913, with little fanfare, they tried out another tactic for boosting productivity: the moving assembly line, a concept borrowed from the meat-packing industry.

At the Ford Motor Company the assembly line was first adopted in the department that built the Model T's magneto, which generated electricity for the ignition system. Previously, one worker had assembled each magneto from start to finish. Under the new approach, however, each worker performed a single task as the unit traveled past his station on a conveyer belt. "The man

An early 1900 advertisement for Oldsmobile's curved-dash runabout

Two Ford workers on either side of the moving assembly line maneuver a car body into position. Someone else—farther down the line—would secure it in place.

Inventor Thomas Edison, shown here in 1928 (opposite), refused to give up his 1914 Model T Ford despite the newer models rolling off the Ford assembly line.

who puts in a bolt does not put on the nut," Ford explained. "The man who puts on the nut does not tighten it."

The savings in time and money were so dramatic that the assembly line approach was soon extended to virtually every phase of the manufacturing process. By 1914 the Ford factory resembled an immense river system, with subassemblies taking shape along tributaries and feeding into the main stream, where the chassis moved continuously along rails at a speed of 6 feet per minute. The time needed for the final stage of assembly dropped from more than 12 hours to just 93 minutes. Eventually, new Model Ts would be rolling off the line at rates as high as one every 10 seconds.

So deep-seated was Henry Ford's belief in the value of simplicity and standardization that the Tin Lizzie was the company's only product for 19 years, and for much of that period it was available only in black because black enamel was the paint that dried the fastest. Since Model Ts accounted for half the cars in the world by 1920, Ford saw no need for fundamental change.

With his son, Edsel, at the wheel, Henry Ford celebrates the Ford Motor Company's 15 millionth automobile in May 1927.

Nonetheless, automotive technology was advancing at a rapid clip. Disk brakes arrived on the scene way back in 1902, patented by British engineer Frederick Lanchester. The catalytic converter was invented in France in 1909, and the V8 engine appeared there a year later. One of the biggest improvements of all, especially in the eyes of women, was the self-starter. It was badly needed. All early internal combustion engines were started by turning over the motor with a hand crank, a procedure that required a good deal of strength and, if the motor happened to backfire, could be wickedly dangerous, breaking many an arm with the kick. In 1911, Charles Kettering, a young Ohio engineer and auto hobbyist, found a better way—a starting system that combined a generator, storage battery, and electric motor. It debuted in the Cadillac the following year and spread rapidly from there.

Even an innovation as useful as the self-starter could meet resistance, however. Henry Ford refused to make Kettering's invention standard in the Model T until 1926, although he offered it as an option before that. Sometimes buyers were the ones who balked at novelty. For example, the first truly streamlined car—the 1934 Chrysler Airflow, designed with the help of aeronautical engineers and wind tunnel testing—was a dud in the marketplace because of its unconventional styling. Power steering, patented in the late 1920s by Francis Davis, chief engineer of the Pierce-Arrow Motor Car Company, didn't find its way into passenger cars until 1951. But hesitantly accepted or not, major improvements in the automobile would keep coming as the decades passed. Among the innovations were balloon tires and safety-glass windshields in the 1920s; front-wheel drive, independent front suspension, and efficient automatic transmissions in the 1930s; tubeless and radial tires in the 1940s; electronic fuel injection in the 1960s; and electronic ignition systems in the 1970s. Engineers outside the United States were often in the vanguard of invention, while Americans continued to excel at all of the unseen details of manufacturing, from glass making and paint drying to the stamping of body panels with giant machines.

Brutal competition was a hallmark of the business throughout the 20th century. In 1926 the United States had no fewer than 43 carmakers, the high point. The fastest rising among them was General Motors, whose marketing strategy was to produce vehicles in a number of distinct styles and price ranges, the exact opposite of Henry Ford's road to riches. GM further energized the market with the concept of an annual model change, and the company grew into a veritable empire, gobbling up prodigious amounts of steel, rubber and other raw materials, and manufacturing components such as spark plugs and gears in corporate subsidiaries.

As the auto giants waged a war of big numbers, some carmakers sold exclusivity. Packard was one. Said a 1930s advertisement: "The Packard owner, however high his station, mentions his car with a certain satisfaction—knowing that his choice proclaims discriminating taste as well as a sound judgment of fine things." Such a car had to be well engineered, of course, and the Packard more than met that standard. So did the lovingly crafted Rolls-Royce from Great Britain and the legendary Maybach Zeppelin of Germany, a 1930s masterpiece that had a huge 12-cylinder engine and a gearbox with eight forward and four reverse gears. (The Maybach marque would be revived by Mercedes seven decades later for a car with a 550-horsepower V12 engine, ultra-advanced audio and video equipment, precious interior veneers, and a price tag over $300,000.)

Designed by automotive engineer Karl Maybach, the incomparable Maybach Zeppelin combined flawless craftsmanship with customized design to give buyers a car created to their specifications, like this 1935 cabriolet.

At the other extreme was the humble, economical Volkswagen—literally, "people's car"—designed by engineer Ferdinand Porsche. World War II delayed its production, but it became a runaway worldwide hit in the 1950s and 1960s, eventually eclipsing the Model T's record of 15 million vehicles sold. Japan, a leader in the development of fuel-efficient engines and an enthusiastic subscriber to advanced manufacturing techniques, also became a major global player, the biggest in the world by 1980.

The automobile's crucial role in shaping the modern world is apparent everywhere. During the 19th century, suburbs tended to grow in a radial pattern dictated by

Delivered by the freighter Ravenstein, *the first complete shipment of Volkswagens arrives from Germany at Baltimore's Pier 11 in March 1956.*

At a Ford plant in 1913 each worker on the flywheel magneto assembly line (right) has a specific job—to put magneto magnets on the flywheel, or put in bolts, or fasten them down—before shoving the flywheel along to the next worker. With a moving conveyor belt, assembly time was cut from 18 minutes per magneto to 5 minutes.

In the engine testing room (left), completed engines are "run in" before going to the assembly line. Each engine is hooked up to an electric motor and turned over for several minutes to wear in the new parts, reducing the power required to drive each engine.

By 1939 the assembly line is chain driven and workers have to keep up with it. At right, an engine is lowered onto a chassis to which front and rear axles have already been attached. Other workers down the line bolt the engine to the frame, after which the radiator is installed. Then the body itself is dropped onto the chassis.

By the mid-1950s, even as other parts of the assembly line are becoming more mechanized, some jobs, such as installing the instruments and wiring that go into the dashboard, still require assembly with simple tools by hand, as seen at right with dashboards for the Ford Thunderbird, introduced in 1955.

MAKING AN AUTOMOBILE

FINE-TUNING THE ASSEMBLY LINE

To Henry Ford, the incentive for building Model Ts on an assembly line boiled down to simple arithmetic. "Save ten steps a day for each of 12,000 employees," he declared, "and you will have saved 50 miles of wasted motion and misspent energy." What moved on an assembly line was the work, not the worker: as shown opposite in photographs from the Ford archives, each employee stayed in one spot and repeatedly performed a single task as some object-in-the-making traveled past. The basic idea was borrowed from the meat-packing industry, but Ford and his engineers carried it to lengths never envisioned before, carefully dividing up tasks for maximum efficiency, creating specialized new tools, designing parts so that they could be fitted together easily, and addressing a host of other details to keep the line flowing smoothly.

Automakers have continued to push the frontiers of mass-production since then. Today, computerized robots (below) have lifted from human shoulders some of the most exacting and strenuous tasks. And although Henry Ford might not have approved, assembly lines have become flexible enough to mass-produce multiple versions of a car without a pause.

Mechanical arms, guided by computerized brains, can do the most demanding and repetitive tasks without growing either weary or bored. They can also take over jobs that are potentially hazardous to human health, operating with impunity amidst a hurricane of welding sparks (left) or the toxic fumes of the spray-painting room (below). And their gloved appendages even handle the more delicate task of installing windshields (bottom).

By the end of the 20th century, robots and computers are integral to the automobile manufacturing process and the minivan is deemed a necessity in the lives of busy families.

trolley lines; the car has allowed them to spring up anywhere within commuting distance of the workplace—frequently another suburb. Malls, factories, schools, fast-food restaurants, gas stations, motels, and a thousand other sorts of waystops and destinations have spread out across the land with the ever-expanding road network. Taxis, synchronized traffic lights, and parking lots sustain modern cities. Today's version of daily life would be unthinkable without the personal mobility afforded by wheels and the internal combustion engine.

Not surprisingly, the automobile remains an engineering work in progress, with action on many fronts, much of it prompted by government regulation and societal pressures. Concerns about safety have put seatbelts and airbags in cars, led to computerized braking systems, and—on the cutting edge of technology—fostered interest in devices that can enhance night vision or warn of impending collisions. Onboard microprocessors reduce polluting emissions and maximize fuel efficiency by controlling the fuel-air ratio. New materials—improved steels, aluminum, plastics, and composites—save weight and may add structural strength.

As for the motive power, engineers are working hard on designs that complement or may someday even supplant the internal combustion engine. One avenue of research involves electric motors whose power is generated by fuel cells that draw electrical energy from an abundant substance such as hydrogen. Already at hand are hybrid cars, powered by both gasoline and electricity. Unlike all-electric cars, hybrids don't have to be plugged in to be recharged; instead, their battery is charged by either the gasoline engine or the electric motor acting as a generator when the car slows. Finally, manufacturing has seen an ongoing revolution that would dazzle even Henry Ford, with computers greatly shortening the time needed to design and test a car, and regiments of industrial robots doing machining and assembly work with a degree of speed, strength, precision, and endurance that no human can match.

Back in 1923 a national magazine declared that the automobile had "outrun the dreamers, confounded the prophets, and amazed the world." True enough—and that was just the beginning.

Perspective

Donald E. Petersen

President and Chairman/CEO
Ford Motor Company
1980–1990

Through continuous improvement and the ingenious application of new technology, the automobile reconfirmed and updated its status as a triumph of engineering throughout the 20th century. I was fortunate to witness and participate in one of the most significant stages of this ongoing transformation. When I joined the industry in 1949, automobiles were still literally just mechanical objects. By the time I retired 40 years later they had become complex electronic devices on wheels.

The first semiconductor computer chip went onboard in the mid-1970s. Before long, microprocessors were improving just about every aspect of the vehicle—emissions, fuel economy, safety, security, engine and transmission performance, ride and handling, even seat positioning. Electronics also transformed cars and trucks into mobile entertainment and communication centers.

During my years in the industry, there were other profound changes that challenged the engineering community. Government regulations in the 1960s mandated cleaner, safer, more fuel-efficient vehicles in a rapid time frame. In the 1970s increasing global competition brought a surge of high-quality, low-cost competitive products from overseas into the United States.

American manufacturers were painfully reminded of the fundamental importance of quality and took on the challenge of making our vehicles world class once again. We had to relearn some of the lessons of manufacturing excellence, such as the critical need for standardized, precision-made parts, that we had taught the world at the beginning of the century.

Shortly after I became president of Ford Motor Company I saw a television program—*If Japan Can, Why Can't We?*—that described Toyota's success in improving quality and gave W. Edwards Deming major credit for Toyota's success. I met with Ed Deming and liked his ideas for improving quality and his emphasis on the importance of people. Peter Drucker also was involved in the Japanese resurgence and emphasized people. For me personally these two men were a major help in forming the ways we worked together to improve product quality.

We began engaging people at all levels and in all functions in what became known as the employee involvement movement in the 1980s. Encouraging everyone to participate and channeling individual and team efforts toward well-defined common goals produced remarkable results. As measured by owner-reported "things gone wrong," vehicle quality improved more than 60 percent from 1980 to 1987 models. Breakthrough products such as the radically aerodynamic 1986 Ford Taurus helped convince consumers that American manufacturers could not only decrease defects but also increase design and engineering attributes that maximized product appeal.

Today the automobile remains the most voracious consumer of new technology of any product in the marketplace. And promising new technological developments, such as the use of fuel cells as a power source, will undoubtedly keep the automobile on the leading edge of technology in the 21st century. But whatever shape the technology takes and wherever it leads us, we would do well to remember the lesson we learned in the 1980s to honor and encourage the people behind new ideas.

Henry Ford shows off one of his late model cars in front of the Ford Motor Company.

Automobiles may not have been born in the 20th century, but they were not yet out of diapers when it began. Even after Gottlieb Daimler and Karl Benz introduced their improved four-stroke internal combustion engine, autos in both the United States and Europe were still poking along at a few miles an hour (a sizeable proportion of them still running on electricity or steam). They could boast no battery starter, roof, or windows and were priced only for the rich. Then Henry Ford fine-tuned the mass production of his Tin Lizzie and the world drove off into the age of affordable transportation—forever altering our notions of place, distance, and community.

1901 The Olds automobile factory starts production in Detroit. Ransom E. Olds contracts with outside companies for parts, thus helping to originate mass production techniques. Olds produces 425 cars in its first year of operation, introducing the three-horsepower "curved-dash" Oldsmobile at $650. The car is a success; Olds is selling 5,000 units a year by 1905.

C. L. Horock designs the "telescope" shock absorber, using a piston and cylinder fitted inside a metal sleeve, with a one-way valve built into the piston. As air or oil moves through the valve into the cylinder, the piston moves freely in one direction but is resisted in the other direction by the air or oil. The result is a smoother ride and less lingering bounce. The telescope shock absorber is still used today.

1902 Standard drum brakes are invented by Louis Renault. His brakes work by using a cam to force apart two hinged shoes. Drum brakes are improved in many ways over the years, but the basic principle remains in cars for the entire 20th century; even with the advent of disk brakes in the 1970s, drum brakes remain the standard for rear wheels.

1908 Henry Ford begins making the Model T. First-year production is 10,660 cars.

Cadillac is awarded the Dewar Trophy by Britain's Royal Automobile Club for a demonstration of the precision and interchangeability of the parts from which the car is assembled. Mass production thus makes more headway in the industry.

1908 William Durant forms General Motors. His combination of car producers and auto parts makers eventually becomes the largest corporation in the world.

1911 Charles Kettering introduces the electric starter. Until this time engines had to be started by hand cranking. Critics believed no one could make an electric starter small enough to fit under a car's hood yet powerful enough to start the engine. His starters first saw service in 1912 Cadillacs.

1913 Ford Motor Company develops the first moving assembly line for automobiles. It brings the cars to the workers rather than having workers walk around factories gathering parts and tools and performing tasks. Under the Ford assembly line process, workers perform a single task rather than master whole portions of automobile assembly. The Highland Park, Michigan, plant produces 300,000 cars in 1914. Ford's process allows it to drop the price of its Model T continually over the next 14 years, transforming cars from unaffordable luxuries into transportation for the masses.

1914 Dodge introduces the first car body made entirely of steel, fabricated by the Budd Company. The Dodge touring car is made in Hamtramck, Michigan, a suburb of Detroit.

1919 The Hispano-Suiza H6B, a French luxury car, demonstrates the first single foot pedal to operate coupled four-wheel brakes. Previously drivers had to apply a hand brake and a foot brake simultaneously.

1922 The Duesenberg, made in Indianapolis, Indiana, is the first American car with four-wheel hydraulic brakes, replacing ones that relied on the pressure

William Durant (left) and Charles Kettering played key roles in the auto industry's early years.

of the driver's foot alone. Hydraulic brakes use a master cylinder in a hydraulic system to keep pressure evenly applied to each wheel of the car as the driver presses on the brake pedal.

1926 Francis Wright Davis uses a Pierce-Arrow to introduce the first power steering system. It works by integrating the steering linkage with a hydraulics system.

1931 Mercedes-Benz introduces the first modern independent front suspension system, giving cars a smoother ride and better handling. By making each front wheel virtually independent of the other though attached to a single axle, independent front suspension minimizes the transfer of road shock from one wheel to the other.

1934 The French automobile Citroën Traction Avant is the first successful mass-produced front-wheel-drive car. Citroën also pioneers the all-steel unitized body-frame structure (chassis and body are welded together). Audi in Germany and Cord in the United States offer front-wheel drive.

1935 A Delaware company uses a thermal interrupter switch to create flashing turn signals. Electricity flowing through a wire expands it, completing a circuit and allowing current to reach the lightbulb. This short-circuits the wire, which then shrinks and terminates contact with the bulb but is then ready for another cycle. Transistor circuits begin taking over the task of thermal interrupters in the 1960s.

1939 The Nash Motor Company adds the first air conditioning system to cars.

1940 Karl Pabst designs the Jeep, workhorse of WWII. More than 360,000 are made for the Allied armed forces.

Oldsmobile introduces the first mass-produced, fully automatic transmission.

1950s Ralph Teeter, a blind man, senses by ear that cars on the Pennsylvania Turnpike travel at uneven speeds, which he believes leads to accidents. Through the 1940s he develops a cruise control mechanism that a driver can set to hold the car at a steady speed. Unpopular when generally introduced in the 1950s, cruise control is now standard on more than 70 percent of today's automobiles.

1960s Automakers begin efforts to reduce harmful emissions, starting with the introduction of positive crankcase ventilation in 1963. PCV valves route gases back to the cylinders for further combustion. With the introduction of catalytic converters in the 1970s, hydrocarbon emissions are reduced 95 percent by the end of the century compared to emissions in 1967.

1966 An electronic fuel injection system is developed in Britain. Fuel injection delivers carefully controlled fuel and air to the cylinders to keep a car's engine running at its most efficient.

1970s Fuel prices escalate, driving a demand for fuel-efficient cars, which increases the sale of small Japanese cars.

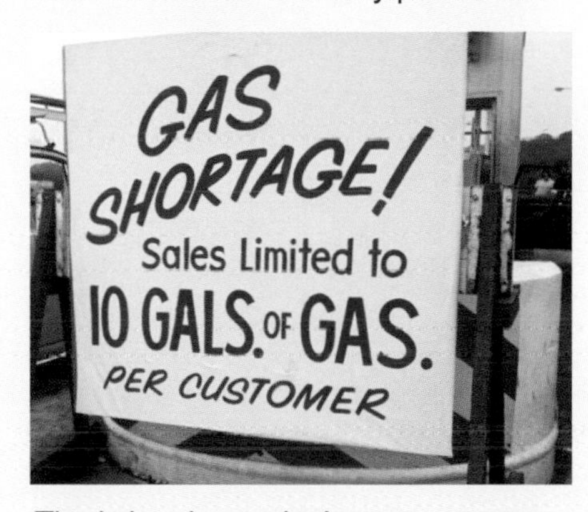

This helps elevate the Japanese automobile industry to one of the greatest in the world.

1980s The Japanese popularize "just in time" delivery of auto parts to factory floors, thus reducing warehousing costs. They also popularize statistical process control, a method developed but not applied in the United States until the Japanese demonstrate how it improves quality.

Airbags, introduced in some models in the 1970s, become standard in more cars.

Originally installed only on the driver's side, they begin to appear on the front passenger side as well.

1985 The Lincoln becomes the first American car to offer an antilock braking system (ABS), which is made by Teves of Germany. ABS uses computerized sensing of wheel movement and hydraulic pressure to each wheel to adjust pressure so that the wheels continue to move somewhat rather than "locking up" during emergency braking.

1992 Passage of the federal Energy Policy Act of 1992 encourages alternative-fuel vehicles. These include automobiles run with mixtures of alcohols and gasoline, with natural gas, or by some combination of conventional fuel and battery power.

1997 Cadillac is the first American carmaker to offer automatic stability control, increasing safety in emergency handling situations.

By 2000, auto manufacturers were introducing fuel efficient and environmentally friendly hybrid vehicles, like this Volkswagen Passat, with both a fuel engine and an electric engine.

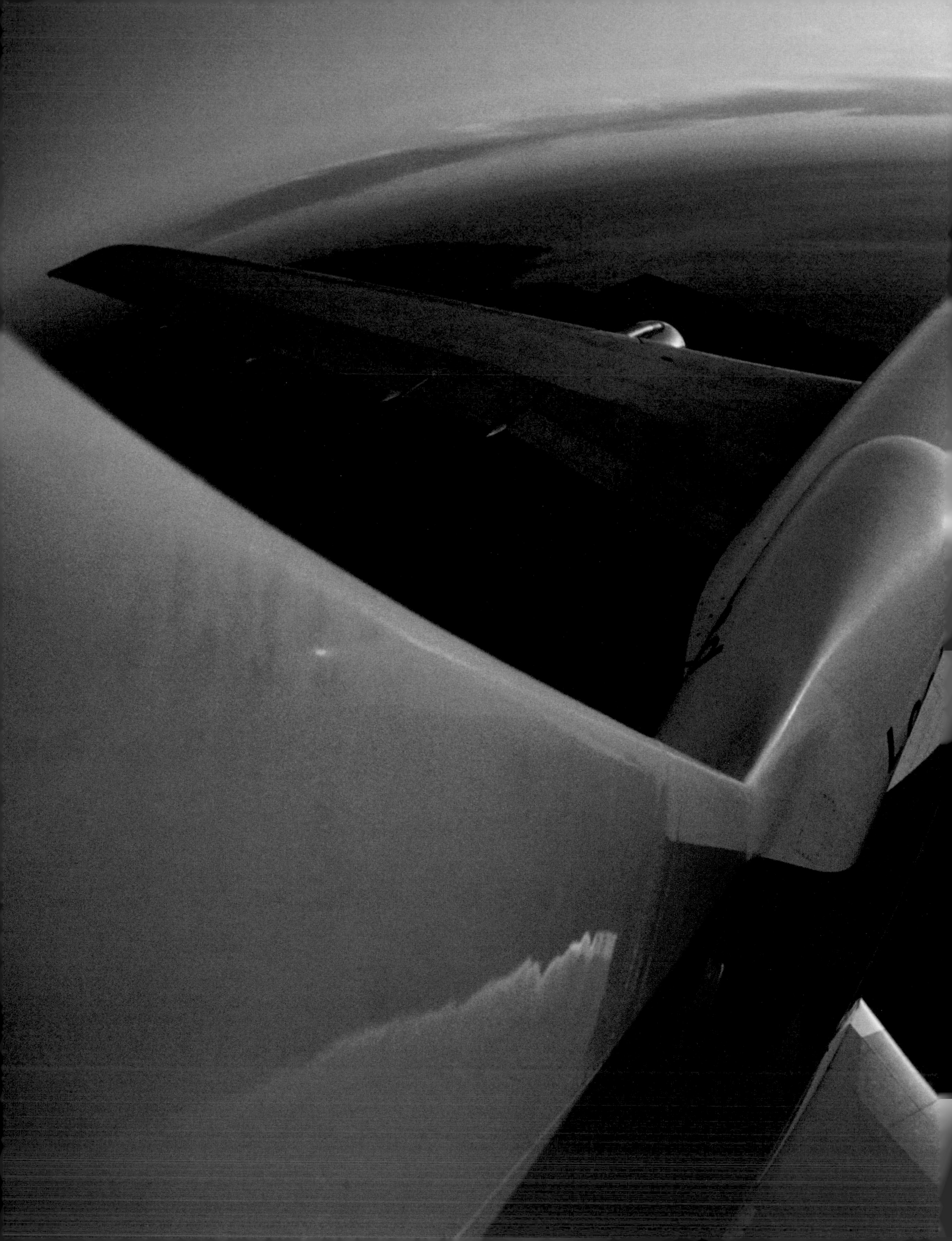

AIRPLANE

Not a single human being had ever flown a powered aircraft when the 20th century began. By century's end, flying had become relatively common for millions of people, and some were even flying through space. The first piloted, powered, controlled flight lasted 12 seconds and carried one man 120 feet. Today, nonstop commercial flights lasting as long as 15 hours carry hundreds of passengers halfway around the world.

Bonfires and beacons showed the way for early tentative transcontinental flights in the 1920s. Now complex computerized systems of navigation and air traffic control manage skies filled with as many as 50,000 planes a day over the United States. Thanks to the airplane, much about the world has changed forever, not only its commerce and wars but also its dimensions. Now that it takes only a few hours to cross a continent or an ocean, the globe has grown small indeed. And propelling virtually every one of aviation's great leaps—from the first flight to the fastest jet—has been the solving of complex engineering problems.

The first of aviation's hurdles—getting an airplane off the ground with a human controlling it in a sustained flight—presented a number of distinct engineering problems: structural, aerodynamic, control, and propulsion. As the 19th century came to a close, researchers on both sides of the Atlantic were tinkering their way to solutions. But it was a fraternal pair of bicycle builders from Ohio who achieved the final breakthrough.

Orville and Wilbur Wright learned much from the early pioneers, including Paris-born Chicago engineer Octave Chanute. In 1894, Chanute had compiled existing information on aerodynamic experiments and suggested the next steps. The brothers also benefited from the work during the 1890s of Otto Lilienthal, a German inventor who had designed and flown several different glider models. Lilienthal, and some others, had crafted wings that were curved, or cambered, on top and flat underneath, a shape that created lift by decreasing the air pressure over the top of the wing and increasing the

With Wilbur Wright looking on, Orville Wright handles the controls on the first sustained, powered, and piloted flight (opposite) on December 17, 1903. A century later, vast fleets of metal-skinned aircraft, carrying hundreds of passengers, take off and land at busy airports all around the world (left).

air pressure on the bottom of the wing. By experimenting with models in a wind tunnel, the Wrights gathered more accurate data on cambered wings than the figures they inherited from Lilienthal, and then studied such factors as wing aspect ratios and wingtip shapes.

Lilienthal and others had also added horizontal surfaces behind each wing, called elevators, that controlled the glider's pitch up and down, and Lilienthal used a vertical rudder that could turn his glider right or left. But the third axis through which a glider could rotate—rolling to either left or right—remained problematic. Most experimenters of the day thought roll was something to be avoided and worked to offset it, but Wilbur Wright, the older of the brothers, disagreed. Wilbur's experience with bicycles had taught him that a controlled roll could be a good thing. Wilbur knew that when cyclists turned to the right, they also leaned to the right, in effect "rolling" the bicycle and thereby achieving an efficient, controlled turn. Wilbur realized that creating a proper turn in a flying machine would require combining the action of the rudder and some kind of roll control. While observing the flight of turkey vultures gliding on the wind, Wilbur decided that by twisting the wings—having the left wing twist upward and the right wing twist downward, or vice versa—he would be able to control the roll. He rigged a system that linked the twisting, called wing warping, to the rudder control. This coordination of control proved key. By 1902 the Wrights were flying gliders with relative ease, and a year later, having added an engine they built themselves, Orville made that historic first powered flight—on December 17, 1903.

As happens so often in engineering, however, the first solution turned out not to be the best one. A crucial improvement soon emerged from a group of aviation enthusiasts headed by famed inventor Alexander Graham Bell. The Wrights had shared ideas with Bell's group, including a young engine builder named Glenn Curtiss, who was soon designing his own airplanes. One of the concepts was a control system that replaced wing warping with a pair of horizontal flaps called ailerons, positioned on each wing's trailing edge. Curtiss used ailerons, which made rolls and banking turns mechanically simpler; indeed, aileron control eventually became the standard. But the Wrights were furious with Curtiss, claiming patent infringement on

On July 4, 1908, in Hammondsport, New York, Glenn Curtiss makes the first public, powered flight of more than 1 kilometer (3,261 feet), winning a trophy offered by Scientific American *magazine.*

his part. The ensuing legal battle dragged on for years, with the Wrights winning judgments but ultimately getting out of the business and leaving it open to Curtiss and others.

Then a more brutal war entered the picture, and the major powers were soon vying for control of the air. World War I's flying machines, which served at first only for reconnaissance, were soon turned into offensive weapons, shooting at each other and dropping bombs on enemy positions. The fighting in the skies was matched by a fierce competition among aviation engineers on both sides. When one side built more powerful engines, the other countered with sleeker streamlining; the development of bigger planes that could drop heavier bombs was countered by improved maneuverability to get the upper hand in dogfights. With each adjustment that worked, aviation took another step forward.

Some of the most significant developments involved the airframe itself. The standard construction of fabric stretched over a wood frame and wings externally braced with wire was notoriously vulnerable in the heat of battle. Some designers had experimented with metal sheathing, but the real breakthrough came from the desk of a German professor of mechanics named Hugo Junkers. In 1917 he introduced an all-metal airplane, the Junkers J4, that turned out to be a masterpiece of engineering. Built almost entirely of a relatively lightweight aluminum alloy called duralumin, it also featured steel armor around the fuel tanks, crew, and engine and strong, internally braced cantilevered wings. The J4 was virtually indestructible, but it came along too late in the war to have much effect on the fighting.

Postal service workers transfer mail bags from a mail wagon to a waiting aircraft for the inaugural flight of airmail service between Philadelphia and New York in May 1918.

In the postwar years, however, Junkers and others made further advances based on the J4's features. For one thing, cantilevering made monoplanes—which produce less drag than biplanes—more practical. Using metal also led to what is known as stressed-skin construction, in which the airframe's skin itself supplies structural support, reducing weighty internal frameworking. New, lighter alloys also added to structural efficiency, and wind tunnel experiments led to more streamlined fuselages. Step by step, a more modern-looking airplane was taking shape.

At the same time, a brand-new role was emerging. As early as 1911, airplanes had been used to fly the mail, and it didn't take long for the business world to realize that airplanes could move people as well. The British introduced a cross-channel service in 1919 (as did the French about the same time), but its passengers must have wondered if flying was really worth it. They traveled two to a plane, crammed together facing each other in the converted gunner's cockpit of the De Havilland 4; the engine noise was so loud that they could communicate with each other or with the pilot only by passing notes. Clearly, aircraft designers had to start paying attention to passenger comfort.

The result was a steady accumulation of improvements, fostered by the likes of American businessman Donald Douglas, who founded his own aircraft company in California in 1920. By 1933 he had introduced an airplane of truly revolutionary appeal, the DC-1 (for Douglas Commercial). Its 12-passenger cabin included heaters and soundproofing, and the all-metal airframe was among the strongest ever built. By 1936 Douglas's engineers had produced one of the star performers in the whole history of aviation, the DC-3. This shiny, elegant workhorse incorporated just about every aviation-related engineering advance of the day, including almost completely enclosed engines to reduce drag, new types of wing flaps for better control, and variable-pitch propellers, whose angle could be altered in flight to improve efficiency and thrust. The DC-3 was roomy enough for 21 passengers and could also be configured with sleeping berths for long-distance flights. Passengers came flocking. By 1938, fully 80 percent of U.S. passengers were flying in DC-3s and a dozen foreign airlines had adopted the planes. DC-3s are still in the air today, serving in a variety of capacities, including cargo and medical relief, especially in developing countries.

Improvements in the mechanisms of control and in airframe construction continued, driven by commercial considerations and, with the advent of World War II, military demands for bigger bombers and faster and more maneuverable fighters. Aviation's next great leap forward, however, was all about power and speed. In

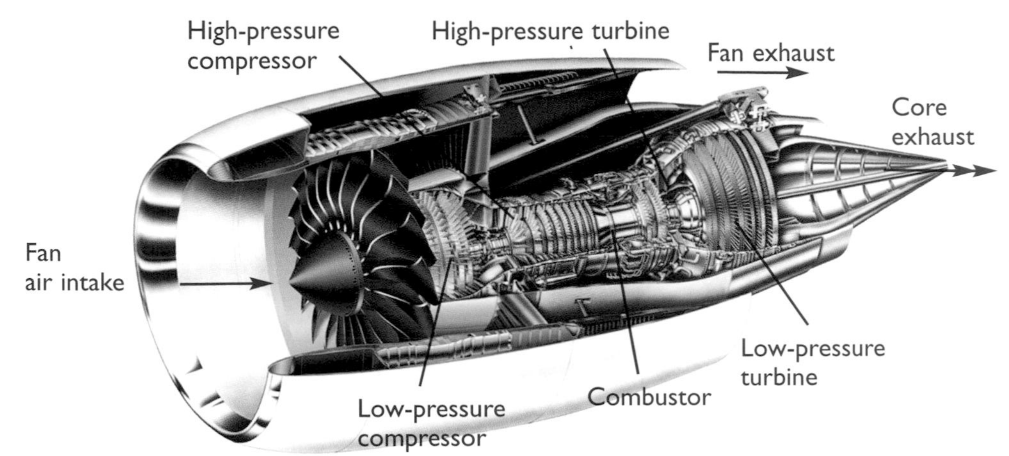

Powerful thrust. Jet engines work on the principle of Newton's third law of physics: for every action there is an equal and opposite reaction. The opposite reaction in this case is called thrust. A fan sucks air into the front of the engine, where the blades of a compressor raise the air pressure. The compressed air is forced into a combustion chamber, sprayed with fuel, and ignited. As the burning gases expand, they pass through a turbine and blast out through the nozzle at the back of the engine and, along with the fan exhaust, thrust the engine—and the aircraft—forward. The turbine, another set of blades, rotates the turbine shaft, which in turn rotates the compressor, bringing in a fresh supply of air. In the turbofan engine shown here, a combination of high- and low-pressure compressors and turbines combine maximum propulsive efficiency with low fuel consumption.

1929 a 21-year-old British engineer named Frank Whittle had drawn up plans for an engine based on jet propulsion, a concept introduced near the beginning of the century by a Frenchman named Rene Lorin. German engineer Hans von Ohain followed with his own design, which was the first to prove practical for flight. In August 1939 he watched as the first aircraft equipped with jet engines, the Heinkel HE 178, took off.

In 1942 Adolf Galland—director general of fighters for the Luftwaffe, veteran of the Battle of Britain, and one of Germany's top aces—flew a prototype of one of the world's first jets, the Messerschmitt ME 262. "For the first time, I was flying by jet propulsion and there was no torque, no thrashing sound of the propeller, and my jet shot through the air," he commented. "It was as though angels were pushing." As Adolf Galland and others soon realized, the angels were pushing with extraordinary speed. The ME 262 that Galland flew raced through the air at 540 miles per hour, some 200 mph faster than its nearest rivals equipped with piston-driven engines. It was the first operational jet to see combat, but came too late to affect the outcome of the war. Shortly after the war, Captain Chuck Yeager of the U.S. Air Force set the bar even higher, pushing an experimental rocket-powered plane, the X-1, past what had once seemed an unbreachable barrier: the speed of sound. This speed varies with air temperature and density but is typically upward of 650 mph. Today's high performance fighter jets can routinely fly at two to three times that rate.

The jet engine had a profound impact on commercial aviation. As late as the 1950s transatlantic flights in propeller-driven planes were still an arduous affair lasting more than 15 hours. But in the 1960s aircraft such as Boeing's classic 707, equipped with four jet engines, cut that time in half. The U.S. airline industry briefly flirted with a plane that could fly faster than sound, and the French and British achieved limited commercial success with their own supersonic bird, the Concorde, which made the run from New York to Paris in a scant three and a half hours. Increases in speed certainly pushed commercial aviation along, but the business of flying was also demanding bigger and bigger airplanes. Introduced in 1969, the world's first jumbo jet, the Boeing 747, still holds the record of carrying 547 passengers and crew.

Like Gulliver among the Lilliputians, the 700,000-pound Boeing 747 dwarfs its admiring public.

AIR TRAFFIC CONTROL

ENGINEERING THE SKIES

It's a marvel of systems engineering that proves itself every day: controlling the traffic flow of tens of thousands of airplanes at takeoff and landing and during flight. The whole process involves not only precision management of the congested skies and runways at airports but also sophisticated coordination of handoffs as aircraft pass from one zone of control to the next en route.

Air traffic controllers and the familiar control towers at airports are only part of the story. The system also depends on automated tracking and communications devices and centers and a carefully plotted division of the airspace equivalent to the roads and highways of earthbound traffic, with one crucial exception: controllers and systems must keep track of three dimensions (bottom).

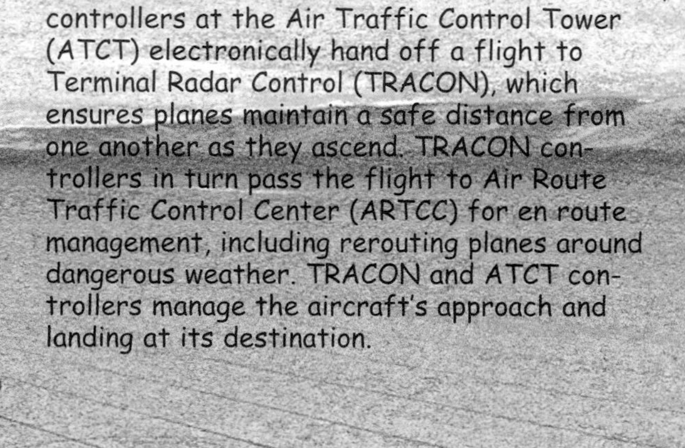

FROM TAKEOFF TO LANDING. After takeoff, controllers at the Air Traffic Control Tower (ATCT) electronically hand off a flight to Terminal Radar Control (TRACON), which ensures planes maintain a safe distance from one another as they ascend. TRACON controllers in turn pass the flight to Air Route Traffic Control Center (ARTCC) for en route management, including rerouting planes around dangerous weather. TRACON and ATCT controllers manage the aircraft's approach and landing at its destination.

MANAGING THE AIRSPACE. The airspace is divided into several different zones, each managed by different sets of controllers. Airport controllers manage takeoffs and landings in each airport's local airspace (orange). The airspace for ascending and descending flights (yellow) typically extends to 26,000 feet. Aircraft at cruising altitude travel in designated flight corridors (blue and green) that are continually adjusted to keep flights safely apart and to route them around turbulence or severe weather.

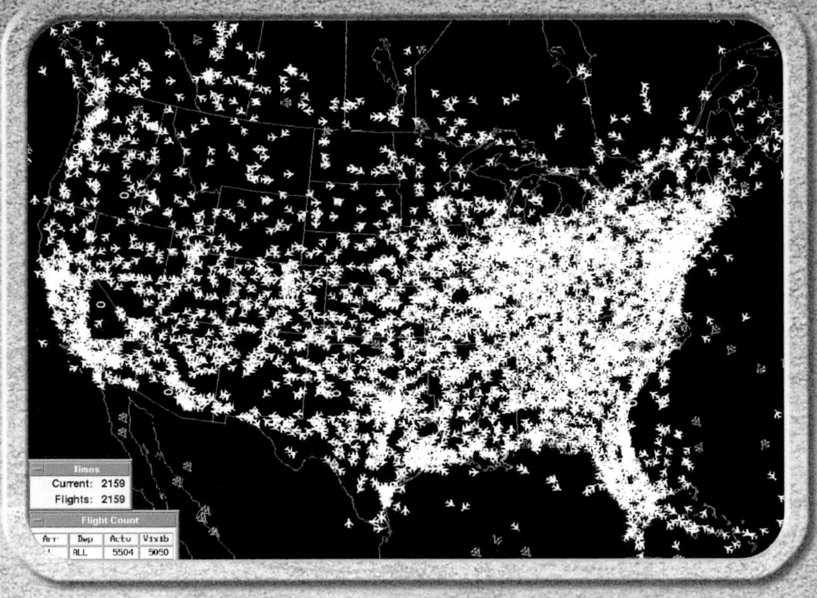

THE CROWDED SKIES. The image at left, compiled from information sent by air traffic radar systems throughout the country, shows some 5,000 aircraft simultaneously in flight across the United States. The National Airspace System oversees and integrates the myriad local air traffic control systems, including more than 250 TRACON centers and 21 ARTCC facilities.

IN THE TOWER. Air traffic controllers in the airport tower monitor weather and flight-plan information and clear individual aircraft for takeoff, coordinating with ground controllers who manage the planes' progress from gate to taxiway. Controllers in the tower also clear incoming planes for landing. The largest U.S. airports handle thousands of takeoffs and landings a day.

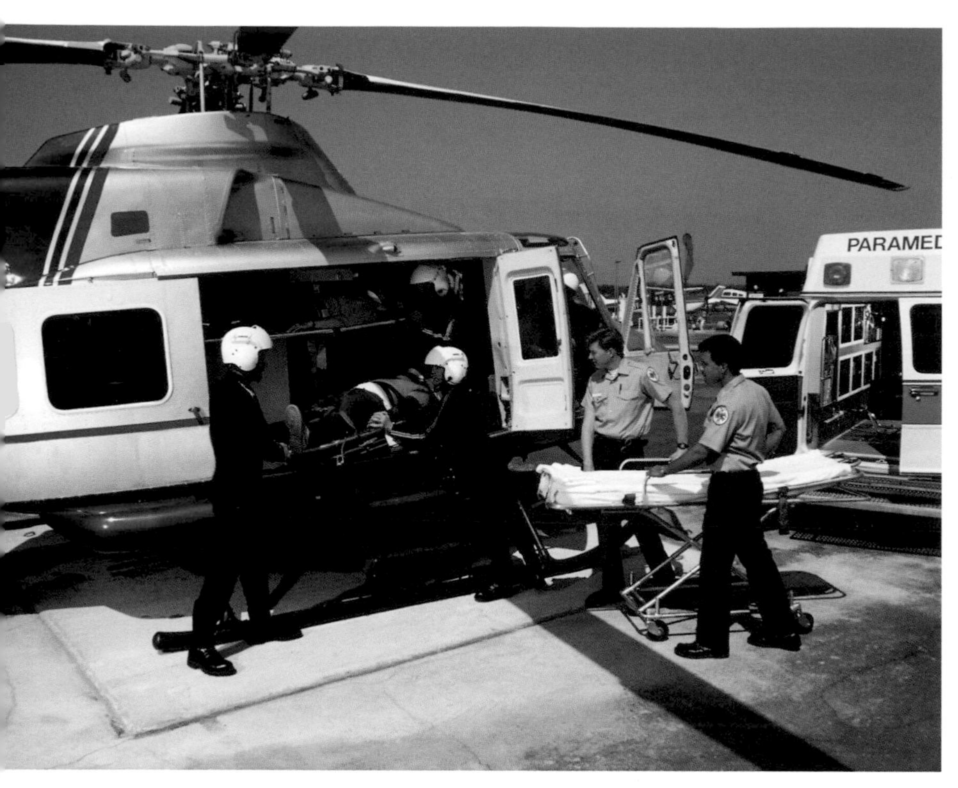

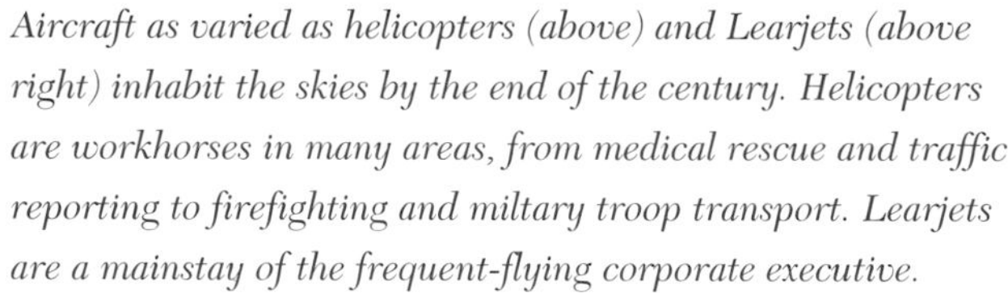

Aircraft as varied as helicopters (above) and Learjets (above right) inhabit the skies by the end of the century. Helicopters are workhorses in many areas, from medical rescue and traffic reporting to firefighting and miltary troop transport. Learjets are a mainstay of the frequent-flying corporate executive.

Building such behemoths presented few major challenges to aviation engineers, but in other areas of flight the engineering innovations have continued. As longer range became more important in commercial aviation, turbojet engines were replaced by turbofan engines, which greatly improved propulsive efficiency by incorporating a many-bladed fan to provide bypass air for thrust along with the hot gases from the turbine. Engines developed in the last quarter of the 20th century further increased efficiency and also cut down on air pollution.

Computers entered the cockpit and began taking a role in every aspect of flight. So-called fly-by-wire control systems, for example, replaced weighty and complicated hydraulic and mechanical connections and actuators with electric motors and wire-borne electrical signals. The smaller, lighter electrical components made it easier to build redundant systems, a significant safety feature. Other innovations also aimed at improving safety. Special collision avoidance warning systems onboard aircraft reduce the risk of midair collisions, and Doppler weather radar on the ground warns of deadly downdrafts known as wind shear, protecting planes at the most vulnerable moments of takeoff and landing.

Another area of flying advanced alongside commercial and military aviation in the last few decades of the century. General aviation, the thousands of private planes and business aircraft flown by more than 650,000 pilots in the United States alone, actually grew to dwarf commercial flight. Of the 19,000 airports registered in the United States, fewer than 500 serve commercial craft. In 1999 general aviation pilots flew 31 million hours compared with 2.7 million for their commercial colleagues. Among the noteworthy developments in this sphere was Bill Lear's Model 23 Learjet, introduced in 1963. It brought the speed and comfort of regular passenger aircraft to business executives, flew them to more airports, and could readily adapt to their schedules instead of the other way around. General aviation is also the stomping ground of innovators such as Burt Rutan, who took full advantage of developments in composite materials (see *High Performance Materials*) to design the sleek *Voyager*, so lightweight and aerodynamic that it became the first aircraft to fly nonstop around the world without refueling.

In today's world, air travel may have lost some of the original glamour that once prompted passengers to dress their best for any flight. But the miracle of flying through the air is still there to be seen, perhaps best in the eyes of a child looking down for the first time on a field of clouds. The dream of flight, a dream turned into reality by the precise work of engineers, continues to enchant.

Perspective

Kent Kresa

Chairman
Northrop Grumman Corporation

I'll never forget my excitement as I watched the maiden flight of the B-2 bomber from a hot tarmac in the desert town of Palmdale, California, in 1989. The flight was the culmination of a dream by the late aviation pioneer Jack Northrop, who had first proposed the flying-wing design more than 50 years earlier. Jack Northrop had developed flying-wing bombers, but none had been widely adopted. With the B-2, however, his dream finally came of age. Watching the bomber go through its maneuvers, I couldn't help but think that this was the future of aviation, a chevron-like structure in which every part of the aircraft contributes to lift. Moreover, the bomber employed stealth characteristics—it was fortunate that the physics of radar reflectivity fit nicely with the physics of flight.

Even more than its aeronautical design, the B-2's electronics pointed toward the future. A triumph of integrated systems and circuitry, the B-2 represented a milestone in the growing dominance of electronics in aerospace engineering. This might seem strange to people not familiar with the industry. In fact, it would have seemed strange to me when I began my career in aerospace engineering in the 1960s, when the focus was on engines and structure. At the time, jet propulsion and wide-body fuselages were in the process of changing aviation dramatically by shortening the time span of air travel and making it affordable for the general public.

But since then most advances in aviation have come about as a result of electronics and its two prodigious offspring—computers and communications. I got a glimpse of the tremendous potential of these areas while I was working at the U.S. government's Defense Advanced Research Projects Agency (DARPA) in the late 1960s and early 1970s. DARPA had developed a prototype system called the ARPANET (Advanced Research Projects Agency Network), which eventually evolved into the Internet. I can't claim to have foreseen how this new medium would blossom into one of the foundations of globalism, but even then I was struck by how a network could make individual computers substantially more useful. It occurred to me that at some future point everything would be "netting," even aircraft.

When I turned to management, "netting" became a major focus of my career. It was also the driving force in many other aerospace careers as well, so that by the end of the 20th century the industry had been reconfigured by advances in electronic systems and networks. Today, for instance, U.S. military aircraft are connected by data links to external sensors and computer processing that guide them to enemy targets. Even the munitions that aircraft launch can be redirected in midflight by networks that feed them continuous real-time information. Commercial aviation grows ever more dependent on electronic networks. With air traffic expected to double by 2015, new air traffic control systems will make greater use of satellite navigation to accommodate the increase. Similarly, air cargo transport is developing radio-based systems that can track individual freight items through every point of the supply chain.

In all of these cases the ability to integrate networked systems into the operation of aircraft is setting new standards for modern-day flight. The military can achieve a more accurate and powerful impact with fewer resources. The commercial system can offer greater transportation and logistical capacity at lower costs.

As we look forward to the next 100 years of aviation, we can expect electronics to continue leading the way in innovation. Certainly there will be additional breakthroughs in aircraft design, such as flying-wing structures developed for commercial transportation and morphing wings that change their shape in flight. But it will be mainly the flight of electrons that pushes the envelope of aerospace engineering in ways we can only dream of today.

AIRPLANE TIMELINE

Efforts to tackle the engineering problems associated with powered flight began well before the Wright brothers' famous trials at Kitty Hawk. In 1804 an English baronet, Sir George Cayley, launched modern aeronautical engineering by studying the behavior of solid surfaces in a fluid stream and flying the first successful winged aircraft of which we have any detailed record. And of course Otto Lilienthal's aerodynamic tests in the closing years of the 19th century influenced a generation of aeronautical experimenters. In the 20th century, advances in aeronautical engineering soon had us soaring in safety and comfort across all the continents and oceans.

1901 Samuel Pierpont Langley builds a gasoline-powered version of his tandem-winged "Aerodromes," the first successful flying model to be propelled by an internal combustion engine. As early as 1896 he launches steam-propelled models with wingspans of up to 15 feet on flights of more than half a mile.

1903 Wilbur and Orville Wright of Dayton, Ohio, complete the first four sustained flights with a powered, controlled airplane at Kill Devil Hills, 4 miles south of Kitty Hawk, North Carolina. On their best flight of the day, Wilbur covers 852 feet over the ground in 59 seconds. In 1905 they introduce the Flyer, the world's first practical airplane.

1904 German professor Ludwig Prandtl presents one of the most important papers in the history of aerodynamics, an eight-page document describing the concept of a fixed "boundary layer," the molecular layer of air on the surface of an aircraft wing. Over the next 20 years Prandtl and his graduate students pioneer theoretical aerodynamics.

1910 Eugene Ely pilots a Curtiss biplane on the first flight to take off from a ship. In November he departs from the deck of a cruiser anchored in Hampton Roads, Virginia, and lands onshore. In January 1911 he takes off from shore and lands on a ship anchored off the coast of California. Hooks attached to the plane's landing gear snag a series of ropes rigged across the deck *(right)*, a primitive version of the system of arresting gear and safety barriers used on modern aircraft carriers.

1914 Lawrence Sperry demonstrates an automatic gyrostabilizer at Lake Keuka, Hammondsport, New York. A gyroscope linked to sensors keeps the craft level and traveling in a straight line without aid from the human pilot. Two years later Sperry and his inventor father, Elmer, add a steering gyroscope to the stabilizer gyro and demonstrate the first "automatic pilot."

1914-1918 During World War I, the requirements of higher speed, higher altitude, and greater maneuverability drive dramatic improvements in aerodynamics, structures, and control and propulsion system design.

1915 Congress charters the National Advisory Committee for Aeronautics, a federal agency to spearhead advanced aeronautical research in the United States.

1917 Hugo Junkers, a German professor of mechanics introduces the Junkers J4, an all-metal airplane built largely of a relatively lightweight aluminum alloy called duralumin.

1918 The U. S. Postal Service inaugurates airmail service from Polo Grounds in Washington, D.C., on May 15. Two years later, on February 22, 1920, the first transcontinental airmail service arrives in New York from San Francisco in 33 hours and 20 minutes, nearly 3 days faster than mail delivery by train.

1919 Britain and France introduce passenger service across the English Channel, flying initially between London and Paris.

U.S. Navy aviators in Curtiss NC-4 flying boats, led by Lt. Cdr. Albert C. Read, make the first airplane crossing of the North Atlantic, flying from Newfoundland to London with stops in the Azores and Lisbon. A few months later British Capt. John Alcock and Lt. Albert Brown make the first nonstop transatlantic flight, from Newfoundland to Ireland.

1925-1926 The introduction of a new generation of lightweight, air-cooled radial engines revolutionizes aeronautics, making bigger, faster planes possible.

1927 On May 21, Charles Lindbergh completes the first nonstop solo flight across the Atlantic, traveling 3,600 miles from New York to Paris in a Ryan monoplane named the *Spirit of St. Louis*. On June 29, Albert Hegenberger and Lester Maitland complete the first flight from Oakland, California, to Honolulu, Hawaii. At 2,400 miles it is the longest open-sea flight to date.

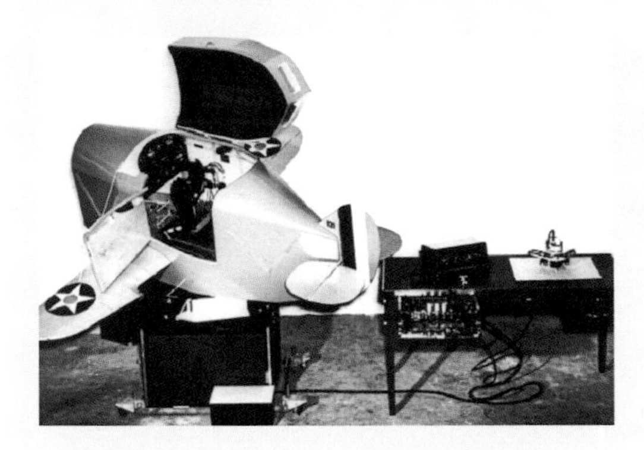

1928 Edwin A. Link introduces the Link Trainer, the first electromechanical flight simulator *(above)*. Mounted on a base that allows the cockpit to pitch, roll, and yaw, these ground-based pilot trainers have closed hoods that force a pilot to rely on instruments. The flight simulator is used for virtually all U.S. pilot training during WWII.

1933 In February, Boeing introduces the 247, a twin-engine 10-passenger monoplane that is the first modern commercial airliner. With variable-pitch propellers, it has an economical cruising speed and excellent takeoff. Retractable landing gear reduces drag during flight.

In that summer Douglas introduces the 12-passenger twin-engine DC-1, designed by aeronautical engineer Arthur Raymond for a contract with TWA. A key requirement is that the plane can take off, fully loaded, if one engine goes out. In September the DC-1 joins the TWA fleet, followed 2 years later by the DC-3, the first passenger airliner capable of making a profit for its operator without a postal subsidy. The DC-3's range of nearly 1,500 miles is more than double that of the Boeing 247. As the C-47 it becomes the workhorse of WWII.

1935 Pan American inaugurates the first transpacific mail service, between San Francisco and Manila, on November 22, and the first transpacific passenger service in October the following year. Four years later, in 1939, Pan Am and Britain's Imperial Airways begin scheduled transatlantic passenger service.

1937 Jet engines designed independently by Britain's Frank Whittle and Germany's Hans von Ohain make their first test runs. (Seven years earlier, Whittle, a young Royal Air Force officer, filed a patent for a gas turbine engine to power an aircraft, but the Royal Air Ministry was not interested in

developing the idea at the time. Meanwhile, German doctoral student Von Ohain was developing his own design.) Two years later, on August 27, the first jet aircraft, the Heinkel HE 178, takes off, powered by von Ohain's HE S-3 engine *(above)*.

1939 Russian emigre Igor Sikorsky develops the VS-300 helicopter for the U.S. Army, one of the first practical single-rotor helicopters.

1939-1945 A world war again spurs innovation. The British develop airplane-detecting radar just in time for the Battle of Britain. At the same time the Germans develop radiowave navigation techniques. Then both sides develop airborne radar, useful for attacking aircraft at night. German engineers produce the first practical jet fighter, the twin-engine ME 262, which flies at 540 miles per hour, and the 600-mph, rocket-powered Messerschmitt 163 Komet. In the United States, the Boeing Company modifies its B-17 into the high-altitude Flying Fortress. Later it makes the 141-foot-wingspan long-range B-29 Superfortress. In Britain the Instrument Landing System (ILS) for landing in bad weather is put into use in 1944.

1947 U.S. Air Force pilot Captain Charles "Chuck" Yeager becomes the fastest man alive when he pilots the Bell X-1 *(above)* faster than sound for the first time on October 14 over the town of Victorville, California.

1949 The prototype De Havilland Comet makes its first flight on July 27. Three years later the Comet starts regular passenger service as the first jet-powered commercial aircraft, flying between London and South Africa.

1950s Boeing makes the B-52 bomber. It has eight turbojet engines, intercontinental range, and a capacity of 500,000 pounds.

1952 Richard Whitcomb, an engineer at Langley Memorial Aeronautical Laboratory, discovers and experimentally verifies an aircraft design concept known as the area rule. A revolutionary method of designing aircraft to reduce drag and increase speed without additional power, the area rule is incorporated into the development of almost every American supersonic aircraft. He later invents winglets, which increase the lift-to-drag ratio of transport airplanes and other vehicles.

1963 The prototype Learjet 23 makes its first flight on October 7. Powered by two GE CJ610 turbojet engines, it is 43 feet long, with a wingspan of 35.5 feet, and can carry seven passengers (including two pilots) in a fully pressurized cabin. It becomes the first small jet aircraft to enter mass production, with more than 100 sold by the end of 1965.

1969 Boeing conducts the first flight of a wide-body, turbofan-powered commercial airliner, the 747, one of the most successful aircraft ever produced.

1976 The Concorde SST is introduced into commercial airline service by both Great Britain and France on January 21. It carries a hundred passengers at 55,000 feet and twice the speed of sound, making the London to New York run in 3.5 hours—half the time of subsonic carriers. But the cost per passenger-mile is high, ensuring that flights remain the privilege of the wealthy. After a Concorde accident kills everyone on board in July 2000, the planes are grounded for more than a year. Flights resume in November 2001, but with passenger revenue falling and maintenance costs rising, British Airways and Air France announce they will decommission the Concorde in October 2003.

1986 Using a carbon-composite material, aircraft designer Burt Rutan crafts *Voyager* for flying around the world nonstop on a single load of fuel. *Voyager* has two centerline engines, one fore and one aft, and weighs less than 2,000 pounds (fuel for the flight adds another 5,000 pounds). It is piloted by Jeana Yeager (no relation to test pilot Chuck Yeager) and Burt's brother Dick Rutan, who circumnavigate the globe (26,000 miles) nonstop in 9 days.

1990s Northrop Grumman develops the B-2 bomber, with a "flying wing" design. Made of composite materials rather than metal, it cannot be detected by conventional radar. At about the same time, Lockheed designs the F-117 stealth fighter, also difficult to detect by radar.

1995 Boeing debuts the twin-engine 777, the biggest two-engine jet ever to fly and the first aircraft produced through computer-aided design and engineering. Only a nose mockup was actually built before the vehicle was assembled—and the assembly was only 0.03 mm out of alignment when a wing was attached.

1996-1998 NASA teams with American and Russian aerospace industries in a joint research program to develop a second-generation supersonic airliner for the 21st century. The centerpiece is the

Tu-144LL, a first-generation Russian supersonic jetliner modified into a flying laboratory *(above)*. It conducts supersonic research comparing flight data with results from wind tunnels and computer modeling.

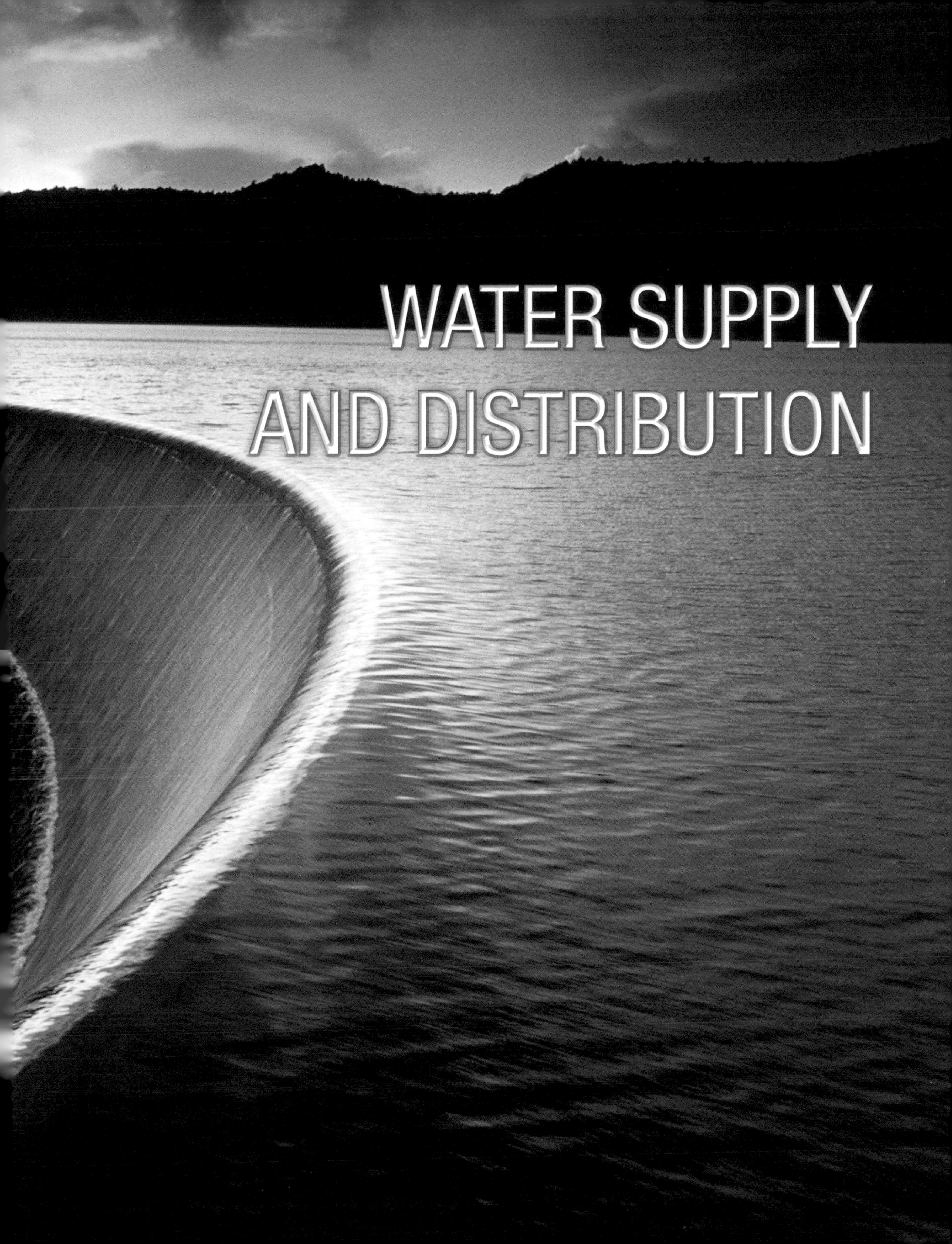

WATER SUPPLY AND DISTRIBUTION

At the beginning of the 20th century, in the United States and in many other countries, water was both greatly in demand and greatly feared. Cities across the nation were clamoring for more of it as their populations grew, and much of the West saw it as the crucial missing ingredient for development. At the same time, the condition of existing water supply systems was abysmal—and a direct threat to public health.

Indoor plumbing was rare, especially in the countryside, and in cities it was inadequate at best. Tenements housing as many as 2,000 people typically had not one bathtub. Raw sewage was often dumped directly into streets and open gutters; untreated industrial waste went straight into rivers and lakes, many of which were sources of drinking water; attempts to purify water consistently fell short, and very few municipalities treated wastewater at all.

As a result, waterborne diseases were rampant. Each year typhoid fever alone killed 25 of every 100,000 people (Wilbur Wright among them in 1912). Dysentery and diarrhea, the most common of the waterborne diseases, were the nation's third leading cause of death. Cholera outbreaks were a constant threat.

These challenges of both quantity and quality—to make sure there was enough water conveniently supplied wherever it was wanted and to make sure that it was safe both before and after use—fell to the nation's civil engineers. The results of their efforts speak for themselves: a deadly handful of waterborne diseases virtually eliminated not only in the United States but throughout the developed world; water distribution systems pumping a clean supply into homes, apartments, businesses, and factories and meeting the needs of tens of millions of people in burgeoning new cities and communities; and the rich potential of western lands realized in acre upon acre of irrigated crops. All told, what 20th-century engineers did to improve the water supply wrought a host of stunning transformations—in public health, in living standards, and in both urban and agricultural development.

As the century began, the most pressing task was to find better ways to make water clean. The impetus came from the discovery only a few years before the turn of the century that diseases such as typhoid and cholera were actually traced to microorganisms living in contaminated water. Treatment systems in place before then had focused on removing particulate matter suspended in water, typically by using various techniques that caused smaller particles to coagulate into heavier clumps that would settle out and by filtering the water through sand and other fine materials. Some harmful microorganisms were indeed removed in this way, but it wasn't good enough. One more step was necessary, and it involved the use of a chemical called chlorine. Known at the time for its bleaching power, chlorine also turned out to be a highly effective disinfectant, and it was just perfect for sterilizing water supplies: It killed a wide range of germs, persisted in residual amounts to provide ongoing protection, and left water free of disease and safe to drink.

In 1908, Jersey City, New Jersey, became the first municipality in the United States to institute chlorination of its water supply, followed that same year by the Bubbly Creek plant in Chicago. As had happened in European cities that had also introduced chlorination and other disinfecting techniques, death rates from waterborne diseases—typhoid in particular—began to plummet. By 1918 more than 1,000 American cities were chlorinating 3 billion gallons of water a day, and by 1923 the typhoid death rate had dropped by more than 90 percent from its level of only a decade before. By the beginning of World War II, typhoid, cholera, and dysentery were, for all practical purposes, nonexistent in the United States and the rest of the developed world.

As the benefits of treatment became apparent, the U.S. Public Health Service set standards for water purity that were continually revised as new contaminants were identified—among them industrial and agricultural chemicals as well as certain natural minerals such as lead, copper, and zinc that could be harmful at high levels. In modern systems, computerized detection devices now monitor water throughout the treatment process for traces of dangerous chemical pollutants and microbes; today's devices are so sophisticated that they can detect contaminants on the order of parts per trillion. More recently, the traditional process of coagulation, sedimentation, and filtration followed by chemical disinfection (see pages 42 and 43) has been complemented by other disinfecting processes, including both ultraviolet radiation and the use of ozone gas (first employed in France in the early 1900s).

One important way to improve water quality, of course, is to reduce the amount of contamination in the first place. As early as 1900, engineers in Chicago accomplished just that with an achievement of biblical proportions: They reversed the flow of the Chicago River. Chicago had suffered more than its fair share of typhoid and cholera outbreaks, a result of the fact that raw sewage and industrial waste were dumped directly into the Chicago River, which flowed into Lake Michigan, the source of the city's drinking

Outhouses like the three-story structure shown opposite contributed to contaminated water and waterborne diseases early in the 20th century, before the advent of indoor plumbing, adequate sewerage, and wastewater treatment plants (left).

By reversing the flow of the Chicago River, seen above in 1945, Chicago stopped contaminating Lake Michigan, the source of the city's drinking water, with industrial waste and raw sewage.

water. In a bold move, Rudolph Hering, chief engineer of the city's water supply system, developed a plan to dig a channel from the Chicago River to rivers that drained not into Lake Michigan but into the Mississippi. When the work was finished, the city's wastewater changed course with the river, and drinking water supplies almost immediately became cleaner.

City fathers in Chicago and elsewhere recognized that wastewater also would have to be treated, and soon engineers were developing procedures for handling wastewater that paralleled those being used for drinking water. It wasn't long before sewage treatment plants became an integrated part of what was fast becoming a complex water supply and distribution system, especially in major metropolitan centers. In addition to treatment facilities, dams, reservoirs, and storage tanks were being constructed to ensure supplies; mammoth tunnel-boring machines were leading the way in the building of major supply pipelines for cities such as New York; networks of water mains and smaller local distribution pipes were planned and laid throughout the country; and pumping stations and water towers were built to provide the needed pressure to support indoor plumbing. Seen in its entirety, it was a highly engineered piece of work.

As the nation's thirst continued to grow, even more was required of water managers—and nowhere more so than in California. The land of the gold rush and sunny skies, of rich alluvial soils and seemingly limitless opportunities, had one major problem—it didn't have nearly enough water. The case was the worst in Los Angeles, where a steadily increasing population and years of uneven rainfall were straining the existing supply from the Los Angeles River. To deal with the problem, the city formed its first official water department in 1902 and put just the right man in the job of superintendent and chief engineer. William Mulholland had moved to Los Angeles in the 1870s as a young man and had worked as a ditch tender on one of the city's main supply channels. In his new capacity he turned first to improving the existing water supply, adding reservoirs, enlarging the entire distribution network, and instituting the use of meters to discourage the wasting of water.

Three vast underground passages of the Catskill Aqueduct, shown here temporarily dry in 1935, suggest the huge volume of water—500 million gallons a day—the aqueduct delivers to Manhattan.

Throngs of people gather in November 1913 for the opening of the Los Angeles–Owens River Aqueduct, designed so that water flows by the force of gravity for 230 miles.

But Mulholland's vision soon reached further, and in 1905 citizens approved a $1.5 billion bond issue that brought his revolutionary plan into being. Work soon began on an aqueduct that would bring the city clear, clean water from the Owens River in the eastern Sierra Nevada, more than 230 miles to the north. Under Mulholland's direction, some 5,000 workers toiled on the project, which was deemed one of the most difficult engineering challenges yet undertaken in America. When it was completed, within the original schedule and budget, commentators marveled at how Mulholland had managed to build the thing so that the water flowed all the way by the power of gravity alone. At a lavish dedication ceremony on November 5, 1913, water finally began to flow. Letting his actions speak for him, Mulholland made one of the shortest speeches on record: "There it is. Take it!"

Los Angeles took what Mulholland had provided, but still the thirst grew. Indeed, throughout the 20th century communities in the American West took dramatic steps to get themselves more water. Most notable is undoubtedly the combined building of the Hoover Dam and the Colorado River Aqueduct in the 1930s and early 1940s. The dam was the essence of multipurposefulness. It created a vast reservoir that could help protect against drought, it allowed for better management of the Colorado River's flow and controlled dangerous flooding, and it provided a great new source of hydroelectric power. The aqueduct brought the bountiful supply of the Colorado nearly 250 miles over and through deserts and mountains to more than 130 communities in Southern California, including the burgeoning metropolis of Los Angeles. Other major aqueduct projects in the state included the California Aqueduct, supplying the rich agricultural lands of the Sacramento and San Joaquin valleys. The unparalleled growth of the entire region quite simply would have been impossible without such efforts.

The American West set the model, and around the world it soon became a mark of progress when a nation would turn to large-scale management of its water resources. Egypt is one of the best examples. The building of the Aswan High Dam in the 1960s created the third-largest reservoir in the world, tamed the disastrous annual flooding of the Nile, and provided controlled irrigation for more than a million acres of arid land. Built a few miles upriver from the original Aswan Dam (built by the British between 1898 and 1902), the Aswan High Dam was a gargantuan project involving its share of engineering challenges as well as the relocation of thousands of people and some of Egypt's most famous ancient monuments. Spanning nearly two miles, the dam increased Egypt's cultivable land by 30 percent and raised the water table for the Sahara Desert as far away as Algeria.

Egypt solved many of its water-related problems with this one grand stroke, but most countries in the developing world don't have the economic resources for such an undertaking. And in many cases, they don't have the water to work with in the first place. In such cases, one solution being adopted more and more widely is desalination—the treatment of seawater to make it drinkable. Once a pipe dream, desalination is now a viable process, and more than 7,500 desalination plants are in operation around the world, the vast majority of them in the desert countries of the Middle East.

Two main processes are used to desalinate seawater. One, called reverse osmosis, involves forcing the water through permeable membranes made of special plastics that let pure water through but filter out salts and any other minerals or contaminants. The other

The arch of Hoover Dam holds back the waters of the Colorado River to form Lake Mead, a 110-mile-long reservoir that meets the water needs of 14 million people in the United States and Mexico. Passing through the dam's four intake towers (above), the river flows through turbines to generate electricity for Nevada, Arizona, and California. When the Aswan High Dam (below) was finished in 1970, it too created a large reservoir—Lake Nasser, the world's third largest with a capacity of 5.97 trillion cubic feet.

TREATING OUR WATER
BEFORE AND AFTER USE

The treatment of a community's water supply—both before and after use—is a multistage affair involving physical, chemical, and in the case of wastewater, biological processes. In most systems, water comes either from surface sources, such as lakes and rivers, or from aquifers, natural underground reservoirs. Before it reaches a treatment plant, the water passes through screens that filter out large pieces of debris. At the plant, chemicals are added that cause fine particles to coagulate into larger particles known as floc. Sedimentation tanks allow these particles to settle out, and then a second round of filtration further clarifies the water. The final treatment stage, disinfection, kills bacteria and other microorganisms, usually with chlorine; some facilities use ultraviolet radiation or other chemicals such as ozone.

Water now safe for drinking is pumped into the distribution system—holding tanks, water towers, and distribution pipes. When demand is low, pump pressure forces water up into water towers; when demand is high, gravity provides the additional necessary pressure to supply all the homes and businesses on the system.

Wastewater is treated similarly to remove contaminants and solid waste. One key additional step is the use of bacteria to digest potentially dangerous organic waste and render it harmless.

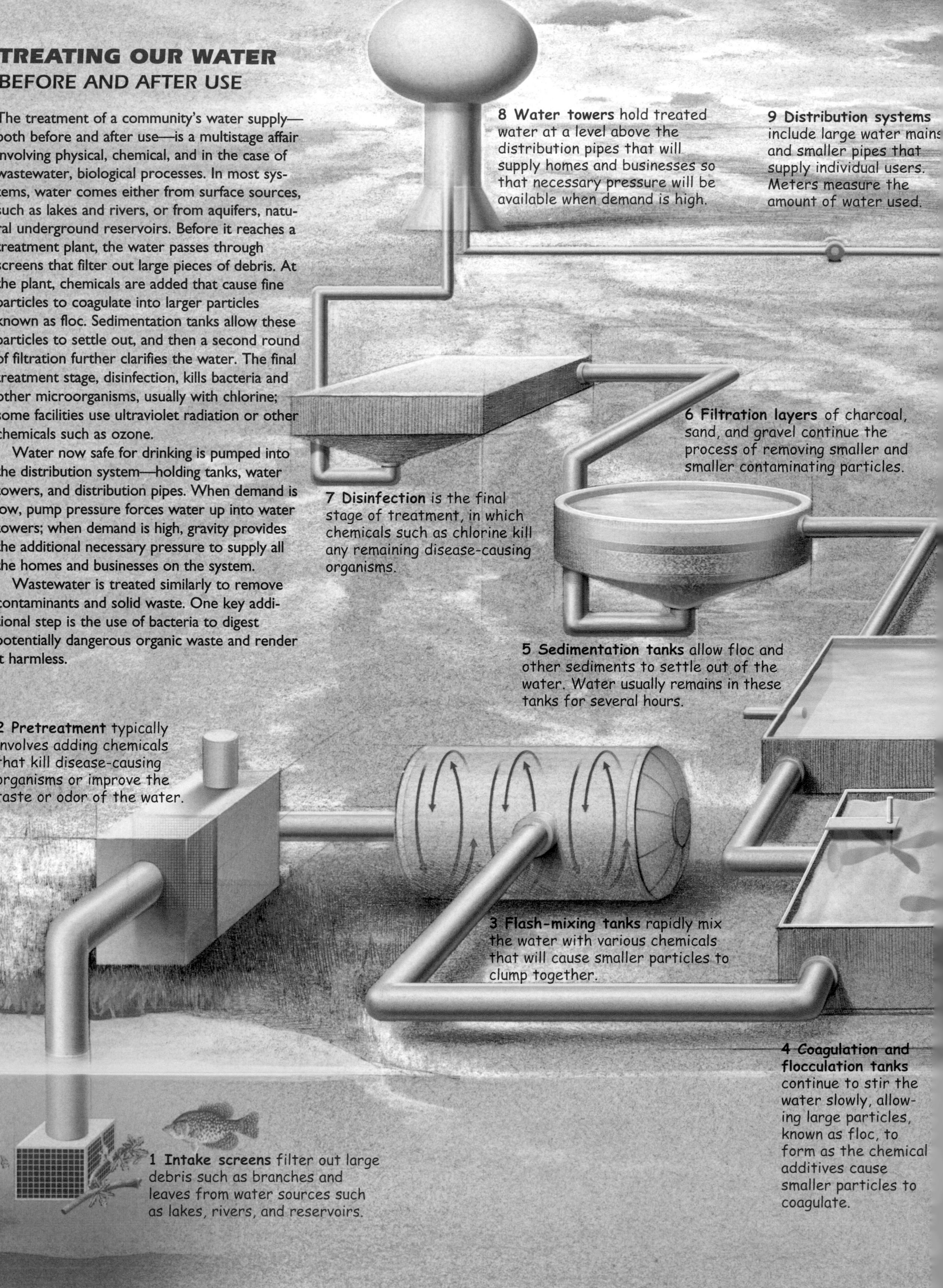

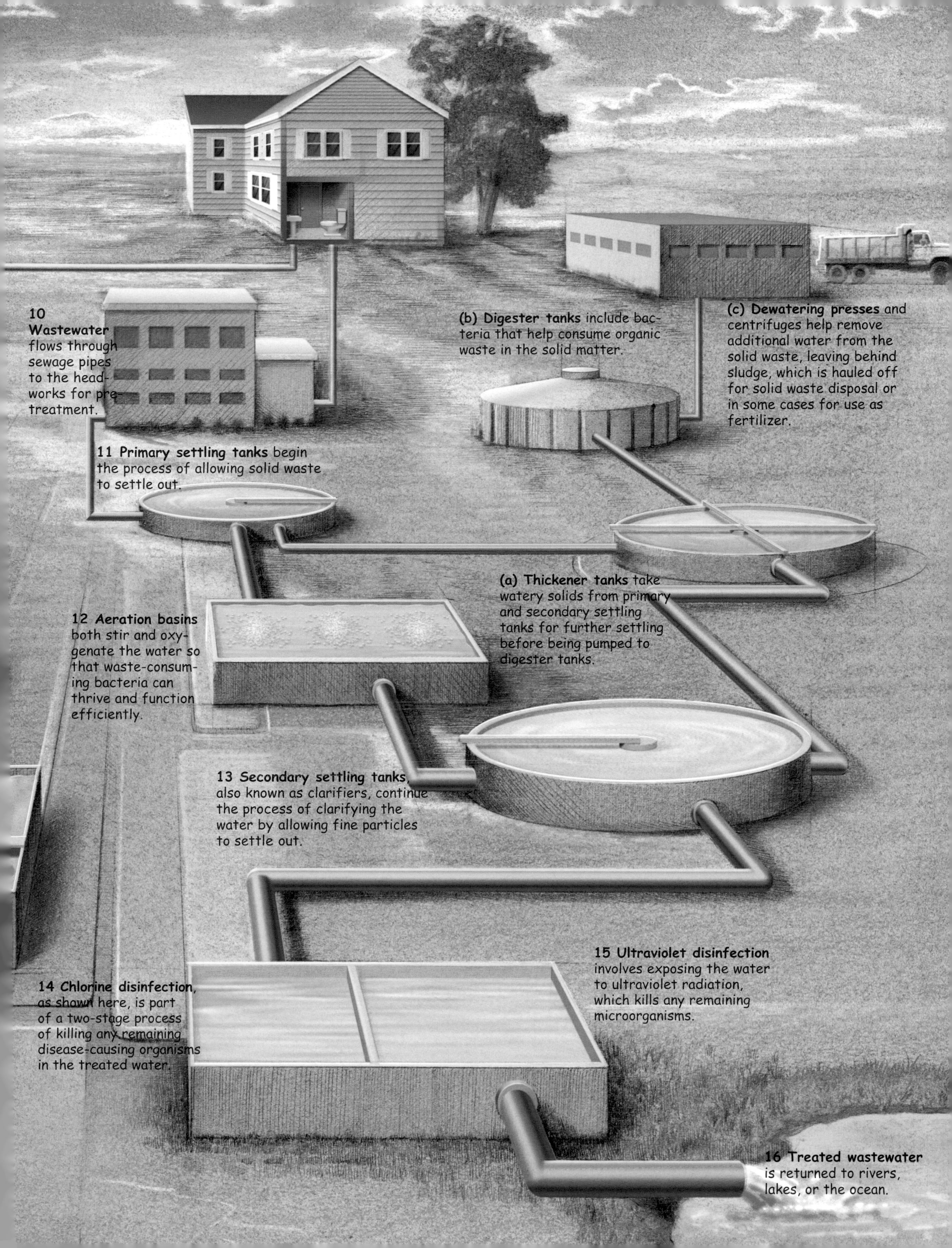

10 Wastewater flows through sewage pipes to the head-works for pre-treatment.
(b) Digester tanks include bacteria that help consume organic waste in the solid matter.
(c) Dewatering presses and centrifuges help remove additional water from the solid waste, leaving behind sludge, which is hauled off for solid waste disposal or in some cases for use as fertilizer.
11 Primary settling tanks begin the process of allowing solid waste to settle out.
(a) Thickener tanks take watery solids from primary and secondary settling tanks for further settling before being pumped to digester tanks.
12 Aeration basins both stir and oxygenate the water so that waste-consuming bacteria can thrive and function efficiently.
13 Secondary settling tanks, also known as clarifiers, continue the process of clarifying the water by allowing fine particles to settle out.
15 Ultraviolet disinfection involves exposing the water to ultraviolet radiation, which kills any remaining microorganisms.
14 Chlorine disinfection, as shown here, is part of a two-stage process of killing any remaining disease-causing organisms in the treated water.
16 Treated wastewater is returned to rivers, lakes, or the ocean.

Once considered a luxury, desalination plants like the one above in Oman are becoming a more viable solution to solving water supply problems in the Middle East. In many other parts of the world, such as Indonesia (above right), people get their water from sources that have not been treated, leading to high rates of disease, illness, and death.

method is distillation, in which the water is heated until it evaporates, then condensed, a process that separates out any dissolved minerals. Although these and other desalination techniques do work and have solved water shortage problems, they are too costly for many countries; distillation, for example, requires a good deal of energy input, and fuel costs can be prohibitively high. In some cases, adequate supplies of fuel aren't even available, at any cost.

The challenge this represents is a mighty one. For a shockingly high proportion of the world's population, clean water is still the rarest of commodities. By some estimates, more than two billion people on the planet have inadequate supplies of safe drinking water. Most of them still get their water from sources outside their homes—water that is for the most part untreated and rife with disease-carrying organisms. In the developing world, more than 400 children die every hour from those old, deadly scourges—cholera, typhoid, and dysentery. In short, the lack of safe water is a global crisis with a lethal toll.

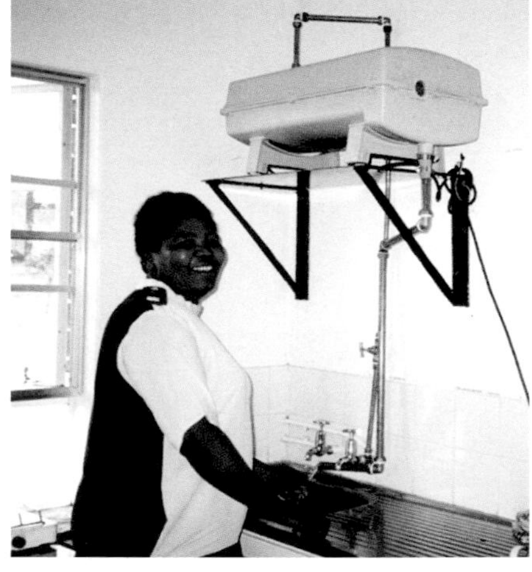

Today's engineers still struggle with the problem, and some of them are coming up with smaller-scale solutions. A case in point is a relatively simple device invented by Ashok Gadgil, an Indian-born research scientist working at the Lawrence Berkeley National Laboratory in California. When a new strain of cholera killed more than 10,000 people in southeastern India and neighboring countries in 1992 and 1993, Gadgil and a graduate student assistant worked to find an effective method for purifying water that wouldn't require the cost-prohibitive infrastructure of treatment plants.

Their device was simplicity itself: a compact box containing an ultraviolet light suspended above a pan of water. Water enters the pan, is exposed to the light, and then passes to a holding tank. At the rate of 4 gallons a minute, the device kills all microorganisms in the water, with the only operating expense being the 40 watts of power needed for the ultraviolet lamp. Dozens of these devices, which can be run off a car battery if need be, are now in use around the world—from Mexico and the Philippines to India and South Africa, where it provides clean drinking water to a rural health clinic (*left*). Regions using the simple treatment have reported dramatic reductions in waterborne diseases and their consequences.

Whatever their scale, from aqueducts and dams to desalination plants and portable ultraviolet devices, the notable successes in water management achieved in the 20th century continue to offer encouragement to a new generation of civil engineers worldwide as they face the challenge of our never-quenched need for clean water.

Perspective

Samuel C. Florman

Chairman
Kreisler Borg Florman
Construction Company

I was born and raised in New York City and have an early memory of a family celebration held at one of Manhattan's more elegant restaurants. I recall the waiter asking my father if he wanted to order a bottle of mineral water with the exotic-sounding name of a European spa. And I recall my father's firm reply: "No thank you, young man. We will all have LaGuardia cocktails." The waiter understood that this reference to our much-beloved mayor meant we wished to be served plain tap water. My father then explained to me that New York City water was the finest, purest beverage one could find anywhere and that it came to us from distant mountains over magnificent aqueducts and through spectacular tunnels carved deep in the earth.

My mother thereupon delivered a lecture on the importance of water to our health and well-being and expressed thanks to providence that many terrible waterborne diseases had recently been conquered, not only because our water came from far away but also because it was filtered and treated with germ-destroying chemicals. After that experience the faucets in our apartment took on for me a fascinating quality they never had before.

A science teacher at school helped nourish my newly awakened interest with a detailed explanation of how the New York City water system was conceived, designed, and built. My father had associated the technological marvel with a popular politician—as did the Romans and many others before and since—and my mother had expressed thanks to providence, surely a benign gesture. But I soon learned that a major part of the credit was due the talented people who had created the marvelous enterprise—the engineers.

I cannot say that this experience, in itself, persuaded me to become an engineer. But I do believe it started me on the way. It prompted me to become an avid sidewalk superintendent, seeking out in our city streets the numerous man-made holes that exposed a fabulous subterranean world of pipes and valves. When, years later, I embarked on my engineering studies, the courses on water supply were among my favorites. The often demanding theoretical work was alleviated by the fun of experimenting with water as it flowed through pipes and channels and poured over weirs. (And the occasional splashing reassured me that engineers are not as totally solemn as they are sometimes said to be.)

Then, as a newly commissioned ensign with the U.S. Navy Seabees, immediately after World War II, I found myself on a small island in the mid-Pacific, assigned to a water supply project. Surrounded by thousands of square miles of salty seas, a supply of fresh water suddenly seemed immensely precious. The elixir we were able to collect from mountain streams, impound behind a small earth-fill dam, then purify and distribute to a military camp reminded me of the water that engineers at home had been able to provide like magic in the midst of large and bustling cities. When work on the island infrastructure was complete and we opened the ceremonial tap, I fleetingly recalled my father's satisfaction in ordering a round of LaGuardia cocktails.

Ultimately, I followed a career in construction engineering and developed a special interest in concrete and steel. Yet each time I see a building rise into the sky, the sight of the plumbing pipes—the final arteries of a marvelous life-sustaining system—evokes a special feeling of wonder and pride.

WATER SUPPLY AND DISTRIBUTION
TIMELINE

In the early 1900s a simple glass of water could quench your thirst—or kill you. The safe drinking water that much of the world takes for granted today did not exist, and deadly waterborne diseases such as cholera, typhoid fever, and dysentery were a constant threat. Thanks to the efforts of scientists and engineers committed to protecting the public health, tap water in most of the world is safe to drink and waterways are guarded against pollution.

1900 In Chicago the Main Channel of the Sanitary and Ship Canal opens, reversing the flow of the Chicago River. The 28-mile, 24-foot-deep, 160-foot-wide drainage canal, built between Chicago and the town of Lockport, Illinois, is designed to bring in water from Lake Michigan to dilute sewage dumped into the river from houses, farms, stockyards, and other industries. Directed by Rudolph Hering, chief engineer of the Commission on Drainage and Water Supply, the project is the largest municipal earth-moving project of the time.

1913 The Los Angeles–Owens River Aqueduct is completed, bringing water 238 miles from the Owens Valley of the Sierra Nevada Mountains into the Los Angeles basin. The project was proposed and designed by William Mulholland *(below)*, an immigrant from Ireland who taught himself geology, hydraulics, and mathematics and worked his way up from a ditch tender on the Los Angeles River to become the superintendent of the Los Angeles Water Department. Mulholland devised a system to transport the water entirely by gravity flow *(below right)* and supervised 5,000 construction workers over 5 years to deliver the aqueduct within original time and cost estimates.

In Birmingham, England, chemists experiment with the biosolids in sewage sludge by bubbling air through wastewater and then letting the mixture settle; once solids had settled out, the water was purified. Three years later, in 1916, this activated sludge process is put into operation in Worcester, England, and in 1923 construction begins on the world's first large-scale activated sludge plant, at Jones Island, on the shore of Lake Michigan.

1914 Boston engineers Leonard Metcalf and Harrison P. Eddy publish *American Sewerage Practice*, Volume I: *Design of Sewers*, which declares that working for "the best interests of the public health" is the key professional obligation of sanitary engineers. The book becomes a standard reference in the field for decades.

1915 In December the new Catskill Aqueduct is completed. The 92-mile-long aqueduct *(one of whose tunnels is shown above)* joins the Old Croton Aqueduct system and brings mountain water from west of the Hudson River to the water distribution system of Manhattan. Flowing at a speed of 4 feet per second, it delivers 500 million gallons of water daily.

1919 Civil engineer Abel Wolman and chemist Linn H. Enslow of the Maryland Department of Health in Baltimore develop a rigorous scientific formula for the chlorination of urban water supplies. (In 1908 Jersey City Water Works, New Jersey, became the first facility to chlorinate, using sodium hypochlorite, but there was uncertainty as to the amount of chlorine to add and no regulation of standards.) To determine the correct dose, Wolman and Enslow analyze the bacteria, acidity, and factors related to taste and purity. Wolman overcomes strong opposition to convince local governments that adding the correct amounts of otherwise poisonous chemicals to the water supply is beneficial—and crucial—to public health. By the 1930s chlorination and filtration of public water supplies eliminates waterborne diseases such as cholera, typhoid, hepatitis A, and dysentery. The formula is still used today by water treatment plants around the world.

1930 Hardy Cross, civil and structural engineer and educator, develops a method for the analysis and design of water flow in simple pipe distribution systems, ensuring consistent water pressure. Cross employs the same principles for the water system problem that he devised for the "Hardy Cross method" of structural analysis, a technique that enables engineers—without benefit of computers—to make the thousands of mathematical calculations necessary to distribute loads and moments in building complex structures such as multi-bent highway bridges and multistory buildings.

1935 In September, President Franklin D. Roosevelt speaks at the dedication of Hoover Dam, which sits astride the Colorado River in Black Canyon, Nevada. Five years in construction, the dam ends destructive flooding in the lower canyon;

provides water for irrigation and municipal water supplies for Nevada, Arizona, and California; and generates electricity for Las Vegas and most of Southern California.

1937 Construction begins on the 115-mile-long Delaware Aqueduct System. Water for the system is impounded in three upstate reservoir systems, including 19 reservoirs and three controlled lakes with a total storage capacity of approximately 580 billion gallons. The deep, gravity-flow construction of the aqueduct allows water to flow from Rondout Reservoir in Sullivan County into New York City's water system at Hillview Reservoir in Westchester County, supplying more than half the city's water. Approximately 95 percent of the total water supply is delivered by gravity with about 5 percent pumped to maintain the desired pressure. As a result, operating costs are relatively insensitive to fluctuations in the cost of power.

1938-1957 The Colorado–Big Thompson Project (C-BT), the first transmountain diversion of water in Colorado, is undertaken during a period of drought and economic depression. The C-BT brings water through the 13-mile Alva B. Adams Tunnel, under the Continental Divide, from a series of reservoirs on the Western Slope of the Rocky Mountains to the East Slope, delivering 230,000 acre-feet of water annually to help irrigate more than 600,000 acres of farmland in northeastern Colorado and to provide municipal water supplies and generate electricity for Colorado's Front Range.

1951 Mining engineer James S. Robbins builds the first hard rock tunnel-boring machine (TBM). Robbins discovers that if a sharp-edged metal wheel is pressed on a rock surface with the correct amount of pressure, the rock shatters. If the wheel, or an array of wheels, continually rolls around on the rock and the pressure is constant, the machine digs deeper with each turn. The engineering industry is at first reluctant to switch from the commonly used drill-and-blast method because Robbins's machine has a $10 million price tag. Today, TBMs are used to excavate circular cross-section tunnels through a wide variety of geology, from soils to hard rock.

1955 Ductile cast-iron pipe, developed in 1948, is used in water distribution systems. It becomes the industry standard for metal due to its superior strength, durability, and reliability over cast iron. The pipe is used to transport potable water, sewage, and fuel, and is also used in fire-fighting systems.

1960s Kuwait is the first state in the Middle East to begin using seawater desalination technology *(below),* providing the dual benefits of fresh water and electric power. Kuwait produces fresh water from seawater with the technology known as multistage flash (MSF) evaporation. The MSF process begins with heating saltwater, which occurs as a byproduct of producing steam for generating electricity, and ends with condensing potable water. Between the heater and condenser stages

are multiple evaporator-heat exchanger subunits, with heat supplied from the power plant external heat source. During repeated distillation cycles cold seawater is used as a heat sink in the condenser.

1970s The Aswan High Dam construction is completed, about 5 kilometers upstream from the original Aswan Dam (1902). Known as Saad el Aali in Arabic, it impounds the waters of the Nile to form Lake Nasser, the world's third-largest reservoir, with a capacity of 5.97 trillion cubic feet. The project requires the relocation of thousands of people and floods some of Egypt's monuments and temples *(above)*, which are later raised. But the new dam controls annual floods along the Nile, supplies water for municipalities and irrigation, and provides Egypt with more than 10 billion kilowatt-hours of electric power every year.

1980s James Barnard, a South African engineer, develops a wastewater treatment process that removes nitrates and phosphates from wastewater without the use of chemicals. Known as the Bardenpho process, it converts the nitrates in activated sludge into nitrogen gas, which is released into the air, removing a high percentage of suspended solids and organic material.

1996 Ashok Gadgil, a scientist at the Lawrence Berkeley National Laboratory in California, invents an effective and inexpensive device for purifying water. UV Waterworks, a portable, low-maintenance, energy-efficient water purifier, uses ultraviolet light to render viruses and bacteria harmless. Operating with hand-pumped or hand-poured water, a single unit can disinfect 4 gallons of water a minute, enough to provide safe drinking water for up to 1,500 people, at a cost of only one cent for every 60 gallons of water—making safe drinking water economically feasible for populations in poor and rural areas all over the world.

ELECTRONICS

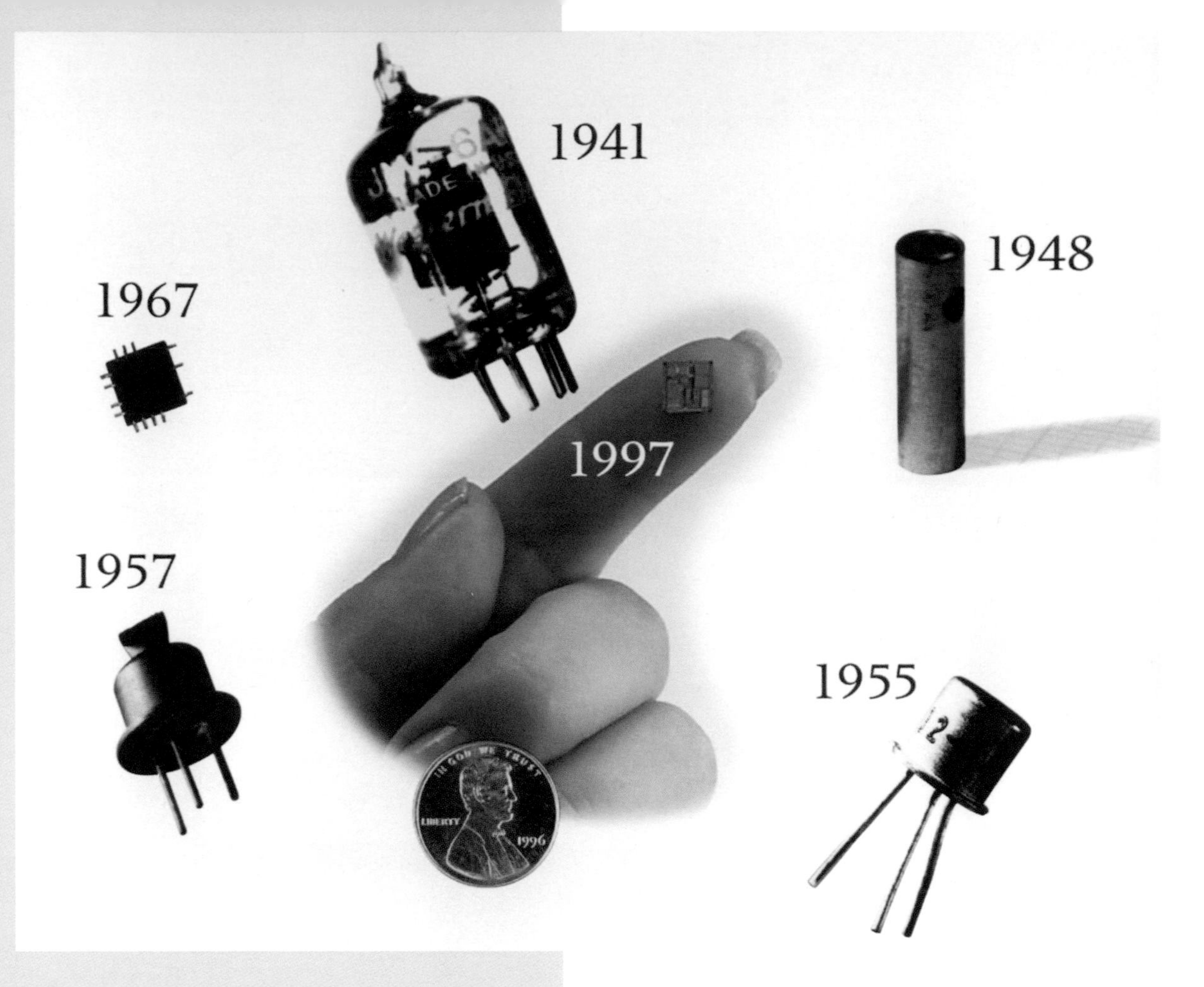

Barely stifled yawns greeted the electronics novelty that was introduced to the public in mid-1948. "A device called a transistor, which has several applications in radio where a vacuum tube ordinarily is employed, was demonstrated for the first time yesterday at Bell Telephone Laboratories," noted an obviously unimpressed New York Times reporter on page 46 of the day's issue.

To be sure, the gadget had pluses. Not only could the transistor amplify electric current like a vacuum tube, it also used little power, didn't need to warm up, and was compact—a thimble-sized cylinder with a couple of protruding wires. But because its main ingredient was an expensive, hard-to-handle element called germanium, the transistor seemed likely to remain a laboratory curiosity.

What happened instead was total technological conquest. Scientists and engineers learned how to make it from abundant silicon, shrink it to microscopic size, and harness it for once-unimaginable powers of digital computing, control, communication, detection, and display. As costs plunged and performance soared, transistor-based circuits found their way into thermostats and sewing machines, power tools, children's toys and greeting cards, cameras and cell phones, fax machines and industrial robots, tractors and missiles. Home, office, factory, farm, and practically every other venue of human activity, within a few decades, became a kind of vast ecosystem of transistorized electronics.

The roots of the triumph reach deep. Germanium and silicon, along with a number of other crystalline materials, are semiconductors, so-called because they neither conduct electricity well, like most metals, nor block it effectively, as do insulators such as glass or rubber. Back in 1874 a German scientist named Ferdinand Braun identified a surprising trait of these on-the-fence substances: Current tends to flow through a semiconductor crystal in only one direction. This phenomenon, called rectification, soon proved valuable in wireless telegraphy, the first form of radio communication.

When electromagnetic radio waves traveling through the atmosphere strike an aerial, they generate an alternating (two-way) electric current. However, earphones or a speaker must be powered by direct (one-way) current. Methods for making the conversion, or rectification, in wireless receivers existed in the closing years of the 19th century, but they were crude. In 1899 Braun patented a superior detector consisting of a semiconductor crystal touched by a single metal wire, affectionately called a "cat's whisker." His device was popular with radio hobbyists for decades, but it was erratic and required much trial-and-error adjustment.

Another route to rectification was soon found, emerging from Thomas Edison's work on the electric lightbulb. Back in 1883 Edison had observed that if he placed a small metal plate in one of his experimental bulbs, it would pick up an electric current that somehow managed to cross the bulb's vacuum from the hot filament. Not long afterward, a British engineer named John Ambrose Fleming noticed that even when the filament carried an alternating current (which Edison hadn't tried), the current passing through the vacuum always traveled from the hot filament to the plate, never the other way around. Early in the new century Fleming devised what he called an "oscillation valve"—a filament and plate in a vacuum bulb. It rectified a wireless signal much more reliably than Braun's crystals.

By then the nature of the invisible current was understood. Experiments in the 1890s by the British physicist Joseph John Thomson had indicated that a flood of infinitesimally small particles—electrons, they would be called—was whizzing through the vacuum at the incredible speed of 20,000 miles per second. Their response to signal oscillations was no less amazing. "So nimble are these little electrons," wrote Fleming, "that however rapidly we change the rectification, the plate current is correspondingly altered, even at the rate of a million times per second."

In 1906 the American inventor Lee De Forest modified Fleming's vacuum tube in a way that opened up broad new vistas for electrical engineers. Between the filament and the plate he inserted a gridlike wire that functioned as a kind of electronic faucet: changes in a voltage applied to the grid produced matching changes in the flow of current between the other two elements. Because a very small voltage controlled a much larger current and the mimicry was exact, the device could serve as an amplifier. Rapidly improved by others, the three-element tube—a triode—made long-distance telephone calls possible, enriched the sound of record players, spawned a host of electronic devices for control or measurement, gave voice to radio by the 1920s, and helped launch the new medium of television in the 1930s. Today, vacuum tubes are essential in high-powered satellite transmitters and a few other applications. Some modern versions are no bigger than a pea.

In addition to amplifying an electric signal, triodes can work as a switch, using the grid voltage to simply turn a current on or off. During the 1930s several researchers identified rapid switching as a way to carry out complex calculations by means of the binary numbering system—a way of counting that uses only ones and zeros rather than, say, the 10 digits of the decimal system. For computing purposes, zeros and ones could be expressed as the two states of a switch, on or off. Moreover, just a few different arrangements of switches would be sufficient to perform any mathematical or logical operation.

Vacuum tubes, being much faster than any mechanical switch, were soon enlisted for the new computing machines. But because a computer, by its nature, requires switches in very large numbers, certain shortcomings of the tubes were glaringly obvious. They were bulky

Building on an early lamp by Thomas Edison (above right), Sir John Ambrose Fleming (above left) developed his "oscillation valve," which evolved over the course of the century into bulky, power hungry vacuum tubes (below) to semiconductor transistors and on to microchips the size of a fingernail (opposite).

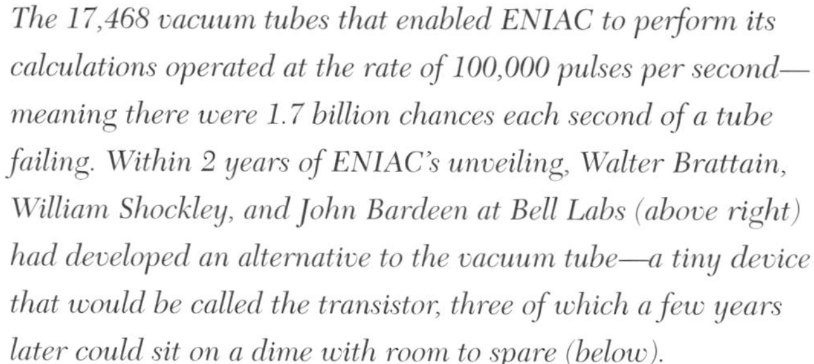

The 17,468 vacuum tubes that enabled ENIAC to perform its calculations operated at the rate of 100,000 pulses per second—meaning there were 1.7 billion chances each second of a tube failing. Within 2 years of ENIAC's unveiling, Walter Brattain, William Shockley, and John Bardeen at Bell Labs (above right) had developed an alternative to the vacuum tube—a tiny device that would be called the transistor, three of which a few years later could sit on a dime with room to spare (below).

and power hungry; they produced a lot of waste heat; and they were prone to failure. The first big, all-electronic computer, a calculating engine known as ENIAC that went to work in 1945, had 17,468 vacuum tubes, weighed 30 tons, consumed enough power to light 10 homes, and required constant maintenance to keep it running.

By that time the search for a semiconductor alternative to vacuum tubes was well under way, aided by the insights of quantum mechanics—the physics of elementary particles. Scientists now knew that conductivity was determined by how tightly electrons were bound to an atom or molecule. In the case of the in-between semiconductors, researchers also saw that electrical behavior was strongly affected by the presence of impurities in the crystal. Some impurities, such as phosphorus, provide a surplus of electrons that are free to wander and contribute to a current. Others, such as boron, create areas of electron deficiency known as holes, and these holes can move about in the lattice structure of the crystal—another sort of electric current. During the years of World War II, researchers at AT&T's Bell Laboratories and elsewhere made great strides in handling semiconductors and various impurities, fashioning them into rectifiers in radar receivers, which had to handle frequencies beyond the reach of vacuum tubes.

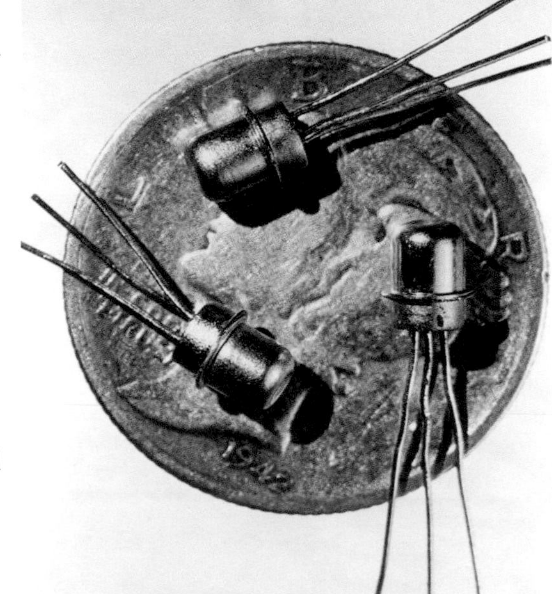

Some investigators were convinced that semiconductors could be given the powers of a triode as well. In late 1947 that goal was met by John Bardeen and Walter Brattain at Bell Labs. Their invention (the little cylinder that provoked a shrug from the *New York Times*) essentially consisted of two "cat's whiskers" placed very close together on the surface of an electrically grounded chunk of germanium. A month later a colleague, William Shockley, came up with a more practical design—a three-layer semiconductor sandwich. The outer layers were doped with an impurity to supply extra electrons, and the very thin inner layer received a different impurity to create holes. By means of complex interactions at the junctions where the layers met, the middle portion of the sandwich functioned like the grid in a triode, with a very small voltage controlling a sizable current flow between the outer layers. Bardeen, Brattain, and Shockley would share a Nobel Prize in physics as inventors of the transistor.

Although Shockley's version was incorporated into a few products where small size and low power consumption were critical—hearing aids, for example—the transistor didn't win widespread acceptance by manufacturers until the mid-1950s. The turning point came when Gordon Teal, a physicist at Texas Instruments, showed that a transistor could be made from highly purified silicon—a component of ordinary sand. Exploiting this abundant material wasn't easy; it depended on growing silicon crystals that were almost totally free of current-disrupting flaws. But once engineers had mastered that difficult art, the way was open to radical new approaches to circuit making.

Raw silicon, component of ordinary sand

Any electronic circuit is an assemblage of several types of components that work together as a unit. Previously, the various circuit elements had always been made separately and then laboriously connected with wires. But in 1958, Jack Kilby, an electrical engineer at Texas Instruments who had been asked to design a transistorized adding machine, came up with a bold unifying strategy. By selective placement of impurities, he realized, a crystalline wafer of silicon could be endowed with all the elements necessary to function as a circuit. As he saw it, the elements would still have to be wired together, but they would take up much less space. In his laboratory notebook, he wrote: "Extreme miniaturization of many electrical circuits could be achieved by making resistors, capacitors and transistors & diodes on a single slice of silicon."

The following year, Robert Noyce, then at Fairchild Semiconductor, independently arrived at the idea of an integrated circuit and added a major improvement. His approach involved overlaying the slice of silicon with a thin coating of silicon oxide, the semiconductor's version of rust. From seminal work done a few years earlier by John Moll and Carl Frosch at Bell Labs, as well as by Fairchild colleague Jean Hoerni, Noyce knew the oxide would protect transistor junctions because of its excellent insulating properties. It also lent itself to a much easier way of connecting the circuit elements. Delicate lines of metal could simply be printed on the coating; they would reach down to the underlying components via small holes etched in the oxide. By 1965 integrated circuits—chips as they were called—embraced as many as 50 elements. That year a physical chemist named Gordon Moore, cofounder of the Intel Corporation with Robert Noyce, wrote in a magazine article: "The future of integrated electronics is the future of electronics itself." He predicted that the number of components on a chip would continue to double every year, an estimate that, in the amended form of a doubling every year and a half or so, would become known in the industry as Moore's Law. While the forecast was regarded as wild-eyed in some quarters, it proved remarkably accurate. The densest chips of 1970 held about 1,000 components. Chips of the mid-1980s contained as many as several hundred thousand. By the mid-1990s some chips the size of a baby's fingernail embraced 20 million components.

Price dropped with size. In the early 1950s a transistor about as big as an eraser cost several dollars. By the mid-1970s, when transistors were approaching the

Robert Noyce of Fairchild Semiconductor (below left) and Jack Kilby at Texas Instruments (center) independently came up with the idea of the integrated circuit. In 1965 Gordon Moore (below right), cofounder with Noyce of Intel Corporation, predicted an annual doubling of the numbers of components integrated on a single chip.

THE TINY ENGINES OF THE DIGITAL WORLD

MAKING MICROPROCESSOR CHIPS

When integrated circuits became the driving force of electronics in the 1960s, Gordon Moore, cofounder of the chip-making company Intel, famously forecast a doubling of their complexity every 18 months or so. Such a pace seemed impossible to sustain for long, yet Moore's Law, as the prediction was dubbed, has held true ever since, leading to chips of staggering intricacy. Shown here, in highly simplified form, are four transistors; today's microprocessors might have nearly 200 million of them, plus many additional elements, packed into a space the size of a fingernail.

Chips are made in groups of several hundred at a time on a wafer of ultrapure silicon, with upwards of 500 steps carried out over a period of several weeks to yield a many-layered structure. Central to the fabrication process is a technique called photolithography: each layer of circuitry is patterned with the aid of a stencil-like mask. Light shined through the mask alters the photoresist surface below and allows it to be removed by etching. The etched areas, in turn, can be chemically treated to determine their electrical properties. Wiring is added by etching channels in the insulator surface, coating the surface with copper, and then polishing away the excess. The copper remaining in the channels forms the necessary connections. At many stages along the way, tests are performed to ensure that the Lilliputian architectures will function as intended.

Current manufacturing techniques are now approaching a physical limit to further miniaturization, but by shifting to extremely short wavelengths of ultraviolet light for photolithography, chipmakers expect to meet Moore's Law for at least another decade, auguring a world in which the powers of today's supercomputers are a desktop staple.

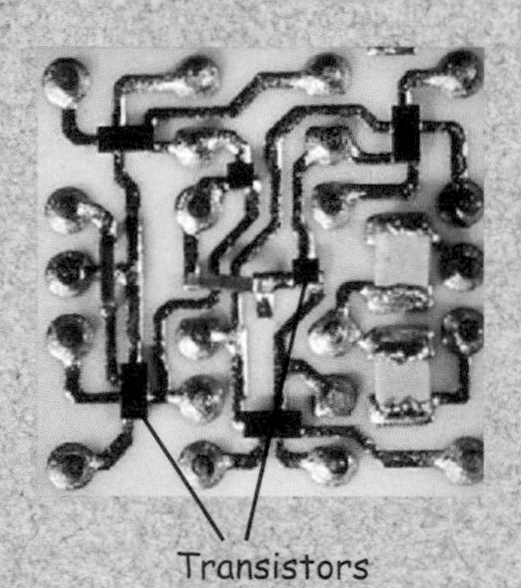

In the early 1960s one-transistor chips were mounted on ceramic to form modules, as in the example at left, with six transistors. An integrated circuit of the same era (right), might contain six transistors on one chip of silicon, a design still simple enough, even with wiring and other components, to be made without machine assistance.

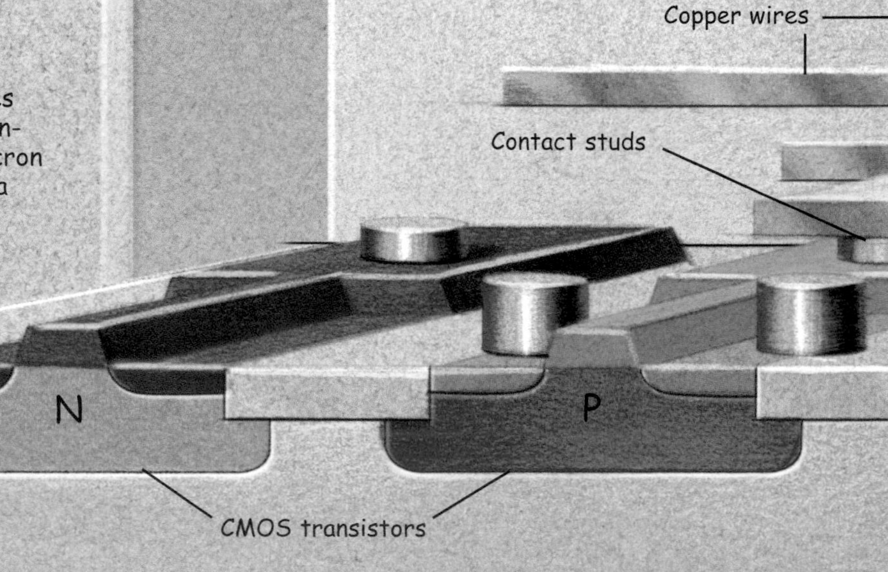

SOLDER CONNECTION. Most communications between a chip's transistors and the outer world pass through solder connections. Each solder bump is comparatively huge by chip standards—about a ten-thousandth of a meter in diameter.

WIRING. The wiring of a microprocessor is aluminum or copper, insulated by glass. Wires are linked to the transistors by tungsten contact studs that measure only about 0.25 micron in diameter—about one-quarter the size of a cell nucleus.

TRANSISTORS. CMOS (complementary metal oxide semiconductor) transistors work in pairs. If their conductive channel is N-type, or negative, material, the current is carried by surplus electrons. With a P-type, or positive, channel, the charge carriers are electron vacancies, called holes, in the silicon crystal lattice.

This recent-vintage microprocessor has 180 million transistors and additional millions of resistors, capacitors, and other components. So intricate is the circuit geometry that computers are needed to fashion the lithographic masks.

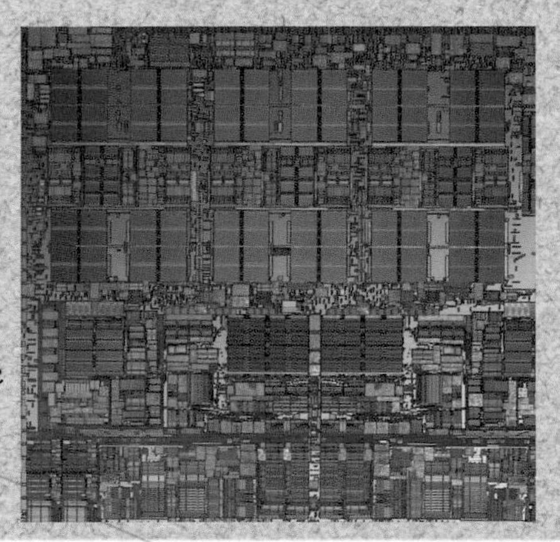

When a modern microprocessor is magnified 1,000 times, seven or eight layers of copper wire are revealed—more than 100 meters' worth in all. To see transistors, a further 1,000-power magnification would be needed, for each transistor is smaller than a blood cell.

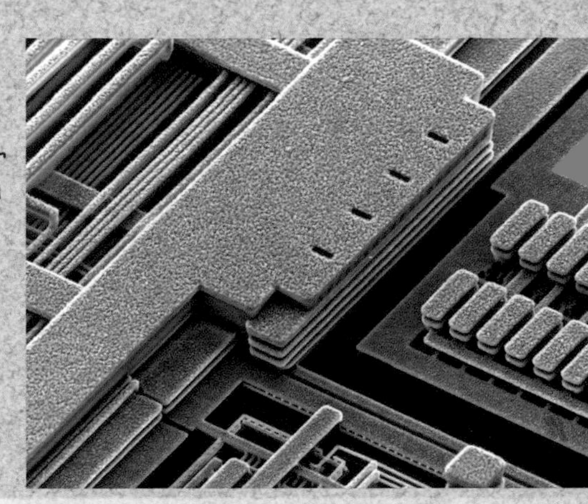

PACKAGING. Plastic sheathing protects the fragile innards of a modern microprocessor. So dense is the circuitry beneath that it puts out as much heat as a 100-watt bulb.

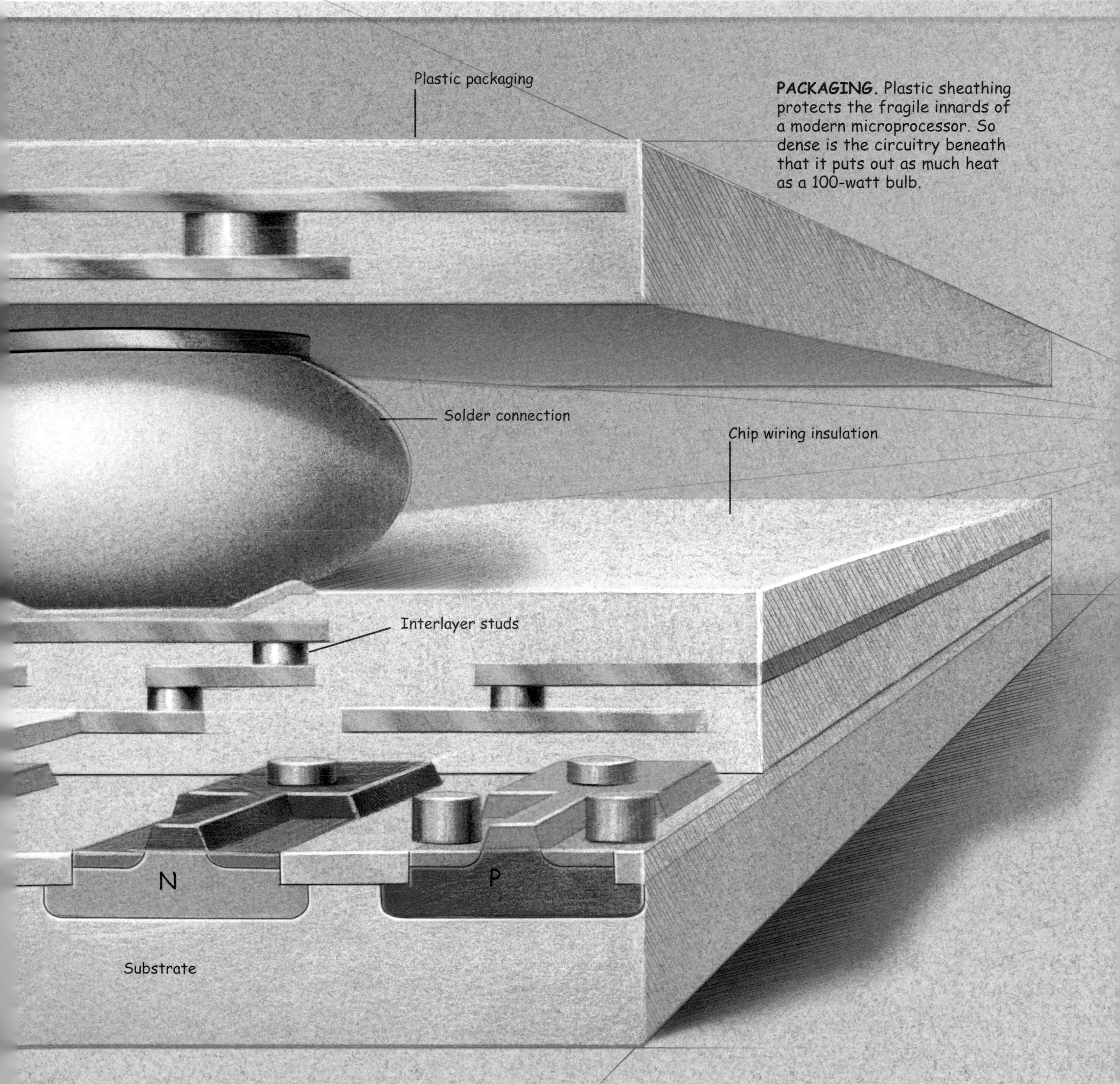

size of a bacterium, they cost mere hundredths of a cent apiece. By the late 1990s the price of a single transistor was less than a hundred-thousandth of a cent—sometimes far less, mere billionths of a cent, depending on the type of chip.

Today's transistors come in a variety of designs and materials and are arrayed in circuits of many degrees of complexity. Some chips provide electronic memory, storing and retrieving binary data. Others are designed to execute particular tasks with maximum efficiency—manipulating audio signals or graphic images, for instance. Still others are general-purpose devices called microprocessors. Instead of being tailored for one job, they do whatever computational work is assigned to them by software instructions.

The first microprocessor was produced by Intel in 1971. Dubbed the 4004, it cost about $1,000 and was as powerful as ENIAC, the vacuum-tube monster of the 1940s. Faster versions soon followed from Intel, and other companies came out with competing microprocessors, with prices dropping rapidly toward $100. The flexibility of the offerings had enormous appeal. If, for instance, the maker of a washing machine or camera wanted to put a chip in the product, it wasn't necessary to commission a special circuit design, await its development, and shoulder the expense of custom manufacturing. An inexpensive, off-the-shelf microprocessor, guided in its work by appropriate software, would often suffice. These devices, popularly known as a computer on a chip, quickly spread far and wide.

Produced in 1971, Intel's 4004 microprocessor (top) was as powerful as ENIAC; its four-bit microprocessor could perform 60,000 operations per second. Today's microprocessors are so complex they must be designed with the assistance of powerful computers and their manufacture requires ultra-sterile conditions (above).

The creation of today's chips is a prodigious challenge. The design stage alone, mapping out the pathways for a forest of interconnected switches, may take months or even years and can be accomplished only with the help of powerful computers. Manufacturing is done in multibillion-dollar plants of unearthly cleanliness, because a single particle of dust, boulder-like in the microworld of transistors, would ruin the circuitry. The tiny electronic creations wrought by all this engineering effort are now everywhere, operating behind the scenes in every household device and every mode of communication, transportation, recreation, and commerce. Most extraordinary of all, the rate of advance shows no signs of slackening. Engineers and scientists are exploring three-dimensional architectures for circuits, seeking organic molecules that may be able to spontaneously assemble themselves into transistors and, on the misty edge of possibility, experimenting with mysterious quantum effects that might be harnessed for computation. Whether we are ready or not, computing power will continue its incredible expansion and change our future in ways yet unimagined.

Perspective

Gordon E. Moore

Chairman Emeritus
Intel Corporation

The discovery of the electron in 1897 set the stage for electronics to develop over the ensuing century. Most of the first half of the 20th century was devoted to controlling electrons in a vacuum with electric and magnetic fields to make amplifiers, oscillators, and switches. These gave us, among other things, radio, television, radar, and the first computers.

The last half of the century saw the rise of solid-state electronics, beginning with the invention of the transistor in 1947. I arrived on the scene in 1956 to join William Shockley, one of the inventors of the transistor, who was establishing the Shockley Semiconductor Laboratory to develop a commercial silicon transistor. By then the advantages of transistors over vacuum tubes were apparent for many applications; it was only necessary to make transistors reliable and cheap.

But Shockley changed his original goal, turning his focus to another semiconductor device he had invented while at Bell Labs—a four-layer diode possibly useful in telephone switches but not much else. A group of us (the Fairchild 8) went off to found a new company, Fairchild Semiconductor, to continue to pursue the silicon transistor. Fortunately we at Fairchild were on the right track technologically when Jack Kilby of Texas Instruments demonstrated a complete circuit made of semiconductor materials. My colleague Bob Noyce saw how the Fairchild technology could be extended to make it practical to manufacture a complete circuit, rather than just individual transistors. Shortly after Bob's inventions he was promoted to general manager and I was left to oversee development of the technology extensions that ultimately led to the computer chips we are all familiar with today.

The new integrated devices did not find a ready market. Users were concerned because the individual transistors, resistors, and other electronic circuit components could not be tested individually to ensure their reliability. Also, early integrated circuits were expensive, and they impinged on the turf that traditionally belonged to the circuit designers at the customer's company. Again, Bob Noyce made a seminal contribution. He offered to sell the complete circuits for less than the customer could purchase individual components to build them. (It was also significantly less than it was costing us to build them!) This step opened the market and helped develop the manufacturing volumes necessary to reduce manufacturing costs to competitive levels. To this day the cost reductions resulting from economies of scale and newer high-density technology are passed on to the user—often before they are actually realized by the circuit manufacturer. As a result, we all know that the high-performance electronic gadget of today will be replaced with one of higher performance and lower cost tomorrow.

The integrated circuit completely changed the economics of electronics. Initially we looked forward to the time when an individual transistor might sell for a dollar. Today that dollar can buy tens of millions of transistors as part of a complex circuit. This cost reduction has made the technology ubiquitous—nearly any application that processes information today can be done most economically electronically. No other technology that I can identify has undergone such a dramatic decrease in cost, let alone the improved performance that comes from making things smaller and smaller. The technology has advanced so fast that I am amazed we can design and manufacture the products in common use today. It is a classic case of lifting ourselves up by our bootstraps—only with today's increasingly powerful computers can we design tomorrow's chips.

ELECTRONICS TIMELINE

Brilliant inventors from the late 19th century to the present day have built on each other's work to launch a revolution in electronics. In recognizing the team of Bardeen, Brattain, and Shockley for their invention of the transistor, the Nobel Prize also paid tribute to their predecessors, the discoverers of electrons, the vacuum tube, purified crystals, and diodes. The transistor spurred experimentation with new materials such as silicon and with a host of manufacturing techniques, leading to electronic devices that have altered every aspect of daily life.

1904 Sir John Ambrose Fleming, a professor of electrical engineering and the first scientific adviser for the Marconi Company, invents the thermionic valve, or diode, a two-electrode rectifier. (A rectifier prevents the flow of current from reversing.) Building on the work of Thomas Edison, Fleming devises an "oscillation valve"—a filament and a small metal plate in a vacuum bulb. He discovers that an electric current passing through the vacuum is always unidirectional.

1907 Lee De Forest *(below)*, an American inventor, files for a patent on a triode, a three-electrode device he calls an Audion. He improves on Fleming's diode by inserting a gridlike wire between the two elements in the vacuum tube, creating a sensitive receiver and amplifier of radio wave signals. The triode is used to improve sound in long-distance phone service, radios, televisions, sound on film, and eventually in modern applications such as computers and satellite transmitters.

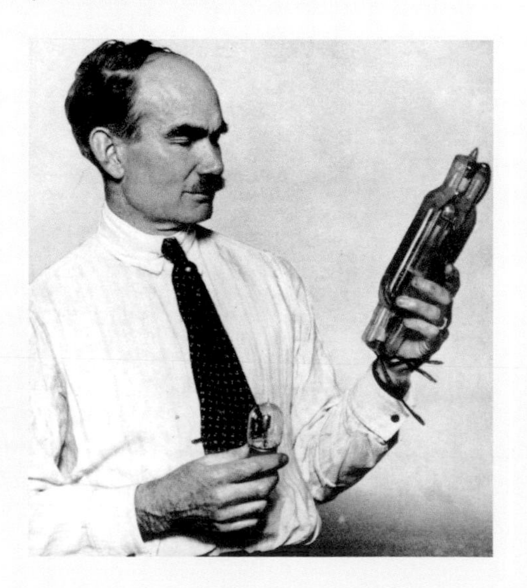

1940 Russell Ohl, a researcher at Bell Labs, discovers that small amounts of impurities in semiconductor crystals create photoelectric and other potentially useful properties. When he shines a light on a silicon crystal with a crack running through it, a voltmeter attached to the crystal registers a half-volt jump. The crack, it turns out, is a natural P-N junction, with impurities on one side that create an excess of negative electrons (N) and impurities on the other side that create a deficit (P). Ohl's crystal is the precursor of modern-day solar cells, which convert sunlight into electricity. It also heralds the coming of transistors.

1947 John Bardeen, Walter H. Brattain, and William B. Shockley of Bell Labs discover the transistor. Brattain and Bardeen build the first point-contact transistor *(right)*, made of two gold foil contacts sitting on a germanium crystal. When electric current is applied to one contact, the germanium boosts the strength of the current flowing through the other contact. Shockley improves on the idea by building the junction transistor—"sandwiches" of N- and P-type germanium. A weak voltage applied to the middle layer modifies a current traveling across the entire "sandwich." In November 1956 the three men are awarded the Nobel Prize in physics.

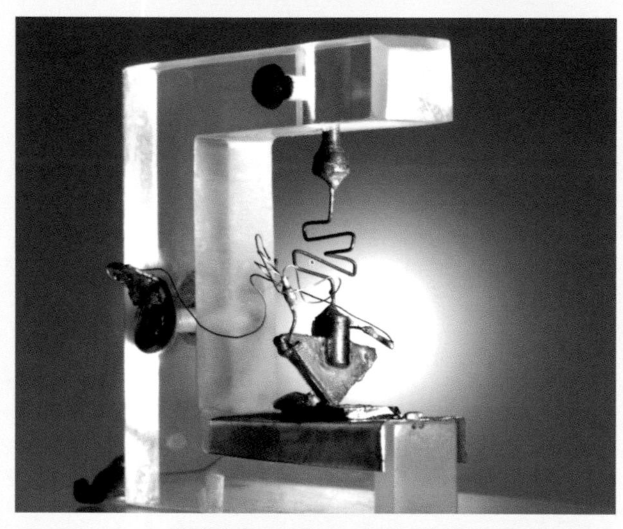

1952 Sonotone markets a $229.50 hearing aid that uses two vacuum tubes and one transistor—the first commercial device to apply Shockley's junction transistor. Replacement batteries for transistorized hearing aids cost only $10, not the nearly $100 of batteries for earlier vaccum tube models.

1954 Gordon Teal, a physical chemist formerly with Bell Labs, shows colleagues at Texas Instruments that transistors can be made from pure silicon—demonstrating the first truly consistent mass-produced transistor. By the late 1950s silicon begins to replace germanium as the semiconductor material out of which almost all modern transistors are made.

Texas Instruments introduces the first transistor radio, the Regency TR1, with radios by Regency Electronics and transistors by Texas Instruments. The transistor replaces De Forest's triode, which was the electrical component that amplified audio signals—making AM (amplitude modulation) radio possible. The door is now open to the transistorization of other mass production devices.

1955 Carl Frosch and Link Derick at Bell Labs discover that silicon dioxide can act as a diffusion mask. That is, when a silicon wafer is heated to about 1200°C in an atmosphere of water vapor or oxygen, a thin skin of silicon dioxide forms on the surface. With selective etching of the oxide layer, they could diffuse impurities into the silicon to create P-N junctions. Bell Labs engineer John Moll then develops the all-diffused silicon transistor, in which impurities are diffused into the wafer while the active elements are protected by the oxide layer. Silicon begins to replace germanium as the preferred semiconductor for electronics.

1958-1959 Jack Kilby, an electrical engineer at Texas Instruments and Robert Noyce of Fairchild Semiconductor independently invent the integrated circuit. In September 1958, Kilby builds an integrated circuit that includes multiple components connected with gold wires on a tiny silicon chip, creating a "solid circuit." (On February 6, 1959, a patent is issued to TI for "miniaturized electronic circuits.") In January 1959, Noyce develops his integrated circuit using the process of planar technology, developed by a colleague, Jean Hoerni. Instead of connecting individual circuits with gold wires, Noyce uses vapor-deposited metal connections, a method that allows for miniaturization and mass production. Noyce files a detailed patent on July 30, 1959.

1962 The metal oxide semiconductor field effect transistor (MOSFET) is invented by engineers Steven Hofstein and Frederic Heiman at RCA's research laboratory in Princeton, New Jersey. Although slower than a bipolar junction transistor, a MOSFET is smaller and cheaper and uses less power, allowing greater numbers of transistors to be crammed together before a heat problem arises. Most microprocessors are made up of MOSFETs, which are also widely used in switching applications.

1965 The automatic adaptive equalizer is invented in 1965 at Bell Laboratories by electrical engineer Robert Lucky. Automatic equalizers correct distorted signals, greatly improving data performance and speed. All modems still use adaptive equalizers.

1967 A Texas Instruments team, led by Jack Kilby, invents the first handheld calculator (*below*) in order to showcase the integrated circuit. Housed in a case made from a solid piece of aluminum, the battery-powered device fits in the palm of a hand and weighs 45 ounces. It accepts six-digit numbers and performs addition, subtraction, multiplication, and division, printing results up to 12 digits on a thermal printer.

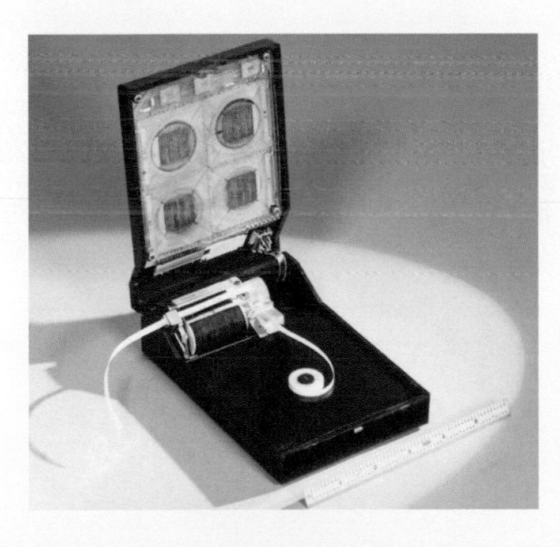

1968 Alfred Y. Cho heads a Bell Labs team that develops molecular beam epitaxy, a process that deposits single-crystal structures one atomic layer at a time, creating materials that cannot be duplicated by any other known technique. This ultra-precise method of growing crystals is now used worldwide for making semiconductor chips and devices such as the semiconductor lasers used in compact disc players. (The term *epitaxy* is derived from the Greek words *epi*, meaning "on," and *taxis*, meaning "arrangement.")

1970 James T. Russell, working at Battelle Memorial Institute's Pacific Northwest Laboratories in Richland, Washington, patents the first system capable of digital-to-optical recording and playback. The CD-ROM (compact disc read-only memory) is years ahead of its time, but in the mid-1980s audio companies purchase licenses to the technology. (See *Computers*, page 85.) Russell goes on to earn dozens of patents for CD-ROM technology and other optical storage systems.

1971 Intel, founded in 1968 by Robert Noyce and Gordon Moore, introduces a "computer on a chip," the 4004 four-bit microprocessor, designed by Federico Faggin, Ted Hoff, and Stan Mazor. It can execute 60,000 operations per second and changes the face of modern electronics by making it possible to include data processing in hundreds of devices. A 4004 provides the computing power for NASA's *Pioneer 10* spacecraft, launched the following year to survey Jupiter.

3M Corporation introduces the ceramic chip carrier, designed to protect integrated circuits when they are attached or removed from circuit boards. The chip is bonded to a gold base inside a cavity in the square ceramic carrier, and the package is then hermetically sealed.

1972 In September, Magnavox ships Odyssey 100 home video game systems to distributors. The system is test marketed in 25 cities, and 9,000 units are sold in Southern California alone during the first month at a price of $99.95.

In November, Nolan Bushnell forms Atari and ships Pong, a coin-operated video arcade game, designed and built by Al Alcorn. The following year Atari introduces its home version of the game, which soon outstrips Odyssey 100.

1974 Texas Instruments introduces the TMS 1000, destined to become the most widely used computer on a chip. Over the next quarter-century, more than 35 different versions of the chip are produced for use in toys and games, calculators, photocopying machines, appliances, burglar alarms, and jukeboxes. (Although TI engineers Michael Cochran and Gary Boone create the first microcomputer, a four-bit microprocessor, at about the same time Intel does in 1971, TI does not put its chip on the market immediately, using it in a calculator introduced in 1972.)

1980 Chuck Stroud, while working at Bell Laboratories, develops and designs 21 different microchips and three different circuit boards—the first to employ built-in self-testing (BIST) technology. BIST results in a significant reduction in the cost, and a significant increase in the quality, of producing electronic components.

1997 IBM announces that it has developed a copper-based chip technology, using copper wires rather than traditional aluminum to connect transistors in chips. Other chip manufacturers are not far behind, as research into copper wires has been going on for about a decade. Copper, the better conductor, offers faster performance, requires less electricity, and runs at lower temperatures. This breakthrough allows up to 200 million transistors to be placed on a single chip.

1998 A team of Bell Labs researchers —Howard Katz, V. Reddy Raju, Ananth Dodabalapur, Andrew Lovinger, and chemist John Rogers—present their latest findings on the first fully "printed" plastic transistor, which uses a process similar to silk screening. Potential uses for plastic transistors include flexible computer screens and "smart" cards, full of vital statistics and buying power, and virtually indestructible.

STEREO

RADIO AND TELEVISION

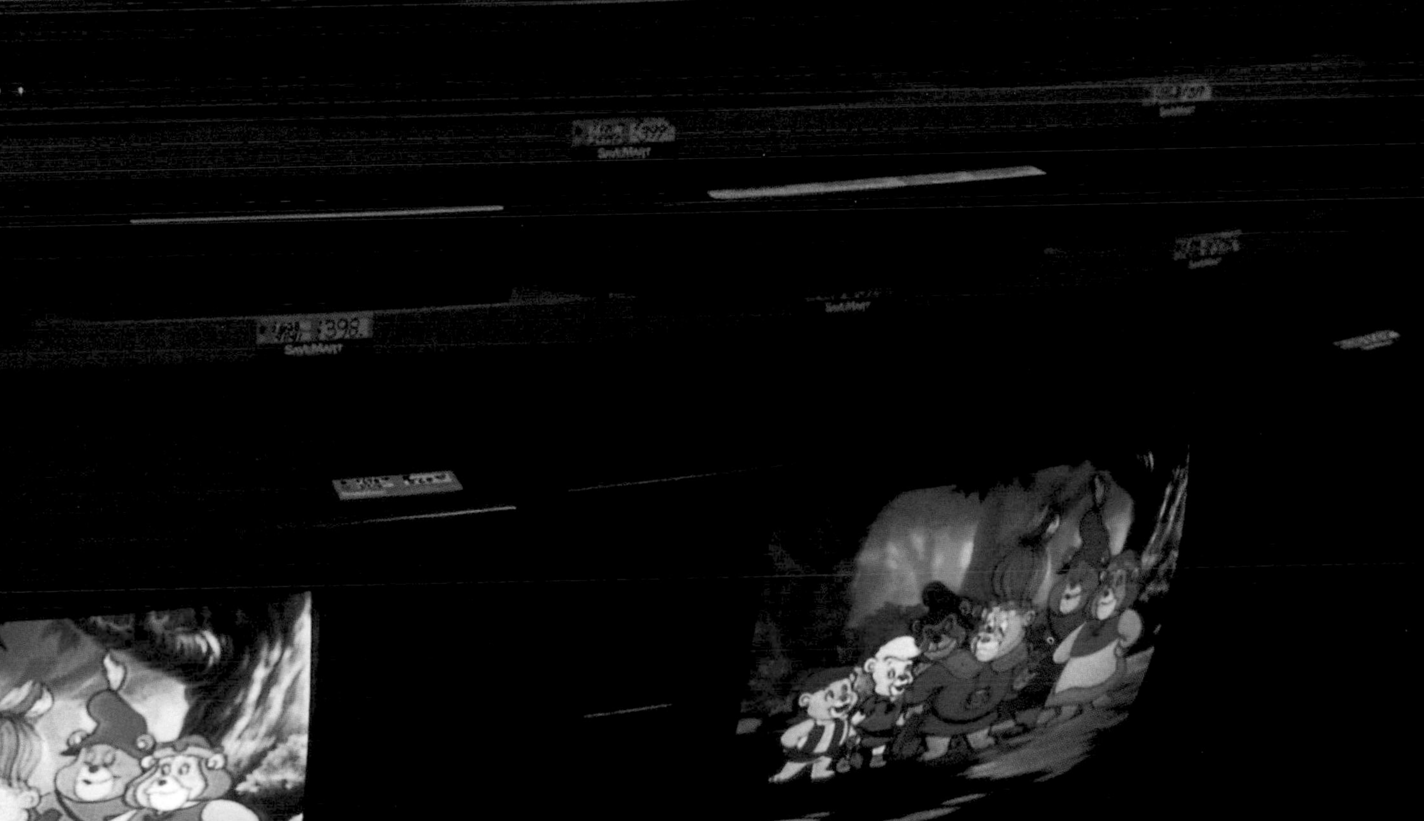

In the autumn of 1899 a new mode of communication wedged its way into the coverage of a hallowed sports event. Outside New York's harbor, two sleek sailboats—*Columbia* of the New York Yacht Club and *Shamrock* of the Ulster Yacht Club in Ireland—were about to compete for the America's Cup, a coveted international trophy. In previous contests the public had no way of knowing what happened on the water until spectators reached shore after the races. This time, however, reports would "come rushing through the air with the simplicity of light," as one newspaper reporter breathlessly put it.

Many people doubted that such a thing was possible, but a young inventor named Guglielmo Marconi proceeded to make good on the promise, using cumbersome sparking devices on observation boats to transmit Morse code messages to land stations a few miles away.

A hundred years later that trickle of dots and dashes had evolved into mighty rivers of information. When another America's Cup competition was held in New Zealand in early 2000, for instance, every detail of the action—the swift maneuvers, straining sails, sunlight winking in spray—was captured by television cameras and then relayed up to a satellite and back down again for distribution to audiences around the world. The imagery rode on the same invisible energy that Marconi had harnessed: radio waves.

Any radio or television signal of today, of course, amounts to only a minuscule fraction of the electromagnetic flow now binding the planet together. Day and night, tens of thousands of radio stations broadcast voice and music to homes, cars, and portable receivers, some that weigh mere ounces. Television pours huge volumes of entertainment, news, sports events, children's programming, and other fare into most households in the developed world. (The household penetration of TV in the United States is 98 percent and average daily viewing time totals 7 hours.) Unrivaled in reach and immediacy, these electronic media bear the main burden of keeping the public informed in times of crisis and provide everyday coverage of the local, regional, and national scenes. But mass communication is only part of the story. Police and fire departments, taxi and delivery companies, jetliner pilots and soldiers all communicate on assigned frequencies. Pagers, cell phones, and wireless links for computers fill additional slices of the spectrum, a now precious realm administered by national and international agencies. As a force for smooth functioning and cohesion of society, radio energy has no equal.

The scientific groundwork for radio and television was laid by the Scottish physicist James Clerk Maxwell, who in 1864 theorized that changes in electrical and magnetic forces send waves spreading through space at 186,000 miles per second. Light consists of such waves, Maxwell said, adding that others might exist at different frequencies. In 1888 a German scientist named Heinrich Hertz confirmed Maxwell's surmise with an apparatus that used sparks to produce an oscillating electric current; the current, in turn, generated electromagnetic energy that caused matching sparks to leap across a gap in a receiving loop of wire a few yards away. And in 1900 brilliant inventor Nikola Tesla was granted two patents for basic radio concepts and devices that inspired others after him.

Fascinated by such findings, Guglielmo Marconi, son of an Irish heiress and Italian aristocrat, began experimenting with electricity as a teenager and soon was in hot pursuit of what he called "wireless telegraphy." In the system he developed, Hertzian sparks created the electromagnetic waves, but Marconi greatly extended their effective range by electrically grounding the transmitter and aerial. At the heart of his receiver was a device called a coherer—a bulb containing iron filings that lost electrical resistance when hit by high-frequency waves. The bulb had to be tapped to separate the filings and restore sensitivity after each pulse was received.

As evidenced by his America's Cup feat in 1899, Marconi was a master of promotion. In 1901 he gained worldwide attention by transmitting the letter "s"—three Morse pips—across the Atlantic. Although his equipment didn't work well over land, he built a successful business by selling wireless telegraphy to shipping companies, maritime insurers, and the world's navies. Telegraphy remained his focus. He didn't see a market beyond point-to-point communication.

Meanwhile, other experimenters were seeking ways to generate radio waves steadily rather than as spark-made pulses. Such continuous waves might be electrically varied—modulated—to convey speech or music. In 1906 that feat was achieved by a Canadian-American professor of electrical engineering, Reginald Fessenden. To create continuous waves, he used an alternator, designed by General Electric engineer Ernst Alexanderson, that rotated at very high speed. Unfortunately, the equipment was expensive and unwieldy, and Fessenden, in any event, was a poor businessman, hatching such unlikely profit schemes as charging by the mile for transmissions.

Fortune also eluded Lee De Forest, another American entrepreneur who tried to commercialize continuous-wave transmissions. In his case the waves were generated with an arc lamp, a method pioneered by Valdemar Poulsen, a Danish scientist. De Forest himself came up with one momentous innovation in 1906—a three-element vacuum tube, or triode, that could amplify an electrical signal. He didn't really understand how it worked or what it might mean for radio, but a young electrical engineer at Columbia University did. In 1912, Edwin Howard Armstrong realized that, by using a feedback circuit to repeatedly pass a signal through a triode, the amplification (hence the sensitivity of a receiver) could be increased a thousandfold. Not only that, but at its highest

Guglielmo Marconi (far left in above photograph) watches as assistants in St. John's, Newfoundland, wrestle with the kite that will be the receiving aerial for the first transatlantic telegraph signals in 1901. A century later, live radio and television coverage could relay the sights and sounds of the America's Cup 2000 race (opposite) from New Zealand to audiences around the world.

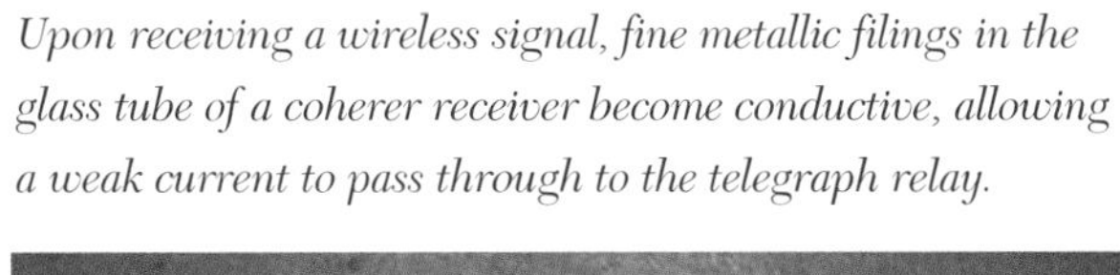
Upon receiving a wireless signal, fine metallic filings in the glass tube of a coherer receiver become conductive, allowing a weak current to pass through to the telegraph relay.

amplification the tube ceased to be a receiving device and became a generator of radio waves. An all-electronic system was at last feasible.

By the early 1920s, after further refinements of transmitters, tuners, amplifiers, and other components, the medium was ready for takeoff. Broadcasting, rather than point-to-point communication, was clearly the future, and the term "wireless" had given way to "radio," suggesting omnidirectional radiation. In the business world, no one saw the possibilities more clearly than David Sarnoff, who started out as a telegrapher in Marconi's company. After the company was folded into the Radio Corporation of America (RCA) in 1919, Sarnoff rose to the pinnacle of the industry. As early as 1915 he wrote a visionary memo proposing the creation of a small, cheap, easily tuned receiver that would make radio a "household utility," with each station transmitting news, lectures, concerts, and baseball games to hundreds of thousands of people simultaneously. World War I delayed matters, but in 1921 Sarnoff demonstrated the market's potential by broadcasting a championship boxing match between heavyweights Jack Dempsey and Georges Carpentier of France. Since radios weren't yet common, receivers in theaters and in New York's Times Square carried the fight—a Dempsey knockout that thrilled the 300,000 gathered listeners. By 1923 RCA and other American companies were producing half a million radios a year.

Radio and television pioneer David Sarnoff (right) began his career as a telegrapher in the Marconi Wireless Telegraph Company and rose to the chairmanship of RCA.

Advertising quickly became the main source of profits, and stations were aggregated into national networks—NBC in 1926, CBS in 1928. At the same time, the U.S. government took control of the spectrum to deal with the increasing problem of signal interference. Elsewhere, some governments chose to go into the broadcasting business themselves, but the American approach was inarguably dynamic. Four out of five U.S. households had radio by the late 1930s. Favorite network shows such as *The Jack Benny Program* drew audiences in the millions and were avidly discussed the next day. During the Depression and the years of war that followed, President Franklin D. Roosevelt regularly spoke to the country by radio, as did other national leaders.

Major advances in radio technology still lay ahead, but many electrical engineers were now focused on the challenge of using electromagnetic waves to transmit moving images. The idea of electrically conveying pictures from one place to another wasn't new. Back in 1884 a German inventor named Paul Nipkow patented a system that did it with two disks, each identically perforated with a spiral pattern of holes and spun at exactly the same rate by motors. The first whirling disk scanned the image, with light passing through the holes and hitting photocells to create an electrical signal. That signal traveled to a receiver (initially by wire) and controlled the output of a neon lamp placed in front of the second disk, whose spinning holes replicated the original scan on a screen. In later, better versions, disk scanning was able to capture and reconstruct images fast enough to be perceived as smooth movement—at least 24 frames per second. The method was used for rudimentary television broadcasts in the United States, Britain, and Germany during the 1920s and 1930s.

But all-electronic television was on the way. A key component was a 19th-century invention, the cathode-ray tube, which generated a beam of electrons and used electrical or magnetic forces to steer the beam across a

A regular family event in the 1930s was gathering around the radio for President Roosevelt's fireside chats.

Philo T. Farnsworth, shown here in 1934 at age 27 demonstrating his television apparatus, patented his invention when he was just 21. By the mid-1950s homes were sprouting TV roof antennas and color television was on the cover of magazines (below).

surface—in a line-by-line scanning pattern if desired. In 1908 a British lighting engineer, Campbell Swinton, proposed using one such tube as a camera, scanning an image that was projected onto a mosaic of photoelectric elements. The resulting electric signal would be sent to a second cathode-ray tube whose scanning beam re-created the image by causing a fluorescent screen to glow. It was a dazzling concept, but constructing such a setup was far beyond the technology of the day. As late as 1920 Swinton gloomily commented: "I think you would have to spend some years in hard work, and then would the result be worth anything financially?"

A young man from Utah, Philo Farnsworth, believed it would. Enamored of all things electrical, he began thinking about a similar scanning system as a teenager. In 1927, when he was just 21, he successfully built and patented his dream. But as he tried to commercialize it he ran afoul of the redoubtable David Sarnoff of RCA, who had long been interested in television. Several years earlier Sarnoff had told his board of directors that he expected every American household to someday have an appliance that "will make it possible for those at home to see as well as hear what is going on at the broadcast station." Sarnoff tried to buy the rights to Farnsworth's designs, but when his offer was rebuffed, he set about creating a proprietary system for RCA, an effort that was led by Vladimir Zworykin, a talented electrical engineer from Russia who had been developing his own electronic TV system. After several years and massive expenditures, Zworykin completed the job, adapting some of Farnsworth's ideas. Sarnoff publicized the product by televising the opening of the 1939 World's Fair in New York, but in the end he had to pay for a license to Farnsworth's patents anyway.

In the ensuing years RCA flooded the market with millions of black-and-white TV sets and also took aim at the next big opportunity—color television. CBS had an electromechanical color system in development, and it was initially chosen as the U.S. standard. However, RCA won the war in 1953 with an all-electronic alternative that, unlike the CBS approach, was compatible with black-and-white sets.

During these years Sarnoff was also locked in a struggle with one of the geniuses of radio technology, Edwin Howard Armstrong, the man

CAMERA LENS. Behind the lens of a video camera, a specially coated prism separates incoming light into its red, green, and blue components and sends them to detectors. These create two electronic signals—luminance, the overall brightness at each point in the picture, and chrominance, the relative strengths of the colors at that spot.

NEWS AT ELEVEN

FROM THE VIDEOCAM IN THE FIELD TO YOUR LIVING ROOM

Television has made seeing-at-a-distance a matter of push-button ease, but remarkable technologies lie behind the feat. In the example shown here, the image of a TV reporter is captured by a portable camera and sent to a distant home via a short microwave hop, a long-haul satellite trip, and a coaxial cable connection for the final leg. The scene that appears on the television screen is an intricate electronic assemblage, a rapid-fire sequence of still images consisting of hundreds of horizontal lines each—525 is the U.S. standard—and displayed at the rate of 30 complete pictures per second to create the illusion of smooth motion. (The pictures are actually double images, successively traced with interlaced lines to eliminate flickering.) To render color, the camera records red, green, and blue versions of the scene; the relative strengths of these primary colors at each point in the image are incorporated into the TV signal and will control three separate electron beams in the picture tube of the receiver. These beams, each replicating the camera's line-by-line scanning in synchrony, cause phosphors on the screen to glow red, green, or blue in proportion to what the camera measured at the start. Various mixes of the three primary colors can create almost all other hues, plus white.

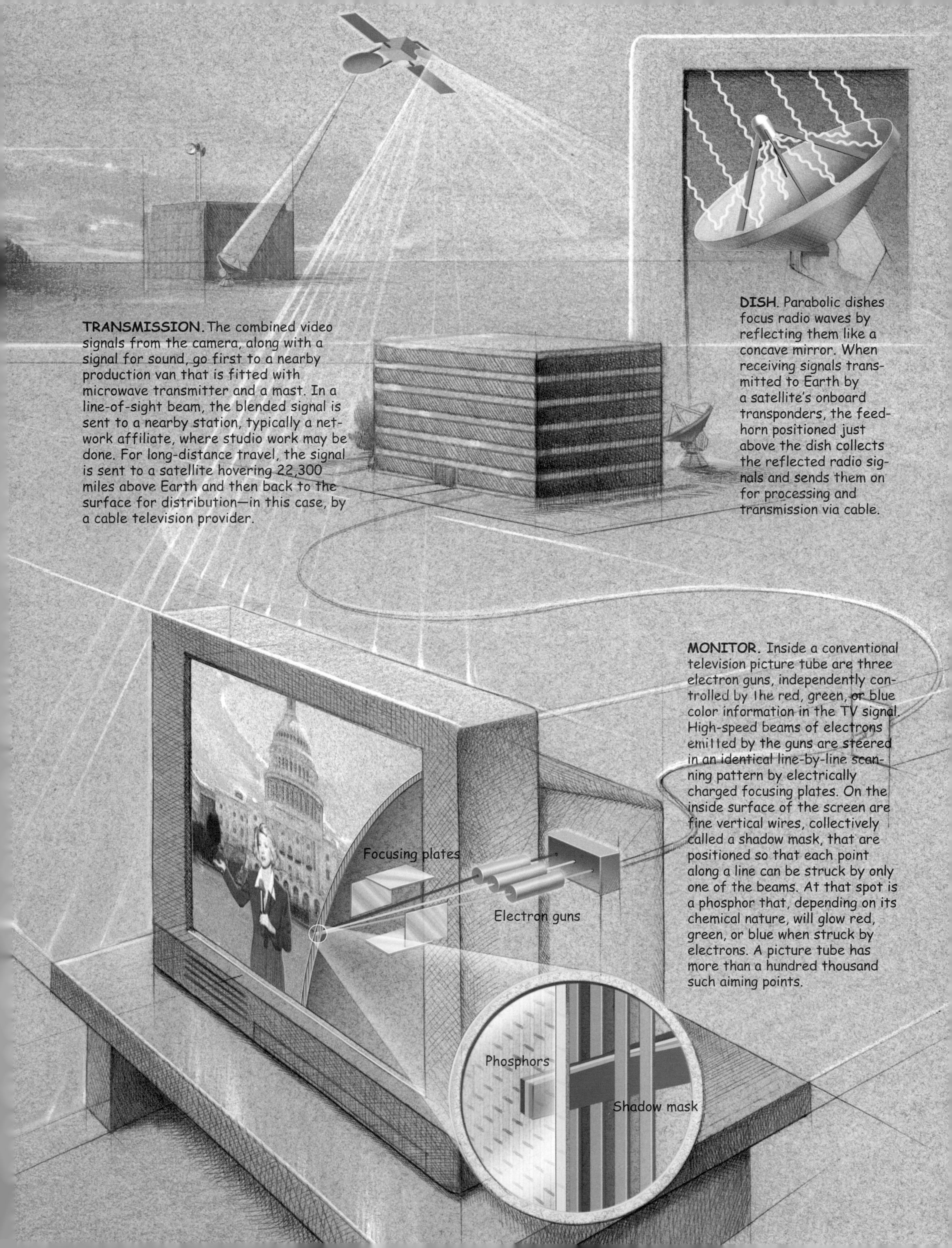
TRANSMISSION. The combined video signals from the camera, along with a signal for sound, go first to a nearby production van that is fitted with microwave transmitter and a mast. In a line-of-sight beam, the blended signal is sent to a nearby station, typically a network affiliate, where studio work may be done. For long-distance travel, the signal is sent to a satellite hovering 22,300 miles above Earth and then back to the surface for distribution—in this case, by a cable television provider.
DISH. Parabolic dishes focus radio waves by reflecting them like a concave mirror. When receiving signals transmitted to Earth by a satellite's onboard transponders, the feedhorn positioned just above the dish collects the reflected radio signals and sends them on for processing and transmission via cable.
MONITOR. Inside a conventional television picture tube are three electron guns, independently controlled by the red, green, or blue color information in the TV signal. High-speed beams of electrons emitted by the guns are steered in an identical line-by-line scanning pattern by electrically charged focusing plates. On the inside surface of the screen are fine vertical wires, collectively called a shadow mask, that are positioned so that each point along a line can be struck by only one of the beams. At that spot is a phosphor that, depending on its chemical nature, will glow red, green, or blue when struck by electrons. A picture tube has more than a hundred thousand such aiming points.
Focusing plates
Electron guns
Phosphors
Shadow mask

The advent in the mid-1960s of commercial telecommunications satellites in geosynchronous orbit transformed spindly TV antennas into concave dish receivers, bringing television to remote areas and additional channels to everyone. By the end of the 20th century, high-definition TV was poised to be the next wave of engineering innovation to tempt the viewing public.

who wrested revolutionary powers from De Forest's vacuum tube. Armstrong had never stopped inventing. In 1918 he devised a method for amplifying extremely weak, high-frequency signals—the superheterodyne circuit. Then in the early 1930s he figured out how to eliminate the lightning-caused static that often plagued radio reception. His solution was a new way of imposing a signal on radio waves. Instead of changing the strength of waves transmitted at a particular frequency (amplitude modulation, or AM), he developed circuitry to keep the amplitude constant and change only the frequency (FM). The result was sound of stunning, static-free clarity.

Once again Sarnoff tried to buy the rights, and once again he failed to reach an agreement. His response this time was to wage a long campaign of corporate and governmental maneuvering that delayed the industry's investment in FM and relegated the technology to low-powered stations and suboptimal frequencies. FM's advantages eventually won it major media roles nonetheless—not only in radio but also as the sound channel for television.

The engineering of radio and television was far from over. The arrival of the transistor in the mid-1950s led to dramatic reductions in the size and cost of circuitry. Videocassette recorders for delayed viewing of TV shows appeared in 1956. Screens grew bigger and more vivid, and some dispensed with cathode-ray technology in favor of new display methods that allowed them to be flat enough to hang on a wall. Cable television—the delivery of signals by coaxial cable rather than through the air—was born in 1949 and gained enormous popularity for its good reception and additional programming. The first commercial telecommunications satellite began service in 1965 and was followed by whole fleets of orbiting transmitters. Satellite television is able to provide far more channels than a conventional TV transmitter because each satellite is allocated a big slice of the electromagnetic spectrum at very high frequencies. With all new wireless technologies, finding room on the radio spectrum—a realm that ranges from waves many miles long to just a millimeter in length—is always a key issue, with conservation growing ever more important.

By century's end the move was toward a future known as high-definition television, or HDTV. The U.S. version, to be phased in over many years, will bring television sets whose digital signals can be electronically processed for superior performance and whose images are formed of more than a thousand scanned lines, yielding much higher resolution than the current 525-line standard. Meanwhile, TV's reach has extended far beyond our world. Television pictures, digitally encoded in radio waves, are streaming to Earth from space probes exploring planets and moons in the far precincts of the solar system. For this most distance dissolving of technologies, no limits are yet in sight.

Perspective

Robert W. Lucky

Retired Corporate Vice President
Applied Research
Telcordia Technologies, Inc.

When I was young there was no television. This was difficult to explain to my children. "Oh no, Dad," they would say, "There was always TV." They can't understand what people did at night in that incomprehensible time when lives were not illuminated by television.

But I remember. My world at night was filled with the magic sounds of radio. I would lie in bed in the darkness, watching the dancing glows of the filaments in my bedside radio. I imagined sometimes that there were little people encased in those tubes and their voices were those I heard. Now in the modern daylight of television it is hard to explain the reality of radio in that long-lost time. I rode with the Lone Ranger. I sent away for the secret decoder ring from Captain Midnight so I could unscramble the coded messages about the next episode. The pictures I drew in my mind may have been more real than the ever-changing, evanescent images from the ubiquitous cathode-ray tubes of today.

I wanted to create this miracle of radio myself. I built crystal radios with "cat whiskers" that touched delicately on little cubes of quartz and listened acutely through earphones as I moved a steel pointer across a coil wound on a cardboard tube. Sadly, I never heard a peep. So I studied a book entitled *Boys' First Book of Radio* and dog-eared a precious copy of the *Amateur Radio Handbook*. From them I learned about superheterodyne receivers. I designed and built one and experienced an unforgettable thrill when I turned the switch and music came from the speaker. That radio made an engineer of me.

The magic of radio lives with me today, but now I see it through the eyes of an experienced engineer. I look out the window at the clear blue sky and think of all the radio waves crossing that seemingly empty space. If those waves had visible color, the sky would be as bright as a laser light show.

It wasn't all that long ago when there were no waves at all. I remember the feeling I had when I visited Marconi's home outside Bologna, Italy, with his daughter, Gioia, who had become a good friend. I looked out the window where he had sent the first radio pulse and wondered what he must have felt like when the iron filings in the glass tube of the coherer detector across the hill jumped at the recognition of his pulse.

Somewhere out there, 100 light-years distant, that first pulse is still traveling among the stars. Its creator, Marconi, must have believed that the heavens had been opened to unlimited communication. As an engineer in the late 20th century, however, I came to realize that the precious spectrum that had seemed free and infinite in Marconi's day had been sold in tiny slivers for billions of dollars.

Today, we again use Marconi's word, "wireless," to describe cellular radio. There has been a renaissance in thinking about the capabilities of that empty sky. New methods of transmission, of processing signals, and of sharing the spectrum have cascaded out of universities and research laboratories. The 20th century saw radio emerge, blossom, and ultimately devour all the capacity that nature had given us. The 21st century may see us reclaim the vastness of Marconi's dream with these new technologies.

RADIO AND TELEVISION TIMELINE

Radio and television were major agents of social change in the 20th century, opening windows to other peoples and places and bringing distant events directly into millions of homes. Although Guglielmo Marconi was the first to put the theory of radio waves into practice, the groundwork for his feat was laid in the 19th century by James Clerk Maxwell, Heinrich Hertz, and Nikola Tesla. Maxwell theorized and Hertz confirmed the feasibility of transmitting electromagnetic signals. Tesla invented a device—the Tesla coil—that converts relatively low-voltage current to high-voltage low current at high frequencies. Some form of the Tesla coil is still used in radio and television sets today.

1900 Nikola Tesla *(above)* is granted a U.S. patent for a "system of transmitting electrical energy" and another patent for "an electrical transmitter"—both the products of his years of development in transmitting and receiving radio signals. These patents would be challenged and upheld (1903), reversed (1904), and finally restored (1943).

1901 Guglielmo Marconi, waiting at a wireless receiver in St. John's, Newfoundland, picks up the first transatlantic radio signal, transmitted some 2,000 miles from a Marconi station in Cornwall, England. To send the signal—the three dots of the Morse letter "s"—Marconi's engineers send a copper wire aerial skyward by hoisting it with a kite. Marconi builds a booming business using radio as a new way to send Morse code.

1904 English physicist John Ambrose Fleming recognizes that the Edison effect—Thomas Edison's discovery in 1883 that an electric current flows through a vacuum from a lighted filament to a metal plate mounted inside a lightbulb—can also be used to detect radio waves and convert them to electricity. Fleming's finding leads to his development of the diode, or two-element vacuum tube, which becomes the first crude radio receiver.

1906 Expanding on Fleming's invention, American entrepreneur Lee De Forest puts a third wire, or grid, into a vacuum tube, creating a sensitive receiver. He calls his invention the "Audion." In later experiments he feeds the Audion output back into its grid and finds that this regenerative circuit can transmit signals.

On Christmas Eve 1906 engineering professor Reginald Fessenden transmits a voice and music program in Massachusetts that is picked up as far away as Virginia.

1912 Columbia University electrical engineering student Edwin Howard Armstrong devises a regenerative circuit for the triode that amplifies radio signals. By pushing the current to the highest level of amplification, he also discovers the key to continuous-wave transmission, which becomes the basis for amplitude modulation (AM) radio. In a long patent suit with Lee De Forest, whose three-element Audion was the basis for Armstrong's work, the courts eventually decide in favor of De Forest, but the scientific community credits Armstrong as the inventor of the regenerative circuit.

Guglielmo Marconi (seen far right) watches a demonstration in 1923 of his wireless telegraphy with a roomful of politicians.

1917 While serving in the U.S. Army Signal Corps during World War I, Edwin Howard Armstrong invents the superheterodyne circuit, an eight-tube receiver that dramatically improves the reception of radio signals by reducing static and increasing selectivity and amplification. He files for a patent the following year.

1920 Station KDKA in Pittsburgh becomes radio's first scheduled commercial programmer with its broadcast of the Harding-Cox presidential election returns, transmitted at 100 watts from a wooden shack atop the Westinghouse Company's East Pittsburgh plant. Throughout the broadcast KDKA intersperses the election returns and occasional music with a message: "Will anyone hearing this broadcast please communicate with us, as we are anxious to know how far the broadcast is reaching and how it is being received?"

1925 Scottish inventor John Logie Baird successfully transmits the first recognizable image—the head of a ventriloquist's dummy *(opposite)*—at a London department store, using a device he calls a Televisor. A mechanical system based on the spinning disk scanner developed in the 1880s by German scientist Paul Nipkow, it requires synchronization of the transmitter and receiver disks. The Televisor images, composed of 30 lines flashing 10 times per second, are so hard to watch they give viewers a headache.

Charles F. Jenkins pioneers his mechanical wireless television system, radiovision, with a public transmission sent from a navy radio station across the Anacostia River to his office in downtown Washington, D.C. Jenkins's radiovisor is a multitube radio set with a special scanning-drum attachment for receiving pictures—cloudy 40- to 48-line images projected on a six-inch-square mirror. Jenkins's system, like Baird's, broadcasts and receives sound and visual images separately. Three years later the Federal Radio Commission grants Charles Jenkins Laboratories the first license for an experimental television station.

1927 Using his all-electronic television system, 21-year-old Utah farm boy and electronic prodigy Philo T. Farnsworth transmits images of a piece of glass painted black, with a center line scratched into the paint. The glass is positioned between a blindingly bright carbon arc lamp and Farnsworth's "image dissector" cathode-ray camera tube. As viewers in the next room watch a cathode-ray tube receiver, someone turns the glass slide 90 degrees—and the line moves. The use of cathode-ray tubes to transmit and receive pictures—a concept first promoted by British lighting engineer A. Campbell Swinton—is the death knell for the mechanical rotating-disk scanner system.

1928 John Logie Baird demonstrates, here with the aid of two ventriloquist's dummies, that his Televisor system can

produce images in crude color by covering three sets of holes in his mechanical scanning disks with gels of the three primary colors. The results, as reported in 1929 following an experimental BBC broadcast, appear "as a soft-tone photograph illuminated by a reddish-orange light."

1929 Vladimir Zworykin, who came to the United States from Russia in 1919, demonstrates the newest version of his iconoscope, a cathode-ray-based television camera that scans images electronically, and a cathode-ray tube receiver called the kinescope. The iconoscope, first developed in 1923, is similar to Philo Farnsworth's "image dissector" camera tube invention, fueling the growing rivalry between the two inventors for the eventual title of "father of modern television."

1933 Edwin Howard Armstrong (*above*) develops frequency modulation, or FM, radio as a solution to the static interference problem that plagues AM radio transmission, especially in summer when electrical storms are prevalent. Rather than increasing the strength or amplitude of his radio waves, Armstrong changes only the frequency on which they are transmitted. However, it will be several years before FM receivers come on the market.

1947 The future of radio and television is forever changed when John Bardeen, Walter Brattain, and William Shockley of Bell Laboratories coinvent the transistor (see *Electronics,* page 52).

1950s Engineers improve the rectangular cathode-ray tube (CRT) for television monitors, eliminating the need for rectangular "masks" over the round picture tubes of earlier monitors. The average price of a television set drops from $500 to $200.

1953 RCA beats out rival CBS when the National Television System Committee adopts RCA's new system for commercial color TV broadcasting. CBS has pioneered color telecasting, but its system is incompatible with existing black-and-white TV monitors throughout the country.

1954 The New Year's Day Tournament of Roses in Pasadena, California, becomes the first coast-to-coast color television transmission, or "colorcast." The parade is broadcast by RCA's NBC network to 21 specially equipped stations and is viewed on newly designed 12-inch RCA Victor receivers set up in selected public venues. Six weeks later NBC's *Camel News Caravan* transmits in color, and the following summer the network launches its first color sitcom, *The Marriage*, starring Hume Cronyn and Jessica Tandy.

Regency Electronics introduces the TR-1, the first all-transistor radio. It operates on a 22-volt battery and works as soon as it is switched on, unlike tube radios, which take several minutes to warm up. The TR-1 sells for $49.95; is available in six colors, including mandarin red, cloud gray and olive green; and is no larger than a package of cigarettes.

1958 Jack S. Kilby of Texas Instruments and Robert Noyce of Fairchild Semiconductor, working independently, create the integrated circuit, a composite semiconductor block in which transistor, resistor, condenser, and other electrical components are manufactured together as one unit. Initially, the revolutionary invention is seen primarily as an advancement for radio and television, which together were then the nation's largest electronics industry (see *Electronics*, page 53).

1962 Communications satellite *Telstar 1*, designed by John Pierce of Bell Labs, is launched by a NASA Delta rocket on July 10, transmitting the first live transatlantic telecast as well as telephone and data signals.

1968 There are 200 million television sets in operation worldwide, up from 100 million in 1960. By 1979 the number reaches 300 million and by 1996 over a billion. In the United States the number grows from 1 million in 1948 to 78 million in 1968. In 1950 only 9 percent of American homes have a TV set; in 1962, 90 percent; and in 1978, 98 percent, with 78 percent owning a color TV.

1988 Sony introduces the first in its "Watchman" series of handheld, battery-operated, transistorized television sets. Model FD-210, with its 1.75-inch screen, is the latest entry in a 30-year competition among manufacturers to produce tiny microtelevisions. The first transistorized TV, Philco's 1959 Safari, stood 15 inches high and weighed 15 pounds.

1990 Following a demonstration by Phillips two years earlier of a high-definition TV (HDTV) system for satellite transmission, the Federal Communications Commission sets a testing schedule for a proposed all-digital HDTV system. Tests begin the next year, and in 1996 Zenith introduces the first HDTV-compatible front-projection television. Also in 1996, broadcasters, TV manufacturers, and PC makers set interindustry standards for digital HDTV. By the end of the century, digital HDTV, which produces better picture and sound than analog television and can transmit more data faster, is on the verge of offering completely interactive TV.

AGRICULTURAL MECHANIZATION

You often see them from the window of a cross-country jet: huge, perfect circles in varying shades of green, gold, or brown laid out in a vast checkerboard stretching to the horizon. Across much of the American Midwest and on farmland throughout the world, these genuine crop circles are the sure sign of an automated irrigation system—and an emblem of a revolution in agriculture, the most ancient of human occupations. At the heart of this transformation is a single concept: mechanization.

When viewed across the span of the 20th century, the effect that mechanization has had on farm productivity—and on society itself—is profound. At the end of the 19th century it took, for example, 35 to 40 hours of planting and harvesting labor to produce 100 bushels of corn. A hundred years later producing the same amount of corn took only 2 hours and 45 minutes—and the farmers could ride in air-conditioned comfort, listening to music while they worked. And as fewer and fewer workers were needed on farms, much of the developed world has experienced a sea-change shift from rural to metropolitan living.

Throughout most of its long history, agriculture—particularly the growing of crops—was a matter of human sweat and draft animal labor. Oxen, horses, and mules pulled plows to prepare the soil for seed and hauled wagons filled with the harvest—up to 20 percent of which went to feed the animals themselves. The rest of the chores required backbreaking manual labor: planting the seed; tilling, or cultivating, to keep down weeds; and ultimately reaping the harvest, itself a complex and arduous task of cutting, collecting, bundling, threshing, and loading. From early on people with an inventive flair—perhaps deserving the title of the first engineers—developed tools to ease farming burdens. Still, even as late as the 19th century, farming and hard labor remained virtually synonymous, and productivity hadn't shifted much across the centuries.

At the turn of the 20th century the introduction of the internal combustion engine set the stage for dramatic changes. Right at the center of that stage was the tractor. It's not just a figure of speech to say that tractors drove the mechanization revolution. Tractors pulled plows. They hauled loads and livestock. Perhaps most importantly, tractors towed and powered the new planters, cultivators, reapers, pickers, threshers, combine harvesters, mowers, and balers that farm equipment companies kept coming out with every season. These vehicles ultimately became so useful and resourceful that farmers took to calling them simply GPs, for general purpose. But they weren't always so highly regarded. Early versions, powered by bulky steam engines, were behemoths, some weighing nearly 20 tons. Lumbering along on steel wheels, they were often mired in wet and muddy fields—practically worthless. Then in 1902 a pair of engineers named Charles Hart and Charles Parr introduced a tractor powered by an internal combustion engine that ran on gasoline. It was smaller and lighter than its steam-driven predecessors, could pull plows and operate threshing machines, and ran all day on a single tank of fuel. Hart and Parr's company was the first devoted exclusively to making tractors, a term they are also credited with introducing. Previously, tractors had been known as "traction engines."

Four Wisconsin farmers show off their labor-saving steam tractor—soon to be replaced by the internal combustion engine.

The Hart-Parr Model 3 tractor was a commercial success, prompting no less a businessman than Henry Ford to get into the picture. In 1917 he introduced the Fordson, weighing as little as one ton and advertised to sell for as little as $395. The Fordson soon ruled the tractor roost, accounting for 75 percent of the U.S. market share and 50 percent of the worldwide share.

Nevertheless, the tractor business remained a competitive field, at least for a few decades, and competition helped foster innovations. Tractors themselves got smaller and more lightweight and were designed with a higher ground clearance, making them capable of such relatively refined tasks as hauling cultivating implements through a standing crop. Another early innovation, introduced by International Harvester in 1922, was the so-called power takeoff. This device consisted of a metal shaft that transmitted the engine power directly to a towed implement such as a reaper through a universal joint or similar mechanism; in other words,

An English farm woman in 1921 plows her field with a Fordson tractor, which dominated the worldwide tractor market.

HIGH-TECH HARVESTING
WITH FINGERTIP CONTROL

Mechanization has revolutionized just about every aspect of farming, but perhaps nowhere is the impact more apparent than in the harvesting of grain. The combine, which has been around in one form or another since the middle of the 19th century, is now a state-of-the-art workhorse. Sophisticated electronics make possible everything from monitoring of crop moisture, yield, and temperature conditions to precision navigating by means of the satellite-based Global Positioning System. The operator can program the combine to cut the grain at different heights depending on field conditions and can even set a desired seed size so that kernels below the desired size will automatically be rejected along with the chaff and straw. Ergonomic seating and climate control in the cab are not just luxuries: along with high-powered halogen lamps, they enable the harvest to continue virtually around-the-clock.

HARVESTING THE CROP. The reel (1) moves the standing grain back over the cutter bar (2). The reel and cutter bar are mounted on a "header" that can be raised and lowered as the combine advances through the field. The cut grain is lifted into the threshing cylinder (3), which separates the grain from the straw through a rubbing action. The threshed grain, straw, and chaff are then conveyed to a set of "walkers" (4) that move the bulk of the straw rearward and out of the machine to the rear (5). The grain and chaff fall through the walkers onto oscillating sieves (6). A fan (7) blows air through the sieves and, since the grain is denser than the chaff, the grain falls down and is collected by an auger (8), while the chaff is blown out of the combine along with the straw. The threshed grain is elevated into a grain storage tank (9). Periodically a grain truck or tractor-drawn trailer pulls alongside the combine and the elevator (10) empties the grain tank on the move (above).

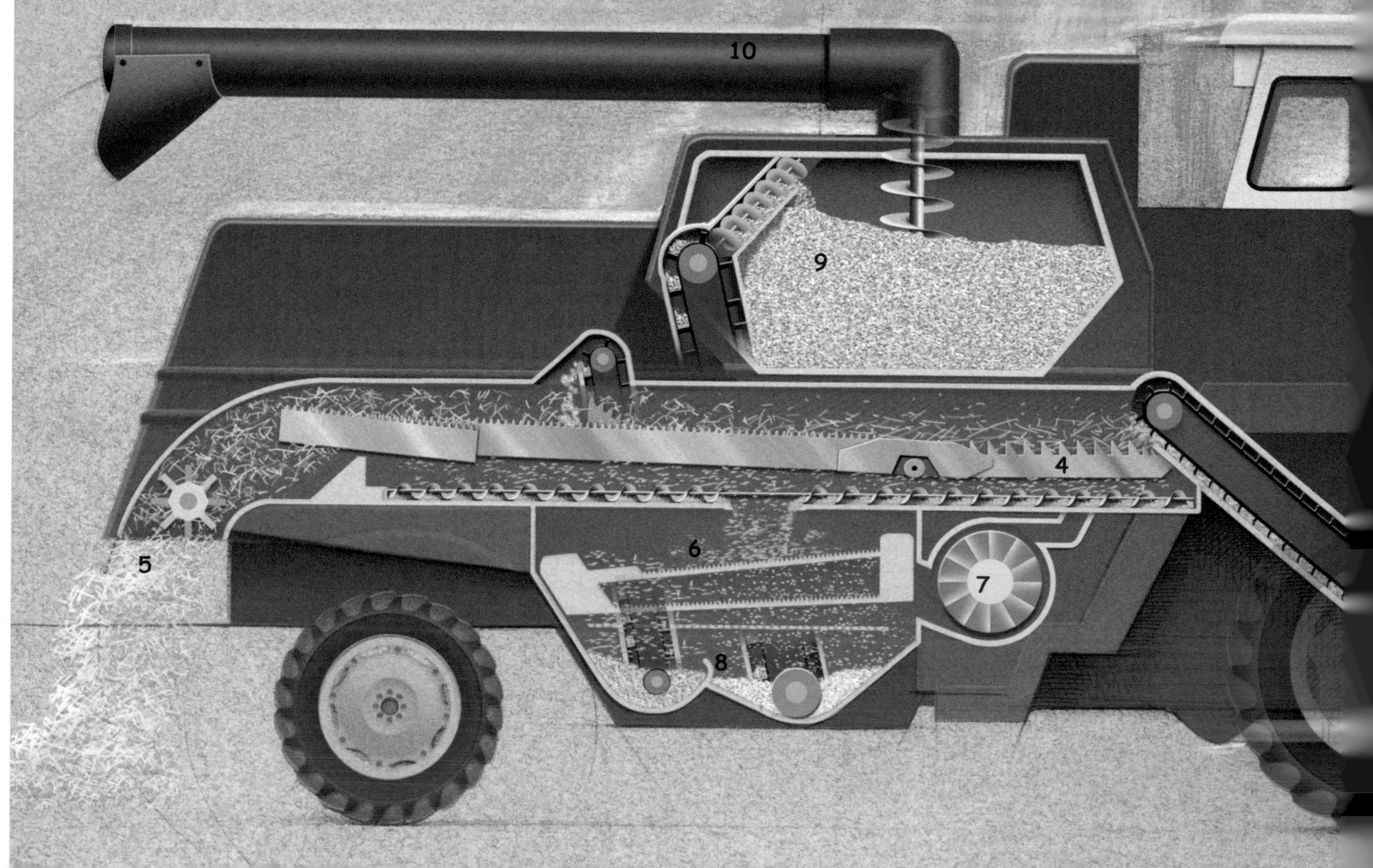

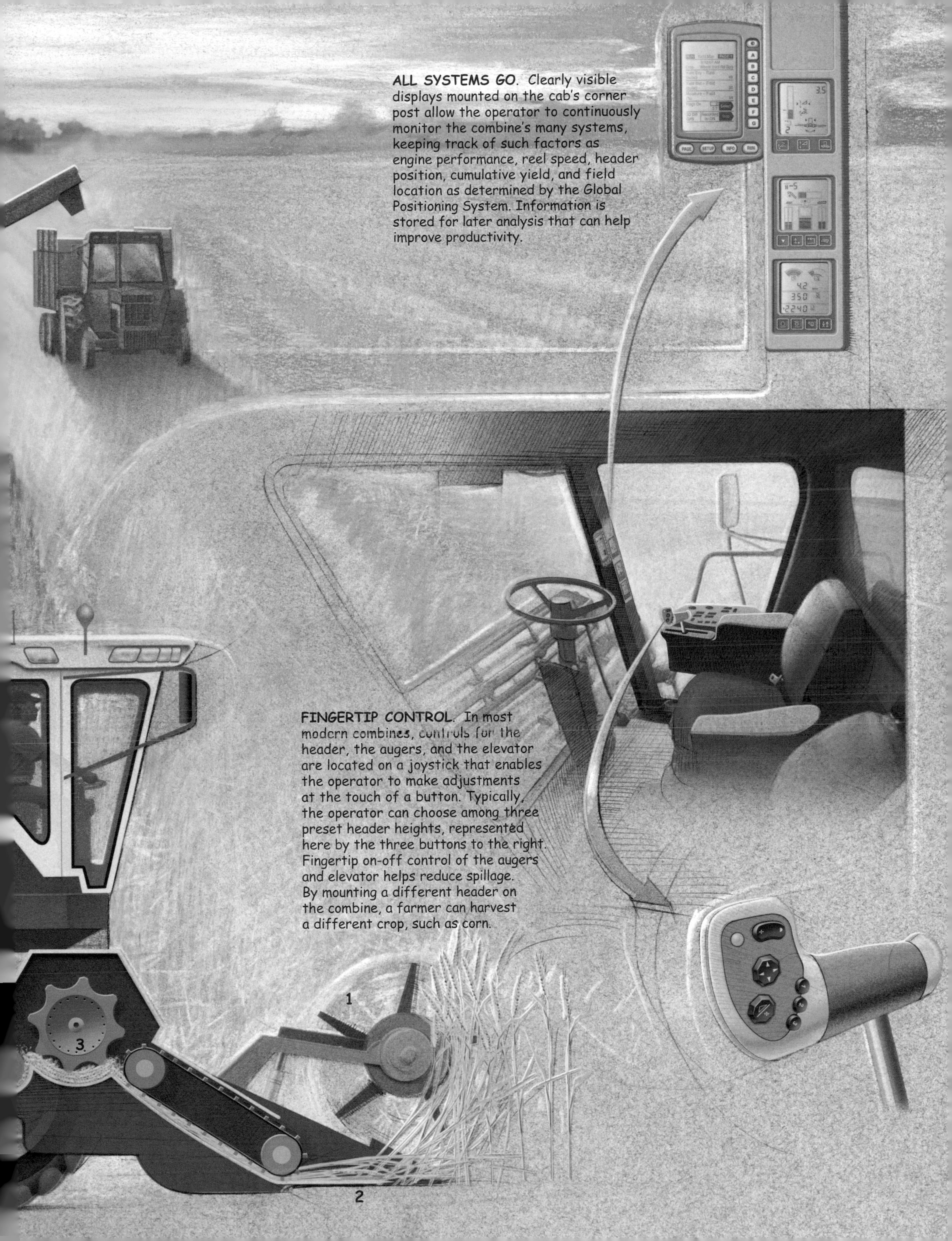

ALL SYSTEMS GO. Clearly visible displays mounted on the cab's corner post allow the operator to continuously monitor the combine's many systems, keeping track of such factors as engine performance, reel speed, header position, cumulative yield, and field location as determined by the Global Positioning System. Information is stored for later analysis that can help improve productivity.

FINGERTIP CONTROL. In most modern combines, controls for the header, the augers, and the elevator are located on a joystick that enables the operator to make adjustments at the touch of a button. Typically, the operator can choose among three preset header heights, represented here by the three buttons to the right. Fingertip on-off control of the augers and elevator helps reduce spillage. By mounting a different header on the combine, a farmer can harvest a different crop, such as corn.

A boon on large farms is center pivot irrigation, which can also be used to apply fertilizer and pesticides.

almost quadrupled irrigated acreage in the United States and has also been used to apply both fertilizers and pesticides.

Mechanization has come to the aid of another critical aspect of agriculture—namely, soil conservation. An approach known as conservation tillage has greatly reduced, or even eliminated, traditional plowing, which can cause soil erosion and loss of nutrients and precious moisture. Conservation tillage includes the use of sweep plows, which undercut wheat stubble but leave it in place above ground to help restrict soil erosion by wind and to conserve moisture. The till plant system is another conservation-oriented approach. Corn stalks are left in place to reduce erosion and loss of moisture, and at planting time the next year the row is opened up, the seeds are planted, and the stalks are turned over beside the row, to be covered up by cultivation. This helps conserve farmland by feeding nutrients back into the soil.

As the century unfolded, everything about farming was changing—not the least its fundamental demographics. In 1900 farmers made up 38 percent of the U.S. labor force; by the end of the century they represented less than 3 percent. With machines doing most of the work, millions of farmers and farm laborers had to look elsewhere for a living—a displacement that helped fuel booms in the manufacturing and service industries, especially after World War II. It also fueled a dramatic shift in the entire culture, as metropolitan and suburban America began to replace the rural way of life. Although some may lament the passing of the agrarian way of life, in much of the developing world these transformations represent hope. The many ways in which agriculture has been changed by engineering—from new methods for land and resource management to more efficient planting and harvesting to the development of better crop varieties—offer the potential solution to the endemic problems of food shortage and economic stagnation.

With precision planting and fertilizing thanks to Global Positioning System technology and computer software that enables farmers to track the productivity of their land virtually acre by acre, one U.S. farmer, on average, can feed 129 people at home and abroad.

Perspective

DONALD JOHNSON

Retired Vice President
Product and Process Technology
Grain Processing Corporation

When I was growing up in north-central Illinois, I watched corn and soybeans being brought to my little rural community's grain elevators in wagons pulled by tractors and sometimes even horses. Talk of 50 bushels per acre of corn was big news as we scrambled to pick up spilled soybeans to arm our peashooters. It was nearly halfway through the 20th century then, and U.S. corn production, the benchmark crop, averaged less than 30 bushels per acre, only slightly higher than at the turn of the century. Over four people resided on and operated an average-sized, 160-acre farm. Farms now average more than twice that size and are commonly operated by only one person.

The big growth in crop yields occurred after the Second World War, increasing more than five-fold through new agricultural practices and hybrid development. Productivity, however, increased by more than 50-fold over the course of the 20th century, due for the most part to mechanization. It is this fantastic productivity that keeps agricultural crops abundantly available at affordable prices as a raw material for industrial products as well as for foodstuffs. Mechanization has made the United States the "breadbasket of the world," and, more than that, it provides the springboard for sustainability on the planet.

My own career encompasses utilizing agricultural crops in the commodity grain-processing industry, an industry that converts oilseeds and grains into millions of pounds per day of foodstuffs, sweeteners, fuels, chemicals, building products, paper adjuvants, and even plastics. This huge industry is but a segment of the broader agricultural commodities industry. The food we eat, the clothes we wear, the houses we live in, and the magazines we read all depend on agricultural crops. They have experienced the least cost inflation of most commodities and a stable supply because of the ingenuity and inventiveness of mechanical and agricultural engineers. Together, the invention and development of implements to plant, cultivate, harvest, and transport agricultural crops efficiently and regardless of unfavorable weather conditions are truly a wondrous and critical achievement.

This fascinating story will continue to unfold throughout the 21st century as we more completely embrace the concept of "sustainability." Fuels, chemicals, and materials will of necessity have to be derived from renewable resources. New crops, new agricultural practices, and new mechanical devices will have to be developed to sustain feeding, clothing, housing, and the quality of life of the growing world population.

AGRICULTURAL MECHANIZATION
TIMELINE

In 1900 farmers represented 38 percent of the U.S. labor force. By the end of the century that number had plunged to 3 percent—dramatic evidence of the revolution in agriculture brought about by mechanization. Beginning with the internal combustion engine and moving on to rubber tires that kept machinery from sinking in muddy soil, mechanization also improved the farm implements designed for planting, harvesting, and reaping. The advent of the combine, for example, introduced an economically efficient way to harvest and separate grain. As the century closed, "precision agriculture" became the practice, combining the farmer's down-to-earth know-how with space-based technology.

1902 Charles Hart and Charles Parr establish the first U.S. factory devoted to manufacturing a traction engine powered by an internal combustion engine. Smaller and lighter than its steam-driven predecessors, it runs all day on one tank of fuel. Hart and Parr are credited with coining the term "tractor" for the traction engine.

1904 Benjamin Holt, a California manufacturer of agricultural equipment, develops the first successful crawler tractor, equipped with a pair of tracks rather than wheels. Dubbed the "caterpillar" tread, the tracks help keep heavy tractors from sinking in soft soil and are the inspiration for the first military tanks. The 1904 version is powered by steam; a gasoline engine is incorporated in 1906. The Caterpillar Tractor Company is formed in 1925, in a merger of the Holt Manufacturing Company and its rival, the C. L. Best Gas Traction Company.

Early tracked farm vehicles—this one built in England in 1905—were the predecessors of WWI armored tanks.

1905 Jay Brownlee Davidson designs the first professional agricultural engineering curriculum at then-Iowa State College. Courses include agricultural machines; agricultural power sources, with an emphasis on design and operation of steam tractors; farm building design; rural road construction; and field drainage. Davidson also becomes the first president of the American Society of Agricultural Engineers in 1907, leading agricultural mechanization missions to the Soviet Union and China.

In the early 1920s International Harvester introduced the Farmall tractor, with its characteristic high-clearance rear axle and closely spaced front wheels to negotiate between rows.

1917 Henry Ford & Son Corporation—a spinoff of the Ford Motor Company—begins production of the Fordson tractor. Originally called the "automobile plow" and designed to work 10- to 12-acre fields, it costs as little as $395 and soon accounts for 50 percent of the worldwide market for tractors.

1918 American Harvester Company of Minneapolis begins manufacturing the horse-drawn Ronning Harvester, a corn silage harvester patented in 1915 by Minnesota farmers Andrean and Adolph Ronning. The Ronning machine uses and improves a harvester developed three years earlier by South Dakotan Joseph Weigel. The first field corn silage harvester was patented in 1892 by Iowan Charles C. Fenno.

1921 U.S. Army pilots and Ohio entomologists conduct the first major aerial dusting of crops, spraying arsenate of lead over 6 acres of catalpa trees in Troy to control the sphinx caterpillar. Stricter regulations on pesticides and herbicides go into effect in the 1960s.

1922 International Harvester introduces a power takeoff feature, a device that allows power from a tractor engine to be transmitted to attached harvesting equipment. This innovation is part of the company's signature Farmall tractor in 1924. The Farmall features a tricycle design with a high-clearance rear axle and closely spaced front wheels that run between crop rows. The four-cylinder tractor can also be mounted with a cultivator guided by the steering wheel.

1931 Caterpillar manufactures a crawler tractor with a diesel engine, which offers more power, reliability, and fuel efficiency than those using low-octane gasoline. Four years later International Harvester introduces a diesel engine for wheeled tractors. Several decades later diesel fuel would still be used for agricultural machinery.

1932 An Allis-Chalmers Model U tractor belonging to Albert Schroeder of Waukesha, Wisconsin, is outfitted with a pair of Firestone 48×12 airplane tires in place of lugged steel wheels. Tests by the University of Nebraska Tractor Test Laboratory find that rubber wheels result in a 25 percent improvement in fuel economy. Rubber wheels also mean smoother, faster driving with less wear and tear on tractor parts and the driver. Minneapolis Marine Power Implement Company even markets a "Comfort Tractor" with road speeds up to 40 mph, making it usable on public roads for hauling grain or transporting equipment.

The Ann Arbor Machine Company of Shelbyville, Illinois, manufactures the first pickup baler, based on a 1929 design by Raymond McDonald. Six years later Edwin Nolt develops and markets a self-tying pickup baler. The baler, attached to a tractor, picks up cut hay in the field, shapes it into a 16–18-inch bale, and knots the twine that holds the bale secure. Self-propelled hay balers soon follow.

1933 Irish mechanic Harry Ferguson develops a tractor that incorporates his innovative hydraulic draft control system, which raises and lowers attached implements—such as tillers, mowers, post-hole diggers, and plows—and automatically sets their needed depth. The David Brown Company in England is the first to build the tractor, but Ferguson also demonstrates it to Henry Ford in the United States. With a handshake agreement, Ford manufactures Ferguson's tractor and implements from 1939 to 1948. A few years later Ferguson's company merges with Canadian company Massey-Harris to form Massey-Ferguson.

1935 Agronomists Frank Duley and Jouette Russell at the University of Nebraska, along with other scientists with the U.S. Soil Conservation Service, begin the first research on conservation tillage. The practice involves various methods of tilling the soil, with stubble mulch and different types of plows and discs, to control wind erosion and manage crop residue. This technology is common on farms by the early 1960s.

Electric cooperatives formed with the help of the Rural Electrification Administration bring electricity that enables farmers to upgrade machinery, operate water pumps, and replace kerosene lamps with modern lighting.

1938 In Australia, Massey-Harris introduces the first self-propelled combine—a thresher and reaper in a single machine—not drawn by a tractor or horse. Welcomed because it replaces the labor-intensive binder, handshocking, and threshing, the new combine becomes increasingly popular. By the end of the century, single-driver combines feature air-conditioned cabins that are lightly pressurized to keep out dirt and debris.

Center pivot irrigation, invented by Frank Zybach, creates circles of green in the Arizona desert.

1943 International Harvester builds "Old Red," the first commercially viable mechanical spindle cotton picker, invented and tested by Texans John and Mack Rust beginning in 1927. The spindle picker features moistened rotating spindles that grab cotton fibers from open bolls while leaving the plant intact. The cotton fibers are then blown into waiting hoppers, free of debris.

1948 Colorado farmer Frank Zybach invents the center pivot irrigation machine, which revolutionizes irrigation technology. The system consists of sprinklers attached to arms that radiate from a water-filled hub out to motorized wheeled towers in the field. Zybach is awarded a patent in 1952 for the "Self-Propelled Sprinkling Irrigating Apparatus."

1954 The John Deere and International Harvester companies introduce corn head attachments for their combines. This attachment rapidly replaces the self-propelled corn picker, which picked the corn and stripped off its husk. The corn head attachment also shells the ears in the field. The attachment allows a farmer to use just one combine, harvesting other grain crops in the summer and corn in the fall.

1956 The Gyral air seeder, which plants seeds through a pneumatic delivery system, is patented in Australia. The technology eventually evolves into large multi-row machines with a trailing seed tank and often a second tank holding fertilizers.

1966 The DICKEY-john Manufacturing Company introduces electronic monitoring devices for farmers that allow them to plant crops more efficiently. Attached to mechanical planters and air seeders, the devices monitor the number and spacing of seeds being planted. The newest devices monitor the planting of up to 96 rows at a time. During the 1990s, similar devices are used at harvest time for yield mapping, or measuring and displaying the quality and quantity of a harvest as the combine moves through the field.

1994 Ushering in the new "precision agriculture," farmers begin using Global Positioning System (GPS) receivers to record precise locations on their farms to determine which areas need particular quantities of water, fertilizer, and pesticides. The information can be stored on a card and transferred to a home computer. Farmers can now combine such data with yield information, weather forecasts, and soil analysis to create spreadsheets. These tools enable even greater efficiency in food production.

Twin combines, each a thresher and reaper in a single machine, work a field at sunset.

P

COMPUTERS

The machine depicted on the cover of the January 1975 issue of Popular Electronics magazine sounded impressive— "World's First Minicomputer Kit to Rival Commercial Models"—and at a price of $397 for the parts, it seemed like quite a bargain. In truth, the Altair 8800 was not a minicomputer, a term normally reserved for machines many times as powerful. Nor was it easy to use. Programming had to be done by adjusting toggle switches, the memory held a meager 256 bytes of data, and output took the form of patterns of flashing lights.

Even so, this was an authentic general-purpose digital computer, a device traditionally associated with air-conditioned sanctums and operation by a technical elite. The Altair's maker, counting on the curiosity of electronics hobbyists, hoped to sell a few hundred. Instead, orders poured in by the thousands, signaling an appetite that, by the end of the century, would put tens of millions of personal computers in homes, offices, and schools around the world. Once again, the greatest productivity tool ever invented would wildly outstrip all expectations.

When the programmable digital computer was born shortly before mid-century, there was little reason to expect that it would someday be used to write letters, keep track of supermarket inventories, run financial networks, make medical diagnoses, help design automobiles, play games, deliver e-mail and photographs across the Internet, orchestrate battles, guide humans to the moon, create special effects for movies, or teach a novice to type. In the dawn years its sole purpose was to reduce mathematical drudgery, and its value for even that role was less than compelling. One of the first of the breed was the Harvard Mark I, conceived in the late 1930s by Harvard mathematician Howard Aiken and built by IBM during World War II to solve difficult ballistics problems. The Mark I was 51 feet long and 8 feet high, had 750,000 parts and 500 miles of wiring, and was fed data in the form of punched cards—an input method used for tabulating equipment since the late 19th century. This enormous machine could do just three additions or subtractions a second.

A route to far greater speeds was at hand, however. It involved basing a computer's processes on the binary numbering system, which uses only zeros and ones instead of the 10 digits of the decimal system. In the mid-19th century the British mathematician George Boole devised a form of algebra that encoded logic in terms of two states—true or false, yes or no, one or zero. If expressed that way, practically any mathematical or logical problem could be solved by just three basic operations, dubbed "and," "or," and "not." During the late 1930s several researchers realized that Boole's operations could be given physical form as arrangements of switches—a switch being a two-state device, on or off. Claude Shannon, a mathematician and engineer at the Massachusetts Institute of Technology (MIT), spelled this out in a masterful paper in 1938. At about the time Shannon was working on his paper, George Stibitz of AT&T's Bell Laboratories built such a device, using strips of tin can, flashlight bulbs, and surplus relays. The K-Model, as Stibitz called it (for kitchen table), could add two bits and display the result. In 1939, John Atanasoff, a physicist at Iowa State College, also constructed a rudimentary binary machine, and unknown to them all, a German engineer named Konrad Zuse created a fully functional general-purpose binary computer (the Z3) in 1941, only to see further progress thwarted by Hitler's lack of interest in long-term scientific research.

The switches used in most early computers were electromechanical relays, developed for the telephone system, but they soon gave way to vacuum tubes, which could turn an electric current on or off much more quickly. The first large-scale, all-electronic computer, ENIAC, took shape late in the war at the University of Pennsylvania's Moore School of Electrical Engineering under the guidance of John Mauchly and John Presper Eckert. Like the Mark I, it was huge—30 tons, 150 feet wide, with 20 banks of flashing lights—and it too was intended for ballistics calculations, but ENIAC could process numbers a thousand times faster. Even before it was finished, Mauchly and Eckert were making plans for a successor machine called EDVAC, conceived with versatility in mind.

Although previous computers could shift from one sort of job to another if given new instructions, this was a tedious process that might involve adjusting hundreds of controls or unplugging and replugging a forest of wires. EDVAC, by contrast, was designed to receive its instructions electronically; moreover, the program,

The K-Model machine (below), assembled in the late 1930s by George Stibitz, could add two bits and display the result. Early electronic computers like ENIAC (above) required a fleet of technicians to connect the wiring needed to program the machine, which could add 5,000 numbers or do fourteen 10-digit multiplications in one second.

coded in zeros and ones, would be kept in the same place that held the numbers the computer would be processing. This approach—letting a program treat its own instructions as data—offered huge advantages. It would accelerate the work of the computer, simplify its circuitry, and make possible much more ambitious programming. The stored-program idea spread rapidly, gaining impetus from a lucid description by one of the most famous mathematicians in the world, John von Neumann, who had taken an interest in EDVAC.

Building such a machine posed considerable engineering challenges, and EDVAC would not be the first to clear the hurdles. That honor was claimed in the spring of 1949 by a 3,000-tube stored-program computer dubbed EDSAC, the creation of British mathematical engineer Maurice Wilkes, of Cambridge University.

Meanwhile, Eckert and Mauchly had left the Moore School and established a company to push computing into the realm of commerce. The product they envisioned, a 5,000-tube machine called UNIVAC, had a breakthrough feature—storing data on magnetic tape

A bank of magnetic tape storage units marks UNIVAC's departure from the punched cards that had long been a staple of computing. Shortly after its completion, UNIVAC correctly predicted the outcome of the 1952 presidential election.

rather than by such unwieldy methods as punched cards. Although a few corporate customers were lined up in advance, development costs ran so high that the two men had to sell their company to the big office equipment maker Remington Rand. Their design proved a marketplace winner, however. Completed in 1951, UNIVAC was rugged, reliable, and able to perform almost 2,000 calculations per second. Its powers were put to a highly public test during the 1952 presidential election, when CBS gave UNIVAC the job of forecasting the outcome from partial voting returns. Early in the evening the computer (represented by a fake bank of blinking lights in the CBS studio) projected a landslide victory by Dwight Eisenhower over Adlai Stevenson. The prediction was made in such unequivocal terms that UNIVAC's operators grew nervous and altered the program to produce a closer result. They later confessed that the initial projection of electoral votes had been right on the mark.

By then several dozen other companies had jumped into the field. The most formidable was International Business Machines (IBM), a leading supplier of office equipment since early in the century. With its deep knowledge of corporate needs and its peerless sales force, IBM soon eclipsed all rivals. Other computer makers often expected customers to write their own applications programs, but IBM was happy to supply software for invoicing, payroll, production forecasts, and other standard corporate tasks. In time the company created extensive suites of software for such business sectors as banking, retailing, and insurance. Most competitors lacked the resources and revenue to keep pace.

Some of the computer projects taken on by IBM were gargantuan in scope. During the 1950s the company had as many as 8,000 employees laboring to computerize the U.S. air defense system. The project, known as SAGE and based on developmental work done at MIT's Lincoln Laboratory, called for a network of 23 powerful computers to process radar information from ships, planes, and ground stations while also analyzing weather, tracking weapons availability, and monitoring a variety of other matters. Each computer had 49,000 tubes and weighed 240 tons—the biggest ever built.

Almost as complex was an airline reservation system, called SABRE, that IBM created for American Airlines in the late 1950s and early 1960s. Using a million lines of program code and two big computers, it linked agents in 50 cities and could handle millions of transactions a year, processing them at the rate of one every 3 seconds. But writing the software for SAGE or SABRE was child's play compared to what IBM went through in the 1960s when it decided to overhaul its increasingly fragmented product line and make future machines compatible—alike in how they read programs, processed data, and dealt with input and output devices. Compatibility required an all-purpose operating system, the software that manages a computer's basic procedures, and it had to be written from scratch. That job took about 5,000 person-years of work and roughly half a billion dollars, but the money was well spent. The new product line, known as System/360, was a smash hit, in good part because it gave customers unprecedented freedom in mixing and matching equipment.

By the early 1970s technology was racing to keep up with the thirst for electronic brainpower in corporations, universities, government agencies, and other such big traffickers in data. Vacuum-tube switches had given way a decade earlier to smaller, cooler, less power-hungry transistors, and now the transistors, along with other electronic components, were being packed together in ever-increasing numbers on silicon chips. In addition to their processing roles, these chips were becoming the technology of choice for memory, the staging area where data and instructions are shuttled in and out of the computer—a job long done by arrays of tiny ferrite doughnuts that registered data magnetically. Storage—the part of a computing system where programs and data are kept in readiness—had gone through punched card, magnetic tape, and magnetic drum phases; now high-speed magnetic disks ruled. High-level programming languages such as FORTRAN (for science applications), COBOL (for business), and BASIC (for beginners) allowed software to be written in English-like commands rather than the abstruse codes of the early days.

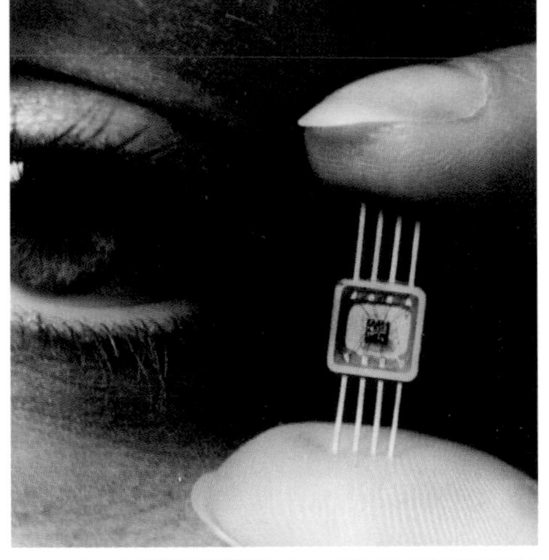

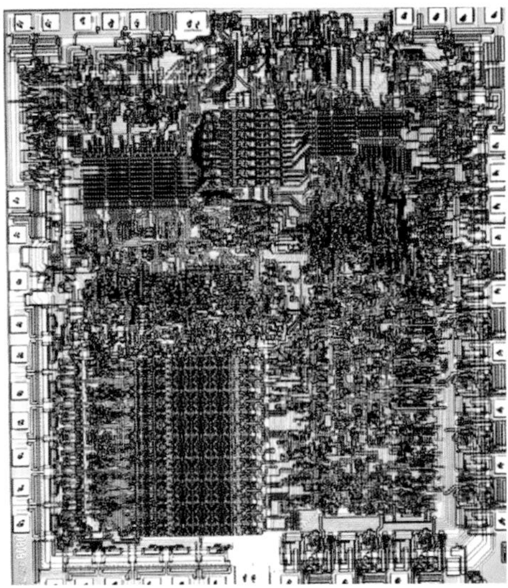

The advent of silicon chips like the device at top by Westinghouse and the Intel 8080 microprocessor (above) dramatically reduced the size of computers by the mid-1970s.

Some computer makers specialized in selling prodigiously powerful machines to such customers as nuclear research facilities or aerospace manufacturers. A category called supercomputers was pioneered in the mid-1960s by Control Data Corporation, whose chief engineer, Seymour Cray, designed the CDC 6600, a 350,000-transistor machine that could execute 3 million instructions per second. The price: $6 million. At the opposite end of the scale, below big mainframe machines like those made by IBM, were minicomputers, swift enough for many scientific or engineering applications but at a cost of tens of thousands rather than hundreds of thousands of dollars. Their development was spearheaded by Kenneth Olsen, an electrical engineer who cofounded Digital Equipment Corporation and had close ties to MIT.

Then, with the arrival of the humble Altair in 1975, the scale suddenly plunged to a level never imagined by industry leaders. What made such a compact, affordable machine possible was the microprocessor, which concentrated all of a computer's arithmetical and logical functions on a single chip—a feat first achieved by an engineer named Ted Hoff at Intel Corporation in 1971. After the Intel 8080 microprocessor was chosen for the Altair, two young computer buffs from Seattle, Bill Gates and Paul Allen, won the job of writing software that would allow it to be programmed in BASIC. By the end of the century the company they formed for that project, Microsoft, had annual sales greater than many national economies.

Nowhere was interest in personal computing more intense than in the vicinity of Palo Alto, California, a place known as Silicon Valley because of the presence of many big semiconductor firms. Electronics hobbyists abounded there, and two of them—Steve Jobs and Steve Wozniak—turned their tinkering into a highly appealing

Microsoft cofounders Paul Allen (below left) and Bill Gates, shown here in 1981 after signing a contract with IBM, and Apple cofounders Steve Jobs and Steve Wozniak, flanking Apple president John Sculley (below right) at the 1984 introduction of the compact Apple IIc, helped make the personal computer a reality.

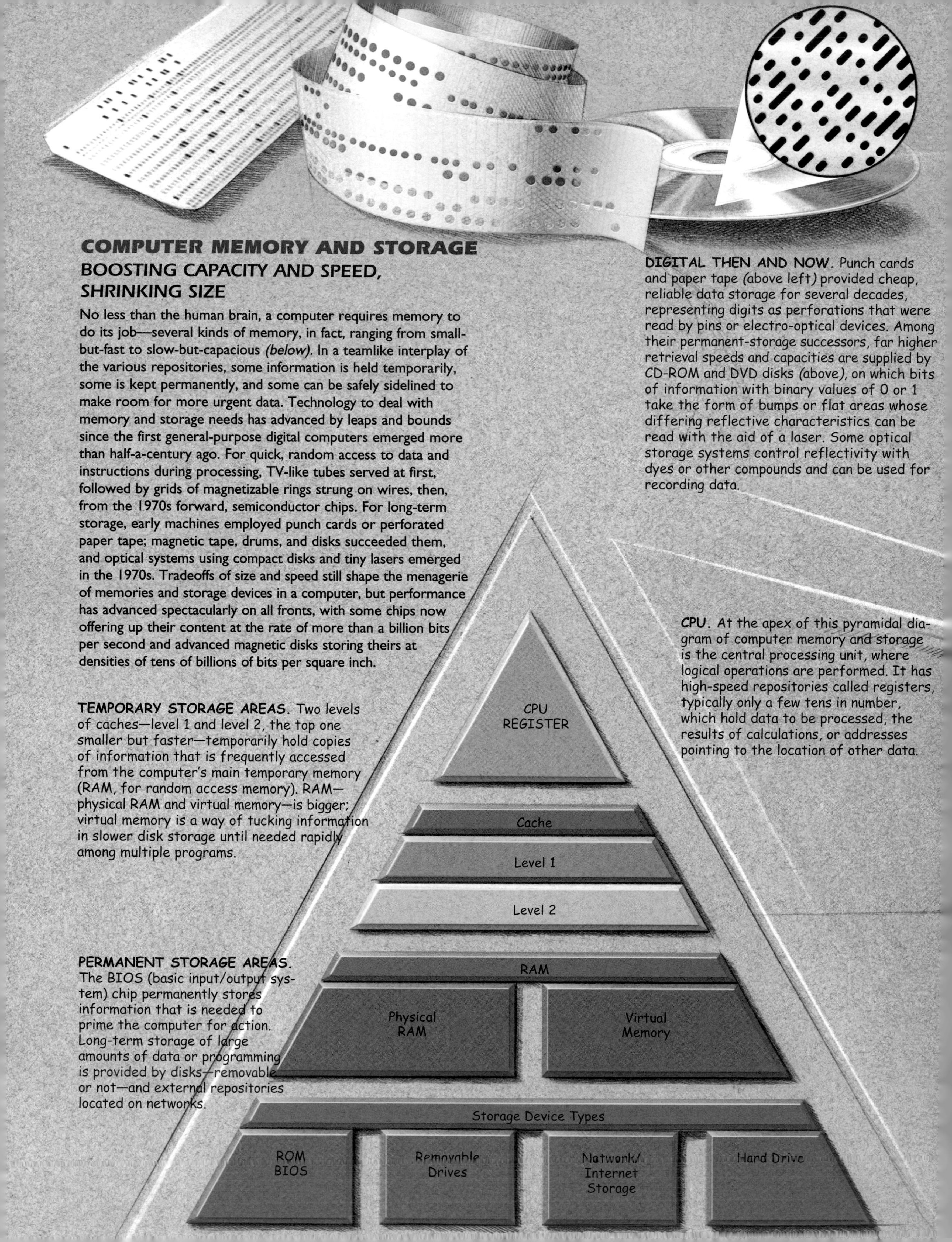

COMPUTER MEMORY AND STORAGE

BOOSTING CAPACITY AND SPEED, SHRINKING SIZE

No less than the human brain, a computer requires memory to do its job—several kinds of memory, in fact, ranging from small-but-fast to slow-but-capacious *(below)*. In a teamlike interplay of the various repositories, some information is held temporarily, some is kept permanently, and some can be safely sidelined to make room for more urgent data. Technology to deal with memory and storage needs has advanced by leaps and bounds since the first general-purpose digital computers emerged more than half-a-century ago. For quick, random access to data and instructions during processing, TV-like tubes served at first, followed by grids of magnetizable rings strung on wires, then, from the 1970s forward, semiconductor chips. For long-term storage, early machines employed punch cards or perforated paper tape; magnetic tape, drums, and disks succeeded them, and optical systems using compact disks and tiny lasers emerged in the 1970s. Tradeoffs of size and speed still shape the menagerie of memories and storage devices in a computer, but performance has advanced spectacularly on all fronts, with some chips now offering up their content at the rate of more than a billion bits per second and advanced magnetic disks storing theirs at densities of tens of billions of bits per square inch.

DIGITAL THEN AND NOW. Punch cards and paper tape *(above left)* provided cheap, reliable data storage for several decades, representing digits as perforations that were read by pins or electro-optical devices. Among their permanent-storage successors, far higher retrieval speeds and capacities are supplied by CD-ROM and DVD disks *(above)*, on which bits of information with binary values of 0 or 1 take the form of bumps or flat areas whose differing reflective characteristics can be read with the aid of a laser. Some optical storage systems control reflectivity with dyes or other compounds and can be used for recording data.

CPU. At the apex of this pyramidal diagram of computer memory and storage is the central processing unit, where logical operations are performed. It has high-speed repositories called registers, typically only a few tens in number, which hold data to be processed, the results of calculations, or addresses pointing to the location of other data.

TEMPORARY STORAGE AREAS. Two levels of caches—level 1 and level 2, the top one smaller but faster—temporarily hold copies of information that is frequently accessed from the computer's main temporary memory (RAM, for random access memory). RAM—physical RAM and virtual memory—is bigger; virtual memory is a way of tucking information in slower disk storage until needed rapidly among multiple programs.

PERMANENT STORAGE AREAS. The BIOS (basic input/output system) chip permanently stores information that is needed to prime the computer for action. Long-term storage of large amounts of data or programming is provided by disks—removable or not—and external repositories located on networks.

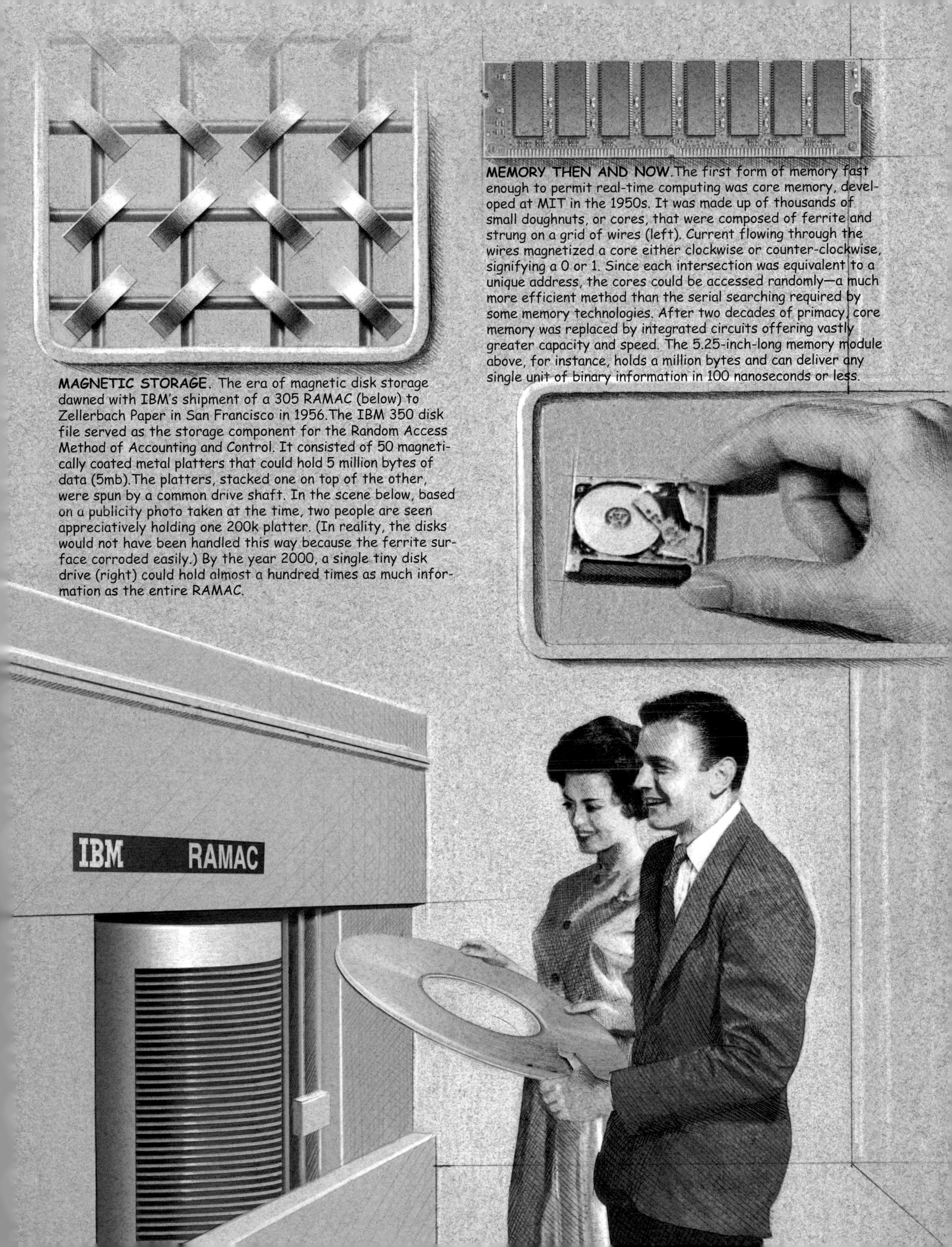

MEMORY THEN AND NOW. The first form of memory fast enough to permit real-time computing was core memory, developed at MIT in the 1950s. It was made up of thousands of small doughnuts, or cores, that were composed of ferrite and strung on a grid of wires (left). Current flowing through the wires magnetized a core either clockwise or counter-clockwise, signifying a 0 or 1. Since each intersection was equivalent to a unique address, the cores could be accessed randomly—a much more efficient method than the serial searching required by some memory technologies. After two decades of primacy, core memory was replaced by integrated circuits offering vastly greater capacity and speed. The 5.25-inch-long memory module above, for instance, holds a million bytes and can deliver any single unit of binary information in 100 nanoseconds or less.

MAGNETIC STORAGE. The era of magnetic disk storage dawned with IBM's shipment of a 305 RAMAC (below) to Zellerbach Paper in San Francisco in 1956. The IBM 350 disk file served as the storage component for the Random Access Method of Accounting and Control. It consisted of 50 magnetically coated metal platters that could hold 5 million bytes of data (5mb). The platters, stacked one on top of the other, were spun by a common drive shaft. In the scene below, based on a publicity photo taken at the time, two people are seen appreciatively holding one 200k platter. (In reality, the disks would not have been handled this way because the ferrite surface corroded easily.) By the year 2000, a single tiny disk drive (right) could hold almost a hundred times as much information as the entire RAMAC.

By the end of the 20th century, personal computers had become so portable and applications software so varied that the machines were used anywhere they were needed, whether at home, at work, or on the road.

consumer product: the Apple II, a plastic-encased computer with a keyboard, screen, and cassette tape for storage. It arrived on the market in 1977, described in its advertising copy as "the home computer that's ready to work, play, and grow with you." Few packaged programs were available at first, but they soon arrived from many quarters. Among them were three kinds of applications that made this desktop device a truly valuable tool for business—word processing, spreadsheets, and databases. The market for personal computers exploded, especially after IBM weighed in with a product in 1981. Its offering used an operating system from Microsoft, MS-DOS, which was quickly adopted by other manufacturers, allowing any given program to run on a wide variety of machines.

The next 2 decades saw computer technology rocketing ahead on every front. Chips doubled in density almost annually, while memory and storage expanded by leaps and bounds. Hardware like the mouse made the computer easier to control; operating systems allowed the screen to be divided into independently managed windows; applications programs steadily widened the range of what computers could do; and processors were lashed together—thousands of them in some cases—in order to solve pieces of a problem in parallel. Meanwhile, new communications standards enabled computers to be joined in private networks or the incomprehensibly intricate global weave of the Internet.

Where it all will lead is unknowable, but the rate of advance is almost certain to be breathtaking. When the Mark I went to work calculating ballistics tables back in 1943, it was described as a "robot superbrain" because of its ability to multiply a pair of 23-digit numbers in 3 seconds. Today, some of its descendants need just 1 second to perform several hundred trillion mathematical operations—a performance that, in a few years, will no doubt seem slow.

Perspective

William H. Gates III

Chairman and Chief Software Architect
Microsoft Corporation

For me the personal computer revolution started in the mid-1970s, when my friend Paul Allen and I saw a magazine article about the MITS Altair 8800. The Altair was the first build-it-yourself computer kit for hobbyists. For a few hundred dollars, MITS would mail you a few bags of parts and some photocopied instructions. After some careful soldering, you had your own computer, roughly the size of a bread box, with rows of switches and blinking lights.

It wasn't much to look at and it wasn't terribly useful, but it felt like the start of a revolution. Until then computers were used mostly by technicians in air-conditioned rooms. Few people had the opportunity even to see a computer and even fewer got to use one. But the Altair was a computer that people could put on their desks, and what they could do with it was limited only by their imagination—and the modest capabilities of Intel's 8080 microprocessor.

We knew that microprocessors would become cheaper and more powerful, making personal computers increasingly capable. We also knew those computers would need software to make them do useful things. So Paul and I founded a company we called Microsoft that we hoped would meet this need.

Our first product was a version of the BASIC programming language that could run on the Altair. Unlike many other languages available at the time, BASIC was relatively simple to use. After a few minutes of instruction, even a nontechnical person could start writing simple programs. Actually developing this product, however, was not very simple. First, it was challenging to come up with a BASIC that could run in the Altair's limited memory and still leave room for people to write programs. Second, we didn't have an Altair to work with. Only a few prototypes were available at the time.

After writing software that would mimic the Altair's functions on another computer and spending nearly all our spare time writing code—some of it on paper notepads—we managed to create a BASIC that worked. For its time the Altair was a huge success, and thousands of programmers used our software to make it do interesting and useful things. Since then the PC has evolved from a hobbyist's toy into a powerful tool that has transformed how we work, learn, play, and keep in touch. And it has created an industry that employs millions of people and plays a leading role in our global economy.

Computing has made many evolutionary leaps over the decades—from the command line to the graphical user interface, from stand-alone PCs to a globally connected Internet. But we're now seeing an even more fundamental change. We're in what I call the "digital decade," a time when computers are moving beyond being merely useful to becoming an essential part of our everyday lives. Today we use computers for discrete tasks—like doing e-mail and paying bills—but in the years ahead they'll play a key role in almost everything we do. We'll rely on them to run our lives and businesses. We'll want them to keep us informed and entertained. We'll expect them to be wherever we need them. It will be an era of truly personal computing.

Many of our early dreams for the PC have already come true. They can recognize speech and handwriting, create realistic animation, and enable people to collaborate, communicate, and find information around the world. But we've barely scratched the surface of the PC's potential, and I'm incredibly excited about the amazing innovations that are just over the horizon.

COMPUTERS TIMELINE

The 20th century was nearly into its fourth decade before the first electronic computer came along, and those early machines were behemoths capable of only the most basic tasks. Today, tiny "handhelds" are used for word processing and storage, delivery of documents and images, inventory management, and remote access by workers to central offices. Programmable electronic devices of all sorts have come to pervade modern society to such a degree that future generations may well designate the 20th century as the Computer Age.

1936 Electrical engineer and mathematician Claude Shannon, in his master's thesis, "A Symbolic Analysis of Relay and Switching Circuits," uses Boolean algebra to establish a working model for digital circuits. This paper, as well as later research by Shannon, lays the groundwork for the future telecommunications and computer industries.

1939 John Atanasoff and Clifford Berry at Iowa State College design the first electronic computer. The obscure project, called the Atanasoff-Berry Computer (ABC), incorporates binary arithmetic and electronic switching. Before the computer is perfected, Atanasoff is recruited by the Naval Ordnance Laboratory and never resumes its research and development. However, in the summer of 1941, at Atanasoff's invitation, computer pioneer John Mauchly of the University of Pennsylvania, visits Atanasoff in Iowa and sees the ABC demonstrated.

The first binary digital computers are developed. Bell Labs's George Stibitz designs the Complex Number Calculator, which performs mathematical operations in binary form using on-off relays, and finds the quotient of two 8-digit numbers in 30 seconds. In Germany, Konrad Zuse develops the first programmable calculator, the Z2, using binary numbers and Boolean algebra—programmed with punched tape.

1943 Colossus, the world's first vacuum-tube programmable logic calculator, is built in Britain for the purpose of breaking Nazi codes. On average, Colossus deciphers a coded message in two hours.

1945 Two mathematicians, Briton Alan Turing and Hungarian John von Neumann, work independently on the specifications of a stored-program computer. Von Neumann writes a document describing a computer on which data and programs can be stored. Turing publishes a paper on an Automatic Computing Engine, based on the principles of speed and memory.

1946 The first electronic computer put into operation is developed late in World War II by John Mauchly and John Presper Eckert at the University of Pennsylvania's Moore School of Electrical Engineering. The Electronic Numerical Integrator and Computer (ENIAC), used for ballistics computations, weighs 30 tons and includes 18,000 vacuum tubes, 6,000 switches, and 1,500 relays.

1947 John Bardeen, Walter H. Brattain, and William B. Shockley of Bell Telephone Laboratories invent the transistor. (See *Electronics*, page 52.)

1949 The Electronic Delay Storage Automatic Calculator (EDSAC), the first stored-program computer, is built and programmed by British mathematical engineer Maurice Wilkes.

1951 Eckert and Mauchly, now with their own company (later sold to Remington Rand), design UNIVAC (UNIVersal Automatic Computer)—the first computer for U.S. business. Its breakthrough feature: magnetic tape storage to replace punched cards. First developed for the Bureau of the Census to aid in census data collection, UNIVAC passes a highly public test by correctly predicting Dwight Eisenhower's victory over Adlai Stevenson in the 1952 presidential race. But months before UNIVAC is completed, the British firm J. Lyons & Company unveils the first computer for business use, the LEO (Lyons Electronic Office), which eventually calculated the company's weekly payroll.

1952 Grace Murray Hopper *(below)*, a senior mathematician at Eckert-Mauchly Computer Corporation and a programmer for Harvard's Mark I computer, develops the first computer compiler, a program that translates computer instructions from English into machine language. She later creates Flow-Matic, the first programming language to use English words and the key influence for COBOL (Common Business Oriented Language). Attaining the rank of rear admiral in a navy career that brackets her work at Harvard and Eckert-Mauchly, Hopper eventually becomes the driving force behind many advanced automated programming technologies.

1955 IBM engineers led by Reynold Johnson design the first disk drive for random-access storage of data, offering more surface area for magnetization and storage than earlier drums. In later drives a protective "boundary layer" of air between the heads and the disk surface would be provided by the spinning disk itself. The Model 305 Disk Storage unit, later called the Random Access Method of Accounting and Control, is released in 1956 with a stack of fifty 24-inch aluminum disks storing 5 million bytes of data.

1957 FORTRAN (for FORmula TRANslation), a high-level programming language developed by an IBM team led by John Backus, becomes commercially available. FORTRAN is a way to express scientific and mathematical computations with a programming language similar to mathematical formulas. Backus and his team claim that the FORTRAN compiler produces machine code as efficient as any produced directly by a human programmer. Other programming languages quickly follow, including ALGOL, intended as a universal computer language, in 1958 and COBOL in 1959. ALGOL has a profound impact on future languages such as Simula (the first object-oriented programming language), Pascal, and C/C++. FORTRAN becomes the standard language for scientific computer applications, and COBOL is developed by the U.S. government to standardize its commercial application programs. Both dominate the computer-language world for the next 2 decades.

1958 Jack Kilby of Texas Instruments and Robert Noyce of Fairchild Semiconductor independently invent the integrated circuit. (See *Electronics*, page 53.)

1960 Digital Equipment Corporation introduces the "compact" PDP-1 for the science and engineering market. Not including software or peripherals, the system costs $125,000, fits in a corner of a room, and doesn't require air conditioning. Operated by one person, it features a cathode-ray tube display and a light pen. In 1962 at MIT a PDP-1 becomes the first computer to run a video game when Steve Russell programs it to play "Spacewar." The PDP-8, released 5 years later, is the first computer to fully use integrated circuits.

1964 Dartmouth professors John Kemeny and Thomas Kurtz develop the BASIC (Beginners All-Purpose Symbolic Instruction Code) programming language specifically for the school's new time-sharing computer system. Designed for non-computer-science students, it is easier to use than FORTRAN. Other schools and universities adopt it, and computer manufacturers begin to provide BASIC translators with their systems.

1968 The computer mouse (*above*) makes its public debut during a demonstration at a computer conference in San Francisco. Its inventor, Douglas Engelbart of the Stanford Research Institute, also demonstrates other user-friendly technologies such as hypermedia with object linking and addressing. Engelbart receives a patent for the mouse 2 years later.

1970 Xerox Corporation assembles a team of researchers in information and physical sciences in Palo Alto, California, with the goal of creating "the architecture of information." Over the next 30 years innovations emerging from the Palo Alto Research Center (PARC) include the concept of windows (1972), the first real personal computer (Alto in 1973), laser printers (1973), the concept of WYSIWYG (what you see is what you get) word processors (1974), and EtherNet (1974). In 2002 Xerox PARC incorporates as an independent company—Palo Alto Research Center, Inc.

1975 The Altair 8800, widely considered the first home computer, is marketed to hobbyists by Micro Instrumentation Telemetry Systems. The build-it-yourself kit doesn't have a keyboard, monitor, or its own programming language; data are input with a series of switches and lights. But it includes an Intel microprocessor and costs less than $400. Seizing an opportunity, fledgling entrepreneurs Bill Gates and Paul Allen propose writing a version of BASIC for the new computer. They start the project by forming a partnership called Microsoft.

1977 Apple Computer, founded by electronics hobbyists Steve Jobs and Steve Wozniak, releases the Apple II, a desktop personal computer for the mass market that features a keyboard, video monitor, mouse, and random-access memory (RAM) that can be expanded by the user. Independent software manufacturers begin to create applications for it.

1979 Harvard MBA student Daniel Bricklin and programmer Bob Frankston launch the VisiCalc spreadsheet for the Apple II, a program that helps drive sales of the personal computer and becomes its first commercially successful business application. VisiCalc owns the spreadsheet market for nearly a decade before being eclipsed by Lotus 1-2-3, a spreadsheet program designed by a former VisiCalc employee.

What is thought to be the first laptop computer is designed by William Moggridge of GRiD Systems Corporation in England. The GRiD Compass 1109 has 340 kilobytes of bubble memory and a folding electroluminescent display screen in a magnesium case. Used by NASA in the early 1980s for its shuttle program, the "portable computer" is patented by GriD in 1982.

1981 IBM introduces the IBM Personal Computer with an Intel 8088 microprocessor and an operating system—MS-DOS—designed by Microsoft. Fully equipped with 64 kilobytes of memory and a floppy disk drive, it costs under $3,000.

1984 Apple introduces the Macintosh, a low-cost, plug-and-play personal computer whose central processor fits on a single circuit board. Although it doesn't offer enough power for business applications, its easy-to-use graphic interface finds fans in education and publishing.

Phillips and Sony combine efforts to introduce the CD-ROM (compact disc read-only memory), patented in 1970 by James

Steve Jobs (left) and John Sculley of Apple Computers show off the new Macintosh in 1984. An ad heralding the computer debuted during the broadcast of the 1984 Super Bowl.

T. Russell (see *Electronics,* page 59). With the advent of the CD, data storage and retrieval shift from magnetic to optical technology. The CD can store more than 300,000 pages worth of information—more than the capacity of 450 floppy disks—meaning it can hold digital text, video, and audio files. Advances in the 1990s allow users not only to read prerecorded CDs but also to download, write, and record information onto their own disks.

1985 Microsoft releases Windows 1.0, operating system software that features a Macintosh-like graphical user interface (GUI) with drop-down menus, windows, and mouse support. Because the program runs slowly on available PCs, most users stick to MS-DOS. Higher-powered microprocessors beginning in the late 1980s make the next attempts—Windows 3.0 and Windows 95—more successful.

1991 The World Wide Web becomes available to the general public (see *Internet,* page 152).

1992 Apple chairman John Sculley coins the term "personal digital assistant" to refer to handheld computers. One of the first on the market is Apple's Newton, which has a liquid crystal display operated with a stylus. The more successful Palm Pilot is released by 3Com in 1996.

1999 Responding to a more mobile workforce, handheld computer technology leaps forward with the Palm VII connected organizer, the combination of a computer with 2 megabytes of RAM and a port for a wireless phone. At less than $600, the computer weighs 6.7 ounces and operates for up to 3 weeks on two AAA batteries. Later versions offer 8 megabytes of RAM, Internet connectivity, and color screens for less than $500.

7 pqrs
8

TELEPHONY

"The telephone," wrote Alexander Graham Bell in an 1877 prospectus drumming up support for his new invention, "may be briefly described as an electrical contrivance for reproducing in distant places the tones and articulations of a speaker's voice." As for connecting one such contrivance to another, he suggested possibilities that admittedly sounded utopian: "It is conceivable that cables of telephone wires could be laid underground, or suspended overhead, communicating by branch wires with private dwellings, country houses, shops, manufactories, etc."

It was indeed conceivable. The enterprise he helped launch that year—the forerunner of the American Telephone and Telegraph Company—would grow into one of the biggest corporations ever seen. At its peak in the early 1980s, just before it was split apart to settle an antitrust suit by the Justice Department, AT&T owned and operated hundreds of billions of dollars worth of equipment, harvested annual revenues amounting to almost 2 percent of the gross domestic product of the United States, and employed about a million people. AT&T's breakup altered the business landscape drastically, but the telephone's primacy in personal communications has only deepened since then, with technology giving Bell's invention a host of new powers.

Linked not just by wires but also by microwaves, communications satellites, optical fibers, networks of cellular towers, and computerized switching systems that can connect any two callers on the planet almost instantaneously, the telephone now mediates billions of distance-dissolving conversations every day—eight per person, on average, in the United States. As Bell foresaw in his prospectus, it is "utilized for nearly every purpose for which speech is employed," from idle chat to emergency calls. In addition, streams of digital data such as text messages and pictures now often travel the same routes as talk. Modern life and the telephone are inextricably intertwined.

At the outset, people weren't quite sure how to use this newfangled device, but they knew they wanted one—or more accurately, two, because telephones were initially sold in pairs. (The first customer, a Boston banker, leased a pair for his office and home, plus a private line to join them.) Telephony quickly found a more flexible form, however. The year 1878 saw the creation of the first commercial exchange, a manual switching device that could form pathways between any of 21 subscribers. Soon that exchange was handling 50 subscribers, and bigger exchanges, with operators handling a maze of plugs and cords to open and close circuits, quickly began popping up in communities all across America. Although the Bell system would stick with operators and plugboards for a while, an automated switchboard became available in the 1890s, invented by an Indiana undertaker named Almon Strowger, who suspected that local telephone operators were favoring his competitors. His apparatus made a connection when a caller pressed two buttons on a telephone a certain number of times to specify the other party. Soon, a 10-digit dialing wheel replaced the buttons, and it would hold sway until Touch-Tone dialing—a faster method that expressed numbers as combinations of two single-frequency tones—arrived on the scene in the 1960s.

When inventor Alexander Graham Bell inaugurated the telephone line between New York and Chicago in 1892 (opposite), callers used a stationary telephone plugged into a wired telephone circuit. Today, people by the hundreds of millions make phone calls from anywhere within range of a cellular phone tower (left).

At the start of the 20th century, many basic features of telephone technology were in place. By then, the human voice was captured by a method that would remain standard for many decades: In the microphone, sound waves pressed carbon granules together, changing their electrical resistance and imposing an analogous pattern of variations on a current passing through them. The signal was carried by a pair of copper wires rather than the single iron or steel wire of the dawn years. Copper's electrical resistance was only a tenth as much, and random noise was dramatically reduced by using two wires instead of completing the circuit through the ground, as telegraphy did. In cities, unsightly webs of wires were minimized by gathering lines in lead pipes about 2 inches in diameter. The early cables, as such bundles were called, held a few dozen pairs; by 1940 about 2,000 could be packed into the pipe.

As other countries caught the telephone contagion, their governments frequently claimed ownership, but the private enterprise model prevailed in the United States, and a multitude of competitors leaped into the business as the original Bell patents expired. At the turn of the 20th century, the Bell system accounted for 856,000 phones, and the so-called independents had 600,000. Corporate warfare raged, with AT&T buying up as many of the upstarts as possible and attempting to derail the rest by refusing to connect their lines to its system. Two decades later AT&T secured its supremacy when the U.S. Senate declared it a "natural monopoly"—one that would have to accept tight governmental regulation.

During these years, all the contenders experimented with sales gimmicks such as wake-up calls and telephone-delivered sermons. Price was a major marketing issue, of course, and it dropped steadily. At the beginning of the century, the Bell system charged $99 per thousand calls in New York City; by the early 1920s a flat monthly residential rate of $3 was typical. As the habit of talking at a distance spread, some social commentators worried that

In the early part of the 20th century, a supervisor of long-distance operators in Minnesota stands ready to assist a long roomful of operators connecting calls with switches and plugboards.

community ties and old forms of civility were fraying, but the telephone had unstoppable momentum. By 1920 more than a third of all U.S. households were connected. Most had party lines, which generally put two to four households on the same circuit and signaled them with distinctive rings. Phone company publications urged party line customers to keep calls brief and not to eavesdrop on their neighbors, but such rules were often honored only in the breach.

Extending the range of telephone calls was a key engineering challenge, much tougher than for telegraphy because the higher frequency of voice-based signals caused them to fade faster as they traveled along a wire. Early on, a device called a loading coil offered a partial cure. Independently invented in 1899 by Michael Pupin of Columbia University and George Campbell of AT&T, it basically consisted of a coil of wire that was placed along a line every 6,000 feet or so, greatly diminishing attenuation in the range of frequencies suitable for voice transmission. Two years later commercial service began between Philadelphia and Chicago, and by 1911 a long-distance line stretched all the way from New York to Denver. But transcontinental service remained out of reach until Bell engineers began experimenting with the triode vacuum tube, patented in 1907 by the radio pioneer Lee De Forest as "A Device for Amplifying Feeble Electrical Currents."

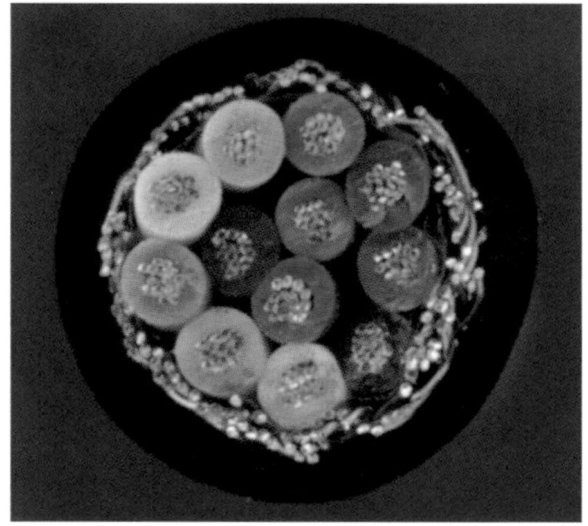

Coaxial cable from 1962 (top) could support 11,000 two-way conversations. Using time-division multiplexing, modern digital coaxial cable can carry more than three times that number (above).

De Forest's tube used a small, varying voltage on a gridlike element to impose matching variations, even at high frequencies, on a much larger flow of electrons between a heated filament and a plate. The inventor's understanding of his device was imperfect, however. He thought that ionized gas in the tube was somehow involved. In 1913 a Bell physicist named H. D. Arnold showed that, on the contrary, the completeness of the vacuum dictated the performance. Arnold and his colleagues designed superior tubes and related circuitry to amplify long-distance telephone transmissions, and service was opened between New York and San Francisco in 1915. Alexander Graham Bell made the first call, speaking to Thomas Watson, who had helped him develop a working telephone four decades earlier. The transcontinental path had 130,000 telephone poles, 2,500 tons of copper wire, and three vacuum-tube devices to strengthen the signals. A 3-minute conversation that year cost $20.70.

By the mid-1920s long distance lines connected every part of the United States. Their capacity was expanded by a technique called frequency multiplexing, which involves electronically shifting the frequencies of speech (about 200 to 3,400 cycles per second) to other frequency bands so that several calls could be sent along a wire simultaneously. After World War II, the Bell system began to use coaxial cable for this kind of multiplexing. Its design—basically a tube of electrically conducting material surrounding an insulated central wire—enabled it to carry a wide range of frequencies.

Stretching coaxial cable beneath oceans posed difficulties so daunting that the first transatlantic link, capable of carrying 36 calls at a time, wasn't established until 1956. But radio had been filling the oceanic gaps for several decades by then while also connecting ships, planes, and cars to the main telephone system or to each other. After mid-century, a previously unexploited form of radio—the microwave frequencies above a billion cycles per second—took over much of the land-based long-distance traffic. Microwaves travel in a straight line rather than following the curvature of the earth like ordinary radio waves, which means that the beam has to be relayed along a chain of towers positioned 26 miles apart on average. But their high frequency permits small antenna size and high volume. Thousands of two-way voice circuits can be crammed into a single microwave channel.

The never-ending need for more capacity brought steady strides in switching technology as well. A simple architecture had been developed early on. Some switching stations handled local circuits, others connected clusters of these local centers, and still others dealt with long-distance traffic. Whenever congestion occurred, the routing was changed according to strict rules. By the 1970s Bell engineers had devised electromechanical switches that could serve more than 30,000 circuits at a time, but an emerging breed of computer-like electronic switches promised speed and flexibility that no electromechanical device could match.

The move to electronic switching began in the 1960s and led to all-digital systems a decade later. Such systems work by converting voice signals into on-off binary pulses and assigning each call to a time slot in a data stream; switching is achieved by simply changing time slot assignments. This so-called time division approach also boosts capacity by packing many signals into the same flow, an efficient vehicle for transmission to and from communications satellites. Today's big digital switches can handle 100,000 or more circuits at a time, maintaining a remarkably clear signal. And like any computer, the digital circuits are versatile. In addition to making connections and generating billing information, their software enables them to provide customers with a whole menu of special services—automatically forwarding calls, identifying a caller before the phone is answered, interrupting one call with an alert of another, providing voice mail, and more.

The first electronic switching system, 1 ESS, installed in Succasunna, New Jersey, in 1965 (below), could store programs and allowed such features as call forwarding and speed dialing. By the end of the century, AT&T's Global Network Operations Center (below right) was thoroughly computerized.

In recent decades, long-distance transmission has undergone a revolution, with such calls migrating from microwave and coaxial cable to threadlike optical fibers that channel laser light. Because light waves have extremely high frequencies, they can be encoded with huge amounts of digital information, a job done by tiny semiconductor lasers that are able to turn on and off billions of times a second. The first fiber-optic telephone links were created in the late 1970s. The latest versions, transmitting several independently encoded streams of light on separate frequencies, are theoretically capable of carrying millions of calls at a time or vast volumes of Internet or video traffic. Today, the world is wrapped in these amazing light pipes, and worries about long-distance capacity are a thing of the past (see *Lasers and Fiber Optics*).

Another technological triumph is the cell phone, a radio-linked device that is taking the world by storm. Old-style mobile telephones received their signals from a single powerful transmitter that covered an area about 50 miles in diameter, an interference-prone method that provided enough channels to connect only a couple of dozen customers at a time. Cellular technology, by contrast, uses low-powered base stations that serve "cells" just a few square miles in area. As a customer moves from one cell to another, the phone switches from a weakening signal to a stronger one on a different

CONNECTING THE WORLD

TELEPHONY'S GLOBE-SPANNING NETWORK

Telephony has come a considerable distance since the days when there was just one "Phone Company" and calling overseas always involved at least two operators and long waits while the connection was made. Today the telephone network, also called the Public Switched Telephone Network, or PSTN, is an interlocking network of switching centers that encompass systems maintained by a plethora of communications companies, including radio frequency cellular phone service. Chaos is averted because all the companies follow the protocols of the switching hierarchy (box below) and each local office maintains a database that tells it which long-distance carrier a given phone number has selected. Long-distance operators have largely been replaced by computerized switches that will route a call from Seattle to Miami in the blink of an eye, and improved copper and fiber-optic links make someone overseas sound like they're right next door.

But perhaps nothing has so revolutionized the idea of telephones as the development of mobile phone technology in the latter few decades of the 20th century. As the technology has improved, cost has come down, and millions of people today carry their cell phones with them almost all the time—handy for emergencies large or small, a way for busy families to synchronize schedules, an essential tool for business people on the road. Whatever comes next in telephone technology, the trend seems to be toward giving us the ability to connect whenever we wish and wherever we are.

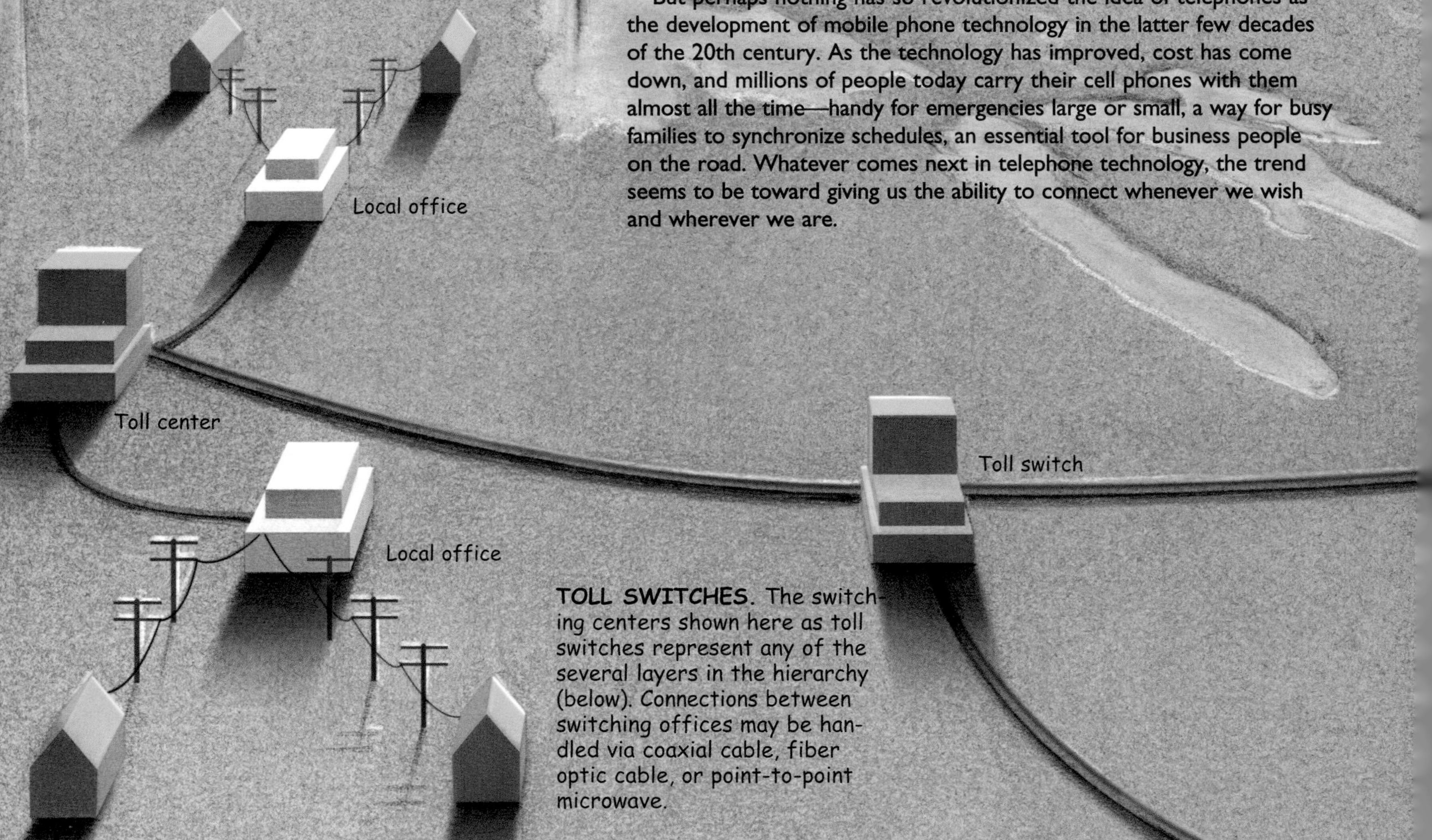

TOLL SWITCHES. The switching centers shown here as toll switches represent any of the several layers in the hierarchy (below). Connections between switching offices may be handled via coaxial cable, fiber optic cable, or point-to-point microwave.

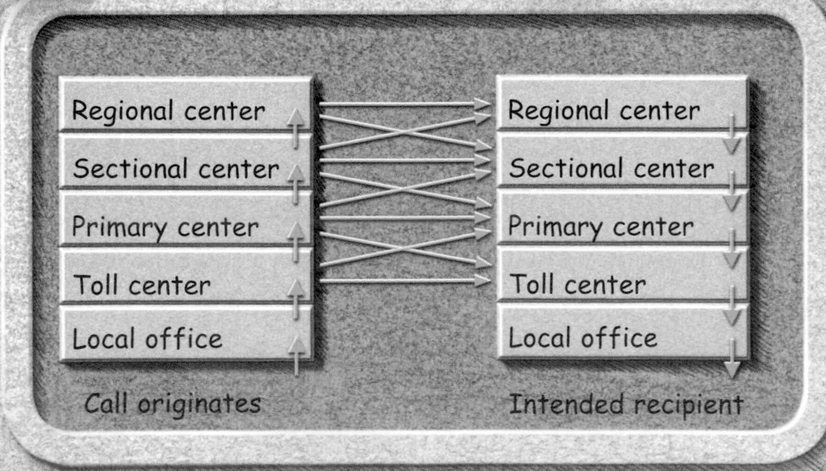

THE SWITCHING HIERARCHY. The hierarchy of telephone switches used by the phone companies to route traffic is also referred to as a homing chain (left). A local office, the lowest level, typically hands off to the toll center, the next level up in its own homing chain. The toll center will try to switch the call to the recipient's toll center, but if that is not possible, it will try the next level up, the recipient's primary center. If that fails, the toll center will switch the call up to its own primary center, where the hierarchy of switching rules begins again. Each succeeding center will try to switch the call to three levels in the recipient's homing chain before passing it up the caller's homing chain. The top level, the regional center, has just two choices: to try the next level down (the sectional center) in the recipient's homing chain, then to switch the call to the recipient's regional center. Once the call is transferred to the recipient's homing chain, it is passed down the chain to the recipient's local office.

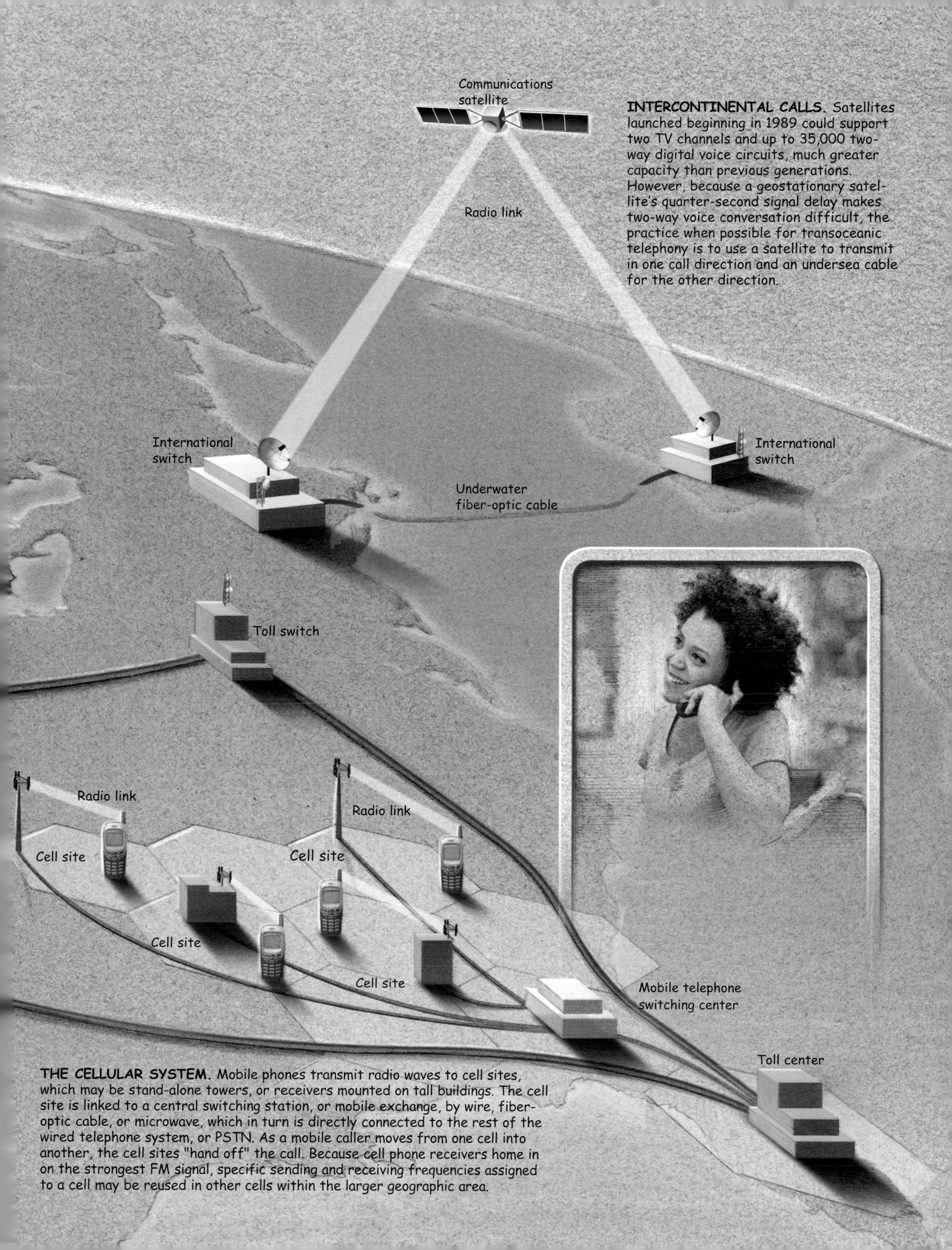

INTERCONTINENTAL CALLS. Satellites launched beginning in 1989 could support two TV channels and up to 35,000 two-way digital voice circuits, much greater capacity than previous generations. However, because a geostationary satellite's quarter-second signal delay makes two-way voice conversation difficult, the practice when possible for transoceanic telephony is to use a satellite to transmit in one call direction and an undersea cable for the other direction.

THE CELLULAR SYSTEM. Mobile phones transmit radio waves to cell sites, which may be stand-alone towers, or receivers mounted on tall buildings. The cell site is linked to a central switching station, or mobile exchange, by wire, fiber-optic cable, or microwave, which in turn is directly connected to the rest of the wired telephone system, or PSTN. As a mobile caller moves from one cell into another, the cell sites "hand off" the call. Because cell phone receivers home in on the strongest FM signal, specific sending and receiving frequencies assigned to a cell may be reused in other cells within the larger geographic area.

Early mobile phone users, such as this British emergency worker, were confined to their cars, where the full-size phones drew power from the car's battery. Today's cell phone users carry phones as small as a deck of cards and rely on a wide distribution of cellular towers (above right) to relay their calls.

frequency, thus maintaining a clear connection. Because transmissions are low powered, frequencies can be reused in nonadjacent cells, accommodating thousands of callers in the same general area.

Although the principles of cellular telephony were worked out at Bell Labs in the 1940s, building such systems had to await the arrival of integrated circuits and other microelectronic components in the 1970s. In the United States, hundreds of companies saw the promise of the business, but government regulators were very slow in making a sufficiently broad band of frequencies available, delaying deployment considerably. As a result, Japan and the Scandinavian countries created the first cellular systems and have remained leaders in the technology. At the start there was plenty of room for improvement. Early cell phones were mainly installed in cars; handheld versions were as big as a brick, cost over a thousand dollars, and had a battery life measured in minutes. But in the 1990s the magic of the microchip drove prices down, shrank the phones to pocket size, reduced their energy needs, and packed them with computational powers.

By the year 2000, 100 million people in the United States and a billion worldwide were using cell phones—not just talking on them but also playing games, getting information off the Internet, and using the keyboard to send short text messages, a favorite

A multitude of cell phones on sale in Tokyo hint at Japan's jump on cellular technology. As recently as 1995 it was illegal for Japanese citizens to own a cell phone. Today, more than 60 million people—half of Japan's population—own cell phones, and more than half of them subscribe to Web phone service.

pastime of Japanese teenagers in particular. In countries where most households still lack a telephone—China and India, for example—the first and only phone for many people is likely to be wireless. Ultimately, Alexander Graham Bell's vision of a wired world may yield to a future in which, for everyone, personal communication is totally portable.

Perspective

Ian M. Ross

President Emeritus
AT&T Bell Laboratories

Shortly after arriving from England to join Bill Shockley's organization at Bell Labs in March 1952, I was asked to arrange a laboratory session for an April symposium on transistor technology. In 1947, Shockley's team had invented the transistor in an effort to replace relays and vacuum tubes in the telephone network with faster, more reliable solid-state devices. My lab session was meant to allow the attendees, most of whom had never seen a transistor, to measure the characteristics of the device touted as having no failure mechanisms and nothing to wear out.

AT&T had long prided itself on its stringent requirements for network reliability. For example, switching machines were to have no more than 2 hours of downtime in 40 years—a policy statement that has had a powerful impact over the years. The requirement may seem somewhat quaint today: Why not 1 hour of downtime in 20 years or 3 minutes per year? I believe the intent was to specify two things—first, the minuscule percentage of time it was acceptable for a machine to be down, and second, that the switch should be designed to serve for 40 years. Two days before the start of the symposium, New Jersey experienced its typical one-day transition from winter directly to the heat and humidity of summer. In those days Bell Labs was not air-conditioned, and when I got to the lab that morning I found that my transistors for the session had lost all their electrical characteristics—the CRT traces were flat! I had discovered for myself what many people already knew: The transistor was sensitive to its environment and particularly to humidity.

The lack of reliability of early transistors was a huge setback and embarrassment to the semiconductor community. The transistor had been lauded as a device with no failure mechanisms, with nothing to wear out. Instead, we had a severe reliability problem—and it took another 14 years to solve the problem completely. Given AT&T's reliability specification, the delay effectively paced the large-scale introduction of semiconductor devices into the telephone network.

Two decades later I was put in charge of the Network Planning Division at Bell Labs. About this time the technology was at hand to permit the conversion of the networks to all-digital operation. Here again AT&T's reliability specification strongly governed the design. The new electronic switching machines were to be controlled by a stored-program processor—in effect a special-purpose computer. The only way we at Bell Labs knew how to meet the downtime requirement was to use dual processors, simultaneously running identical programs. Fortunately, the machines that were introduced in the early 1970s are still providing service today. So far so good!

AIR CONDITIONING
AND REFRIGERATION

Which of the appliances in your home would be the hardest to live without? The most frequent answer to that question in a recent survey was the refrigerator. Over the course of the 20th century, this onetime luxury became an indispensable feature of the American home, a mainstay in more than 99.5 percent of the nation's family kitchens by century's end.

But the engineering principle on which it is based, mechanical refrigeration, has had even more far-reaching effects, through both refrigeration itself and its close cousin, air conditioning. Taken together, these cooling technologies have altered some of our most fundamental patterns of living. Our daily chores are different. What we eat and how we prepare food have both changed. The kinds of buildings we live and work in and even where we choose to live across the whole length and breadth of the United States all changed as a result of 20th-century expertise at keeping things cool.

Look back for a moment to the world before the widespread use of refrigeration and air conditioning—a world that was still very much present well into the first decades of the 20th century. Only fresh foods that could be grown locally were available, and they had to be purchased and used on a daily basis. Meat was bought during the daily trip to the butcher's; the milkman made his rounds every morning. If you could afford weekly deliveries of ice blocks—harvested in the winter from frozen northern lakes—you could keep some perishable foods around for 2 or 3 days in an icebox. As for the nonexistence of air conditioning, it made summers in southern cities—and many northern ones—insufferable. The nation's capital was a virtual ghost town in the summer months. As late as the 1940s, the 60-story Woolworth Building and other skyscrapers in New York City were equipped with window awnings on every floor to keep direct sunlight from raising temperatures even higher than they already were. Inside the skyscrapers, ceiling and table fans kept the humid air from open windows at least moving around. Throughout the country, homes were built with natural cooling in mind. Ceilings were high, porches were deep and shaded, and windows were placed to take every possible advantage of cross-ventilation.

By the end of the century all that had changed. Fresh foods of all kinds were available just about anywhere in the country all year round—and what wasn't available fresh could be had in convenient frozen form, ready to pop into the microwave. The milkman was all but gone and forgotten, and the butcher now did his work behind a counter at the supermarket. Indeed, many families concentrated the entire week's food shopping into one trip to the market, stocking the refrigerator with perishables that would last a week or more. And on the air-conditioning side of the equation, just about every form of indoor space—office buildings, factories, hospitals, and homes—was climate-controlled and comfortable throughout the year, come heat wave or humidity. New homes looked quite different, with lower rooflines and ceilings, porches that were more for ornament than practicality, and architectural features such as large plate glass picture windows and sliding glass doors. Office buildings got a new look as well, with literally acres of glass stretching from street level to the sky-scraping upper floors. Perhaps most significant of all, as a result of air conditioning, people started moving south, reversing a northward demographic trend that had continued through the first half of the century. Since 1940 the nation's fastest-growing states have been in the Southeast and the Southwest, regions that could not have supported large metropolitan communities before air conditioning made the summers tolerable.

Mechanical refrigeration, whether for refrigeration itself or for air conditioning, relies on a closed system in which a refrigerant—basically a compound of elements with a low boiling point—circulates through sets of coils that absorb and dissipate heat as the refrigerant is alternately compressed and allowed to expand (see pages 114-115). In a refrigerator the circulating refrigerant draws heat from the interior of the refrigerator, leaving it cool; in an air conditioner, coils containing refrigerant perform a similar function by drawing heat and moisture from room air.

This may sound simple, but it took the pioneering genius of a number of engineers and inventors to work out the basic principles of cooling and humidity control. Their efforts resulted in air conditioning systems that not only were a real benefit to the average person by the middle of the 20th century but also made possible technologies in fields ranging from medical and scientific research to space travel.

Prominent among air-conditioning pioneers was Willis Haviland Carrier. In 1902, Carrier, a recent graduate of Cornell University's School of Engineering, was working for the Buffalo Forge Company on heating and cooling systems. According to Carrier, one foggy night while waiting on a train platform in Pittsburgh he had a sudden insight into a problem he had been puzzling over for a while—the complex relationship between air temperature, humidity, and dew point. He realized that air could be dried by saturating it with chilled water to induce condensation. After a number of experimental air conditioning installations, he patented Dew Point Control in 1907, a device that, for the first time, allowed for the precise control of temperature and humidity necessary for

Weekly delivery of blocks of ice (opposite) is a thing of the past, now that refrigeration enables us to keep foods cold and fresh in the grocery store (above) as well as at home. In the form of "comfort cooling," refrigeration technology also changed the cityscape, making it possible to work in office buildings (left) whose interiors could be kept at a comfortable temperature no matter the weather outside.

sophisticated industrial processes. Carrier's early air conditioner was put to use right away by a Brooklyn printer who could not produce a good color image because fluctuations of heat and humidity in his plant kept altering the paper's dimensions and misaligning the colored inks. Carrier's system, which had the cooling power of 108,000 pounds of ice a day, solved the problem. That same principle today makes possible the billion-dollar facilities required to produce the microcircuits that are the backbone of the computer industry.

Theaters, among the first public buildings to get air conditioning in the 1920s, lured customers by advertising their cool interiors (right).

Air conditioners were soon being used in a variety of industrial venues. The term itself was coined in 1906 by a man named Stuart Cramer, who had applied for a patent for a device that would add humidity to the air in his textile mill, reducing static electricity and making the textile fibers easier to work with. Air-conditioning systems also benefited a host of other businesses, enumerated by Carrier himself: "lithography, the manufacture of candy, bread, high explosives and photographic films, and the drying and preparing of delicate hygroscopic materials such as macaroni and tobacco." At the same time, it did not go unnoticed that workers in these air-conditioned environments were more productive, with significantly lower absentee rates. Comfort cooling, as it became known, might just be a profitable commodity in itself.

Carrier and others set out to explore the potential. In 1915 he and several partners formed the Carrier Engineering Corporation, which they dedicated to improving the technology of air conditioning. Among the key innovations was a more efficient centrifugal (as opposed to piston-driven) compressor, which Carrier used in the air conditioners he installed in Detroit's J. L. Hudson Department Store in 1924, the first department store so equipped. Office buildings soon followed.

Even as Willis Carrier was pioneering innovations in industrial air conditoners, a number of others were doing the same for comfort cooling. Beginning in 1899, consulting engineer Alfred Wolff designed a number of cooling systems, including prominent installations at the New York Stock Exchange, the Hanover National Bank, and the New York Metropolitan Museum of Art. The public was exposed to air conditioning en masse at the St. Louis World's Fair in 1904, where they enjoyed the air-conditioned Missouri State Building. Dozens of movie theaters were comfort cooled after 1917, the result of innovations in theater air conditioning by Fred Wittenmeier and L. Logan Lewis, with marquees proclaiming" "It's 20 degrees cooler inside." Frigidaire engineers introduced a room cooler in 1929, and they, along with other companies such as Kelvinator, General Electric, and York, pioneered fully air-conditioned homes soon after.

Refrigerators did not represent quite as much of a revolution. Many people at the turn of the century were at least familiar with the concept of a cool space for storing food—the icebox. But true mechanical refrigeration—involving that closed system of circulating refrigerant driven by a compressor—didn't come along in any kind of practical form until 1913. In that year a man named Fred Wolf invented a household refrigerator that ran on electricity (some earlier mechanical refrigerators had run on steam-driven compressors that were so bulky they had to be housed in a separate room). He called it the Domelre, for Domestic Electric Refrigerator, and sold it for $900. It was a quick hit but was still basically an adaptation of the existing icebox, designed to be mounted on top of it. Two years later Alfred Mellowes introduced the first self-contained mechanical refrigerator, which was marketed by the Guardian Refrigerator Company. Mellowes had the right idea, but Guardian didn't make what it could of it. In 2 years the company produced a mere 40 machines.

Into the breach stepped one of the giants of the automotive industry, William Durant, president of General Motors. Realizing the potential of Guardian's product, he bought the company in 1918, renamed it Frigidaire, and put some of GM's best engineering and manufacturing minds to work on mass production. A few years later Frigidaire also bought the Domelre patent and began churning out units, introducing improvements with virtually each new production run.

Other companies, chief among them Kelvinator and General Electric, added their own improvements in a quest for a share of this obviously lucrative new market. By 1923 Kelvinator, which had introduced the first refrigerator with automatic temperature control, held

80 percent of market share, but Frigidaire regained the top in part by cutting the price of its units in half—from $1,000 in 1920 to $500 in 1925. General Electric ended up as industry leader for many years with its Monitor Top model—named because its top-mounted compressor resembled the turret of the Civil War ship—and with innovations such as dual temperature control, which enabled the combining of separate refrigerator and freezer compartments into one unit.

Market forces and other concerns continued to drive innovations. Led by Thomas Midgley, chemical engineers at Frigidaire solved the dangerous problem of toxic, flammable refrigerants—which had been known to leak, with fatal consequences—by synthesizing the world's first chlorofluorocarbon, to which they gave the trademarked name Freon. It was the perfect refrigerant, so safe that at a demonstration before the

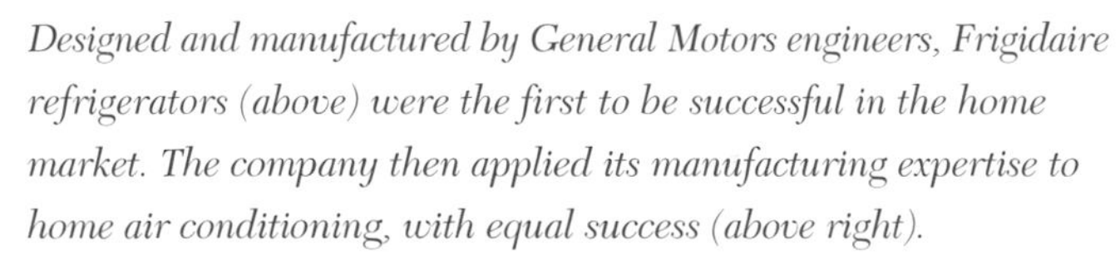

Designed and manufactured by General Motors engineers, Frigidaire refrigerators (above) were the first to be successful in the home market. The company then applied its manufacturing expertise to home air conditioning, with equal success (above right).

American Chemical Society in 1930 Midgley inhaled a lungful of the stuff and then used it to blow out a candle. In the late 1980s, however, chlorofluorocarbons were found to be contributing to the destruction of Earth's protective ozone layer. Production of these chemicals was phased out and the search for a replacement began.

At about the same time Frigidaire was introducing Freon it also turned its attention to the other side of the mechanical refrigeration business: air conditioning. Comfort cooling for the home had been hampered by the fact that air conditioners tended to be bulky affairs that had been designed specifically for large-scale applications such as factories, theaters, and the like. In 1928 Carrier introduced the "Weathermaker," the first practical home air conditioner, but because the company's main business was still commercial, it was slow to turn to the smaller-scale designs that residential applications required. Frigidaire, on the other hand, was ready to apply the same expertise in engineering and manufacturing that had allowed it to mass produce—literally by the millions—the low-cost, small-sized refrigerators that were already a fixture in most American homes. In 1929 the company introduced the first commercially successful "room cooler," and a familiar list of challengers—Kelvinator, GE, and this time Carrier—quickly took up the gauntlet. Window units came first, then central whole-house systems. Without leaving home, Americans could now escape everything from the worst humid summers of the Northeast and Midwest to the year-round thermometer-busting highs of the South and desert Southwest.

At about the same time that both refrigeration and air conditioning were becoming significantly more commonplace, both also went mobile. In 1939 Packard introduced the first automobile air conditioner, a rather awkward affair with no independent shut-off mechanism. To turn it off, the driver had to stop the car and the engine and then open the hood and disconnect a belt connected to the air conditioning compressor. Mechanical engineers weren't long in introducing needed improvements, ultimately making air conditioning on wheels so *de rigueur* that even convertibles had it.

THE BASICS OF COOLING

HOW AIR CONDITIONING AND REFRIGERATION WORK

Both air conditioners and refrigerators rely on the same basic working parts and the same principles of thermodynamics to provide cooling. A refrigerant—a chemical compound with a low boiling point—circulates in a closed system that includes a compressor, a condenser, an expansion valve, and an evaporator. As it circulates, the refrigerant is compressed and then allowed to expand, in the process changing back and forth from a gaseous to a liquid state and alternately absorbing and then dissipating heat as a result. The basic thermodynamic principle is that compression raises the refrigerant's temperature and expansion lowers it. When the refrigerant is colder than the air it is "conditioning," heat is absorbed; when it is hotter, that heat is dissipated and the refrigerant is again capable of absorbing heat.

Early forms of refrigerant, such as methyl chloride, were toxic and highly flammable. Chlorofluorocarbons such as Freon offered a safe alternative until it was discovered that, released into the atmosphere, they played a significant role in destroying Earth's protective ozone layer. In recent years, chemists have developed new, safer forms of refrigerant that are just as effective at keeping indoor air—and the space inside refrigerators and freezers—cool.

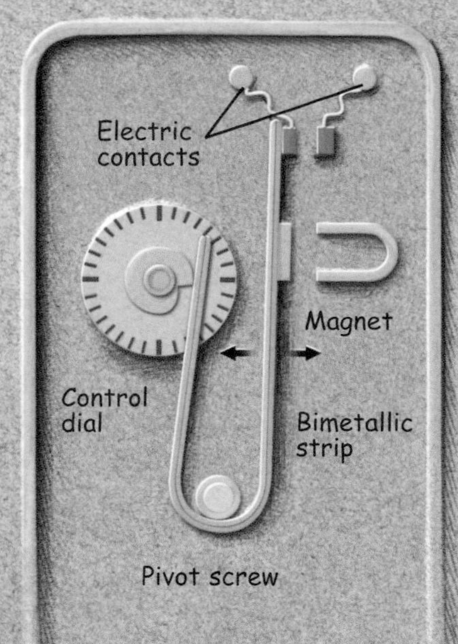

BIMETALLIC THERMOSTAT. The key part of this thermostat is a strip made up of two metals that expand and contract at different rates as the room temperature changes, causing the strip to bend toward or away from a magnet. When the strip connects with the magnet, it closes an electric contact, turning the cooling unit on. As the room cools, the strip bends in the other direction, pulling away from the magnet. At the temperature set by the control dial, the strip breaks the electric contact, turning the unit off.

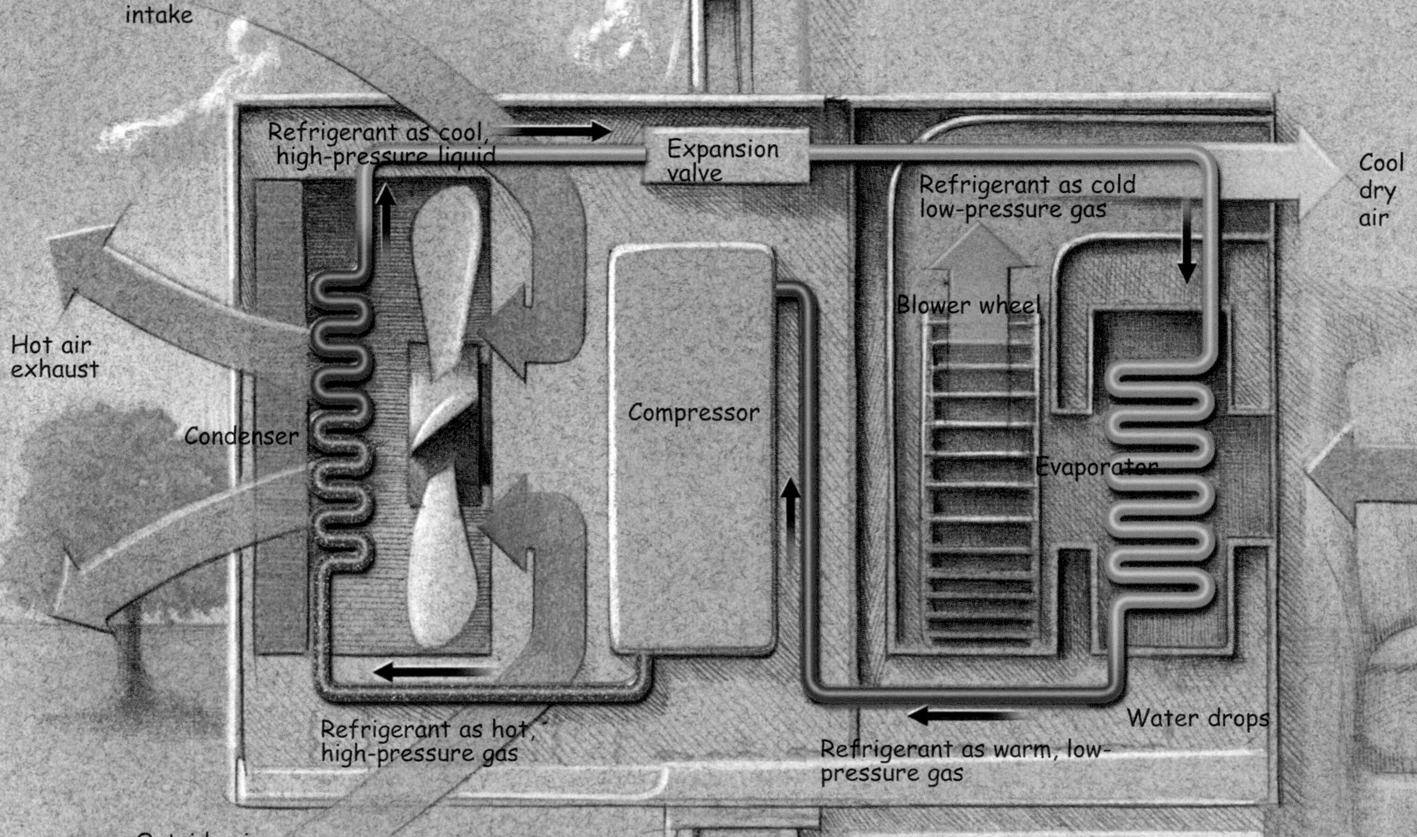

AIR CONDITIONER. In a window air conditioner, warm, humid indoor air passes over coils of cool, low-pressure refrigerant gas (blue) in the evaporator. As the air cools, it releases its moisture, and a fan blows cool dry air back into the room. The warm, low-pressure refrigerant gas is sucked into the compressor, where the gas is compressed. This hot, high-pressure gas (red) runs through a set of coils in the condenser, where a fan and metal fins help to dissipate its heat. As it cools, the gas condenses into a liquid. The cool, high-pressure liquid runs through an expansion valve. The resulting drop in pressure causes the refrigerant to expand back into a cold gas to start the cycle again.

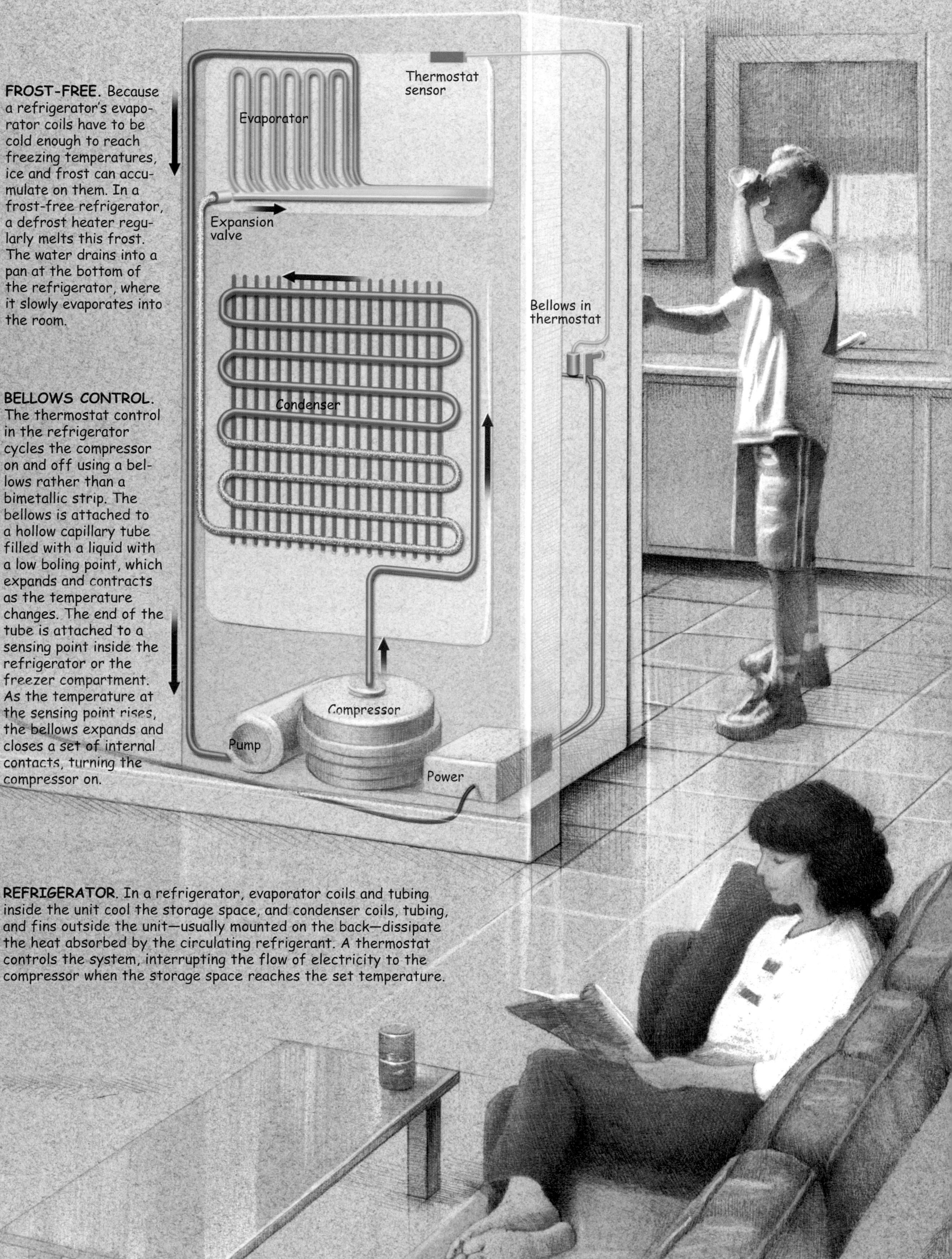

FROST-FREE. Because a refrigerator's evaporator coils have to be cold enough to reach freezing temperatures, ice and frost can accumulate on them. In a frost-free refrigerator, a defrost heater regularly melts this frost. The water drains into a pan at the bottom of the refrigerator, where it slowly evaporates into the room.

BELLOWS CONTROL. The thermostat control in the refrigerator cycles the compressor on and off using a bellows rather than a bimetallic strip. The bellows is attached to a hollow capillary tube filled with a liquid with a low boling point, which expands and contracts as the temperature changes. The end of the tube is attached to a sensing point inside the refrigerator or the freezer compartment. As the temperature at the sensing point rises, the bellows expands and closes a set of internal contacts, turning the compressor on.

REFRIGERATOR. In a refrigerator, evaporator coils and tubing inside the unit cool the storage space, and condenser coils, tubing, and fins outside the unit—usually mounted on the back—dissipate the heat absorbed by the circulating refrigerant. A thermostat controls the system, interrupting the flow of electricity to the compressor when the storage space reaches the set temperature.

With refrigeration at home and at the market, families can now do grocery shopping on a weekly basis, rather than having to buy perishable items daily or every other day.

Keeping cool has been a human preoccupation for millennia, but until the 20th century most efforts were ineffective. People tried everything from draping saturated mats in doorways to the installation of water-powered fans. Even Leonardo da Vinci designed and built a mechanical ventilating fan, the first of its kind. The modern system—involving the exchange of hot, moist air for cool, dry air by way of a circulating refrigerant—was first used in industrial settings. Indeed, a North Carolina textile engineer named Stuart Cramer, impressed with how the latest system of controlling the heat and humidity in his plant improved the cloth fibers, coined the term "air conditioning" in 1906. Since then comfort of cool is no longer considered a luxury but a fact of modern existence.

1902 A 300-ton comfort cooling system designed by Alfred Wolff is installed at the New York Stock Exchange. Using free cooling provided by waste-steam-operated refrigeration systems, Wolff's system functions successfully for 20 years.

The Armour Building in Kansas City, Missouri, becomes the first office building to install an air-conditioning system. Each room is individually controlled with a thermostat that operates dampers in the ductwork, making it also the first office building to incorporate individual "zone" control of separate rooms.

1904 A self-contained mechanical refrigerator is displayed at the St. Louis World's Fair by Brunswick Refrigerating Co., which specializes in designing small refrigerators for residences and butcher shops. The ammonia refrigerating system is mounted on the side of a wooden ice-box-type refrigerator.

Thousands of attendees at the World's Fair also experience the public debut of air conditioning in the Missouri State Building. The system uses 35,000 cubic feet of air per minute to cool a 1,000-seat auditorium, the rotunda, and various other rooms.

1906 Willis Carrier *(above)* files for a patent on his "dew point control" system. Carrier has studied the science of humidity control after designing a rudimentary air-conditioning system for a Brooklyn printing plant in 1902. This and subsequent designs allow him to devise a precise method of controlling humidity using refrigerated water sprays, thereby allowing the manufacture of air-conditioning systems to be standardized.

In Chicago, Frank Lloyd Wright's Larkin Administration Building is the first office building specifically designed for air conditioning. The system uses safe, nonflammable carbon dioxide as a refrigerant.

Boston Floating Hospital becomes the first air-conditioned hospital, using a system designed by Edward Williams to maintain the hospital wards at about 70°F with a relative humidity of 50 percent. The hospital's five wards are individually controlled by thermostats. Williams's system features "reheat" in which cooled air is heated slightly to lower its humidity.

1907 Air-conditioning equipment designed by Frederick Wittenmeier is installed in dining and meeting rooms at Congress Hotel in Chicago. This is one of the first systems designed by Wittenmeier for hotels and movie theaters. His firm, Kroeschell Brothers Ice Machine Company, installs hundreds of cooling plants into the 1930s.

1914 Fred Wolf, Jr., markets an air-cooled, electric, self-contained household refrigerating unit, the Domelre (Domestic Electric Refrigerator), in Chicago. The system is designed to be placed on top of any icebox, operating automatically using a thermostat. The first household refrigerating system to feature ice cubes, the Domelre uses air to cool the condenser, unlike other household refrigerators that need to be hooked up to water.

1916 Clarence Birdseye begins experiments in quick-freezing. Birdseye develops a flash-freezing system that moves food products through a refrigerating system on conveyor belts. This causes the food to be frozen very fast, minimizing ice crystals.

1923 An electrically refrigerated ice cream dipping cabinet is marketed by Nizer and shortly after by Frigidaire. These cabinets use a refrigeration system to chill alcohol-based antifreeze, which surrounds ice cream cans placed in wells in the cabinet. The alcohol is later replaced by salt brine.

1927 Gas-fired household absorption refrigerators that do not require electricity are marketed to rural areas in the United States. One, the Electrolux, marketed in Sweden since 1925, becomes very popular.

General Electric introduces the first refrigerator to be mass produced with a completely sealed refrigerating system. Nicknamed "The Monitor Top" for its distinctive round refrigerating unit, resembling the gun turret of the Civil War ironclad ship *Monitor*, the refrigerator is produced over the next 10 years and is so reliable that thousands are still in use today.

1928 Chlorofluorocarbon (CFC) refrigerants are synthesized for Frigidaire by the General Motors Research Lab team of Thomas Midgley, Albert Henne, and Robert McNary. Announced publicly in 1930 and trademarked as Freon, CFCs are the first nontoxic and nonflammable refrigerating fluids, making it possible for refrigerators and air conditioners to be used with complete safety.

1929 Frigidaire markets the first room cooler. The refrigeration unit, which uses sulfur dioxide refrigerant and has a capacity of one ton (12,000 BTUH), is designed to be located outside the house or in the basement.

1930 With the advent of the centrifugal chiller, smaller air-conditioning units become feasible for trains. In 1930 the Baltimore & Ohio Railroad tests a unit designed by Willis Carrier on the "Martha Washington" *(below)*, the dining car on the *Columbian*, running between Washington, D.C. and New York. To test the system, the car is heated to 93°F. The heat is then turned off and the air conditioner turned on. Within 20 minutes, the temperature in the dining car is a comfortable 73°F.

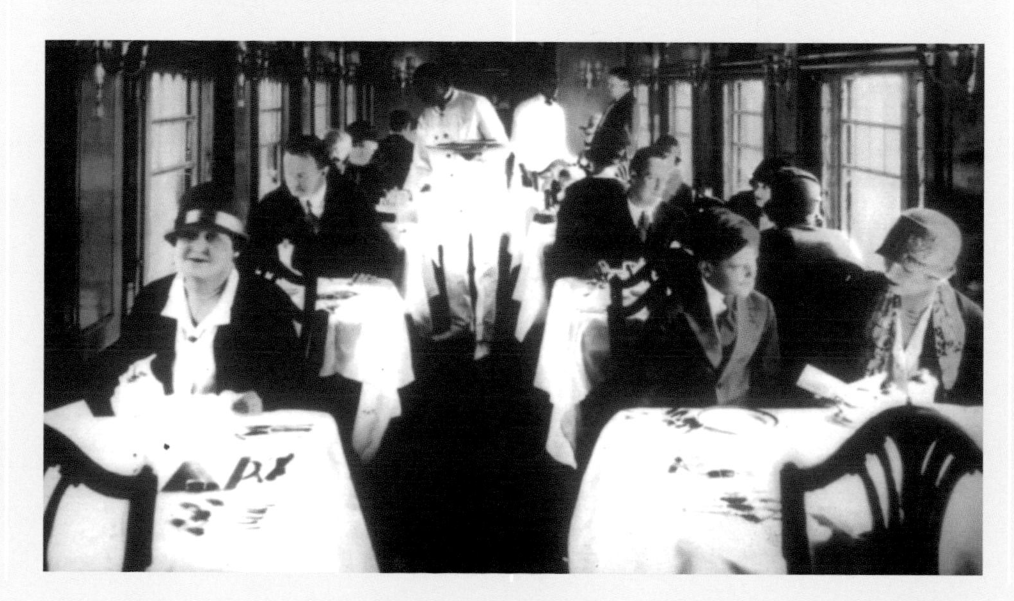

1931 Frigidaire markets the "Hot-Kold" year-round central air-conditioning system for homes. During the early 1930s, a number of manufacturers design central air conditioners for homes, a market that grows slowly until the 1960s, when lower costs make it affordable for many new homes.

Southern California Edison Company installs a heat pump air-conditioning system in its Los Angeles office building. Since a refrigeration system moves heat from one place to another, the same principle can be used to remove heat in summer or add heat in winter by engineering the system to be reversible.

1932 Chesapeake & Ohio Railroad begins running the first overnight train with air conditioning, the *George Washington*, between New York and Washington. Four years later United Air Lines uses air conditioning in its "three miles a minute" passenger planes.

1936 Albert Henne, coinventor of the CFC refrigerants, synthesizes refrigerant R-134a. In the 1980s this refrigerant is hailed as the best nonozone-depleting replacement for CFCs.

1938 A window air conditioner using Freon is marketed by Philco-York. Featuring a beautiful wood front, the Philco air conditioner can simply be plugged into an electrical outlet.

1939 Packard Motor Car Company markets an automobile with air conditioning offered as an option for $274. The refrigeration compressor runs off the engine, and the system has no thermostat. It discharges the cooled air from the back of the car.

1947 Mass-produced, low-cost window air conditioners become possible as a result of innovations by engineer Henry Galson, who sets up production lines for a number of manufacturers. In 1947, 43,000 window air conditioners are sold in the United States. For the first time, many homeowners can enjoy air conditioning without having to buy a new home or renovate their heating system.

1969 More than half of new automobiles (54 percent) are equipped with air conditioning, which is soon a necessity, not only for comfort but also for resale value.

By now, most new homes are built with central air conditioning, and window air conditioners are increasingly affordable.

1987 The National Appliance Energy Conservation Act mandates minimum energy efficiency requirements for refrigerators and freezers as well as room and central air conditioners.

The same year the Montreal Protocol serves as an international agreement to begin phasing out CFC refrigerants, which are suspected of contributing to the thinning of the earth's protective, high-altitude ozone shield.

1992 The U.S. Energy Policy Act mandates minimum energy efficiency standards for commercial buildings, using research and standards developed by the American Society of Heating, Refrigerating, and Air Conditioning Engineers.

HIGHWAYS

Sweeping visions were something of a specialty for William Durant, founder of General Motors, and he ran true to form in a 1922 interview. "Most of us," he said, "will live to see this whole country covered with a network of motor highways built from point to point as the bird flies, the hills cut down, the dales bridged over, the obstacles removed." Given the intensity of America's love affair with the automobile, his prediction wasn't so far-fetched.

At the turn of the century a few thousand people owned cars; in 1922 about 10 million did, and that number more than doubled in the next few years. Sharing their need for decent roads were fast-growing fleets of trucks and buses. And the federal government was thinking big. Congress had just authorized funds to help states create a 200,000-mile web of smooth-surfaced roads that would connect with every county seat in the nation.

It was just a beginning. Ahead lay engineering feats beyond anything Durant could have foreseen: the construction of conduits that can safely handle thousands of cars an hour and endure years of punishment by 18-wheel trucks, expressways and beltways to speed traffic in and around cities, swirling multilevel interchanges, arterial tunnels and mighty suspension bridges. Ahead, as well, lay a host of social and economic changes wrought by roads—among them, spreading suburbs, the birth of shopping malls and fast-food chains, widened horizons for vacationers, a revolution in manufacturing practices, and a general attuning of the rhythms of daily life, from errands to entertainment to the personal mobility offered by the car. Expansion of the network would also bring such indisputable negatives as traffic congestion and air pollution, but the knitting together of the country with highways has barely paused since the first automobiles rolled forth from workshops about a century ago.

Rails ruled the transportation scene then. Like other developed nations, the United States had an intricate system of railroad tracks reaching to almost every sizable community in the land. Virtually all long-distance travel was by train, and electric trolleys running on rails served as the main people movers in cities. The United States also had more than 2 million miles of roads, but practically all were unsurfaced or only lightly layered with broken stone. A "highway census" performed by the federal government in 1904 found a grand total of 141 miles of paved roads outside cities. In rainy weather, travel in the countryside became nightmarish, and even in good conditions, hauling loads over the rough roads was a laborious business; it was cheaper to ship fruit by rail from California to an urban market in the East than to deliver it by wagon from a farm 15 miles away. As for anyone contemplating a lengthy drive in one of the new horseless carriages, an ordeal was in store. The first crossing of the continent by car in 1903 required 44 days of hard driving. By train the trip took just 4 days.

Narrow tracks early in the 20th century needed places for vehicles to pull over, as this Buick Roadster did, to let a horse-drawn carriage pass (left). Today roadways like the scenic drive along California's coast (opposite) give drivers in each direction plenty of room—and gorgeous scenery besides.

But cars had the irresistible advantage of flexibility, allowing drivers to go wherever they wanted, whenever they wanted, provided suitable roads were available. Efforts to accommodate motorized travel were soon launched by all levels of government, with particular emphasis on relieving the isolation of farmers. Beginning in 1907 the federal Office of Public Roads built experimental roads to test concrete, tars, and other surfacing materials. The agency also trained engineers in the arts of road location, grading, and drainage, then sent them out to work with state highway departments, which selected the routes and set the construction standards. Federal-state partnerships became the American way of road building, with the states joining together to harmonize their needs.

When the United States entered World War I in 1917, trucks carrying extra-heavy loads of munitions and other supplies pounded many sections of highway to ruin. Even so, shippers were so impressed by their performance that the trucking industry boomed after the war, and new highways were engineered accordingly. During the 1920s, states increased the recommended thickness of concrete pavement on main roads from 4 inches to at least 6 and set the minimum pavement width at 20 feet. Extensive research was done on soil types to ensure adequate underlying support. Engineers improved old roads by smoothing out right-angle turns and banking the curves. At the same time, much research was done on signs, pavement markings, and other methods of traffic control. The first four-way, three-color traffic light appeared in Detroit in 1920.

Europe provided some compelling lessons in road construction. In Italy, whose heritage included raised paved roads that allowed Roman armies to move swiftly across the empire, private companies began to build toll highways called *autostrade* in the mid-1920s. Although not especially well suited for fast-moving traffic, their limited-access design minimized disruption of the flow, and safety was further enhanced by the elimination of intersections with other roads or with railways. These features were also incorporated into the first true expressways, the national network of autobahns built in Germany between 1929 and 1942. The 1,310-mile system consisted of twin 30-foot-wide roadways separated by a grassy central strip, which significantly boosted capacity while allowing higher speeds.

The United States adopted the four-lane, limited-access scheme for relatively modest highways in Connecticut and California in the late 1930s and then

By the mid-1930s Germany's expressways, or autobahns, were a motorist's dream, as seen below on a stretch between Frankfurt and Mannheim.

produced a true engineering masterpiece, the Pennsylvania Turnpike, whose initial 164-mile section opened in 1940. A model for future high-speed, heavy-duty routes, the turnpike had a 10-foot median strip and a 200-foot total right-of-way. Each lane was 12 feet wide; curves were long and banked; grades were limited to 3 feet in a hundred; feeder and exit lanes merged smoothly with the main traffic streams; and the concrete pavement was surpassingly sturdy—9 inches thick, with a reinforcement of welded steel fabric. Travel time between Philadelphia and Pittsburgh was reduced by as much as 6 hours, but not for free. The Pennsylvania Turnpike was a toll road, and it did such an active business that many other states soon created their own turnpike authorities to construct similar self-financing superhighways.

As the nation's highway network grew, the challenge of leaping over water barriers inspired some structural wonders. One was the George Washington Bridge, which opened in 1931. To connect the island of Manhattan with New Jersey, Swiss-born engineer Othmar Ammann suspended a 3,500-foot, eight-lane roadway—the longest span in the world at the time—between a pair of lattice-steel towers on either side of the Hudson River. Special machinery spun and compressed the 105,000 miles of wires that went into the cables, and everything was made strong enough to support a second deck added later. In 1937, San Francisco was joined to Marin County with an even longer suspension span—4,200 feet. The Golden Gate Bridge, designed by Joseph Strauss, was built to withstand the swift tides and high winds of the Golden Gate strait. One of its tower supports had to be built almost a quarter-mile from shore in water 100 feet deep. A million tons of concrete went into the massive anchors for the cable ends.

Coiling freeway interchanges, like these on the New Jersey Turnpike in the 1950s, are now a familiar part of the landscape throughout the United States.

Bridge construction on both sides of the country boomed during the 1930s, from the George Washington Bridge in New York City (below left) to the Golden Gate Bridge in San Francisco (below right), whose two main towers each rise 746 feet above the water and are strung with 80,000 miles of cable. During construction, a safety net stretched above the level of the water saved the lives of 19 workers.

The United States would eventually need half a million highway bridges, most of them small and unmemorable, some ranking among the loveliest structures ever created. Roads, too, aspired to beauty at times. During the 1920s and 1930s, parkways that meandered through pristine landscapes were laid out around New York City, and the National Park Service constructed scenic highways such as Skyline Drive along Virginia's Blue Ridge Mountains *(right)*. In general, however, highways have done far more to alter the look of America than to celebrate it.

Beginning in the 1920s, residential communities left the old urban trolley lines far behind and spread out ward from cities via roads. Stores, factories, and other businesses followed, sometimes aggregating into mini-metropolises themselves. As roads were improved to serve commuters and local needs, the outward migration accelerated, producing more traffic, which required more roads—almost limitlessly, it often seemed. In the late 1940s, for example, California began building an extensive system of express highways in and around Los Angeles and San Francisco, only to have congestion steadily worsen and a major expansion of the freeway system become necessary just a decade later.

The long-distance links of the nation's road system were under pressure as well, and federal and state highway officials sketched some grand plans to meet the expected demand. One of the boldest dreamers of all was President Franklin D. Roosevelt, a man who loved long drives through the countryside in a car. He suggested that the government buy a 2-mile-wide strip of land stretching from one coast to the other, with some of it to be sold for development and the rest used for a magnificent toll highway. By 1944 this scheme had been brought down to earth by Congress, which passed a bill offering the states 50 percent federal funding of a 40,000-mile arterial network. Little could be done with a war on, however, and progress was fitful in its hectic aftermath. But President Dwight D. Eisenhower turned that basic vision into one of the greatest engineering projects in history.

As a former military man, Eisenhower was keenly interested in transportation. When he was a young lieutenant in 1919, he had traveled from Washington to San Francisco with a caravan of cars and trucks, experiencing delays and breakdowns of every kind and averaging 5 miles an hour on the miserable roads. At the opposite extreme, he had noted the swift movement of German forces on autobahns when he was commanding Allied forces in Europe during World War II. The United States, he was convinced, needed better highways.

Getting legislation—and funding—through Congress was no small task. In 1954 Eisenhower appointed an advisory committee, chaired by his wartime colleague General Lucius Clay, to establish consensus among the many parties to the nation's road-building program. It took 2 years, but the quiet diplomacy and technical expertise of the committee's executive secretary, a Bureau of Public Roads engineer named Frank Turner, ultimately helped steer legislation through the political shoals in both the House and the Senate. In 1956 Eisenhower signed into law the act initiating the epic enterprise known as the National System of Interstate and Defense Highways. It called for the federal government to pay $25 billion—90 percent of the estimated total cost—toward building limited-access expressways that would crisscross the nation and speed traffic through and around cities.

The network, to be completed by 1972, would incorporate some older toll roads, and its length was

In January 1955 President Eisenhower accepted the Clay Commission report on financing the interstate highway system. Standing behind the president are (from left) Gen. Lucius Clay, Frank Turner, Steve Bechtel, Sloan Colt, William Roberts, and Dave Beck.

TUNNELS TO TAKE US FROM HERE TO THERE

MACHINES THAT BORE THROUGH SOLID ROCK

As a barrier-defeating tactic, digging through the earth is at least as old as civilization itself (the Babylonians carved a passageway beneath the Euphrates River in the 3rd millennium B.C.), but nowadays much of the work on major tunneling projects is likely to be automated—carried out by huge, wormlike machines that gnaw their way forward at a typical pace of a dozen-or-so yards a day. At the front of one of these tunnel-borers is a wheel that rotates slowly—perhaps twice a minute—to grind rock or soil apart. Hydraulic jacks behind the wheel supply forward thrust, and other machinery whisks the scrapings away and constructs the tunnel wall. Although such borers are unsuited for very hard rock, they can be configured to cut efficiently though strata ranging from limestone to clay. With such versatility and brute power available, transportation engineers are increasingly opting for subterranean routes beneath busy waterways or cities. Some tunnel links are many miles long. To join Britain and France across the English Channel, for instance, giant borers pierced 31 miles of chalk; automobiles piggybacking on special rail cars can now make the crossing in just 35 minutes.

BORE HEAD. On the slow-spinning wheel at the head of a tunnel-boring machine are cutting rollers and scrapers that vary in number and shape according to the strata being excavated. A gripper ring is pressed tightly against the tunnel wall to keep the borer head stable, and hydraulic rams shove the rotating disk against the rock or soil. An arm mounted behind the bore head lifts and places concrete panels that form the tunnel lining.

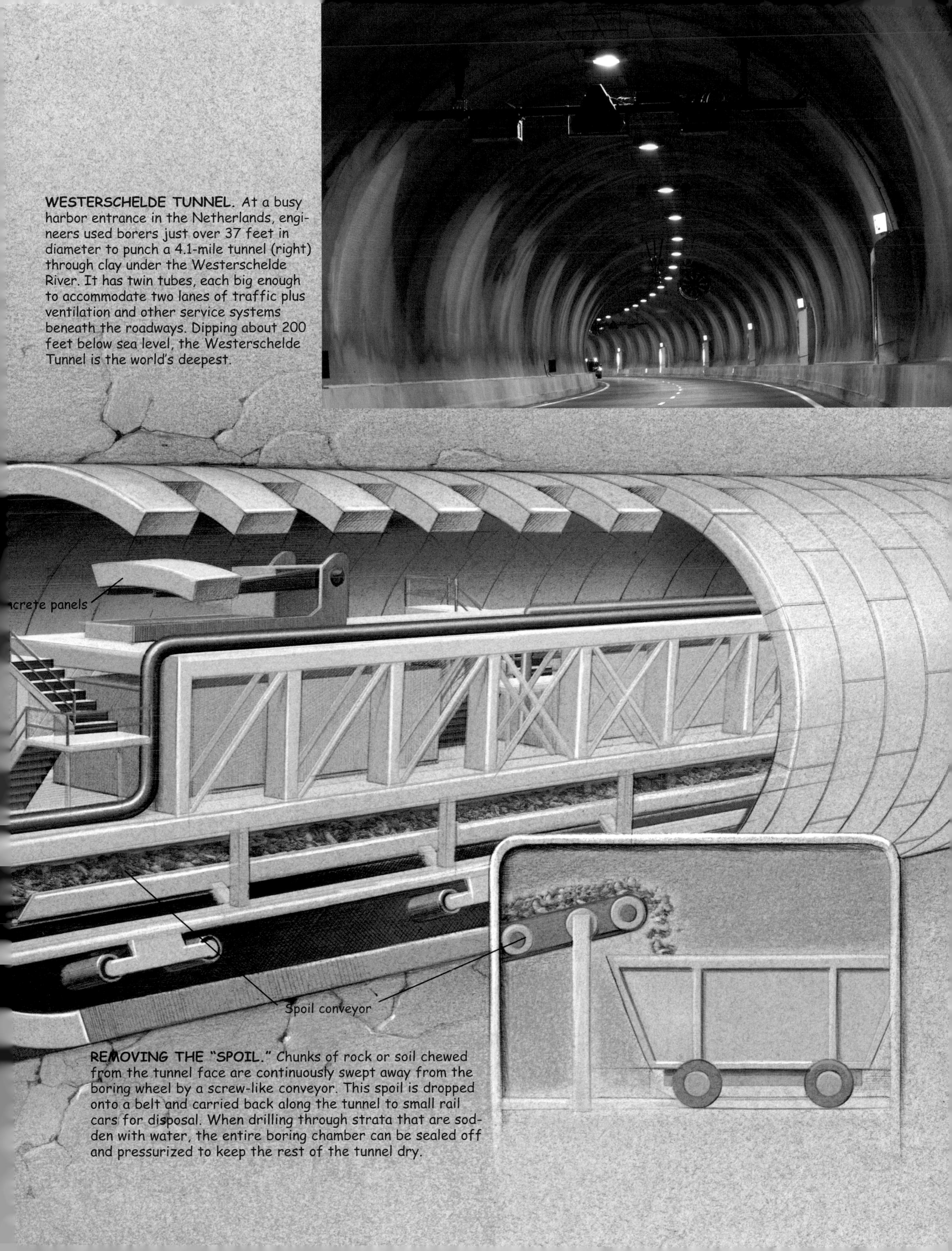

WESTERSCHELDE TUNNEL. At a busy harbor entrance in the Netherlands, engineers used borers just over 37 feet in diameter to punch a 4.1-mile tunnel (right) through clay under the Westerschelde River. It has twin tubes, each big enough to accommodate two lanes of traffic plus ventilation and other service systems beneath the roadways. Dipping about 200 feet below sea level, the Westerschelde Tunnel is the world's deepest.

REMOVING THE "SPOIL." Chunks of rock or soil chewed from the tunnel face are continuously swept away from the boring wheel by a screw-like conveyor. This spoil is dropped onto a belt and carried back along the tunnel to small rail cars for disposal. When drilling through strata that are sodden with water, the entire boring chamber can be sealed off and pressurized to keep the rest of the tunnel dry.

In constructing bridges like the 4.35-mile-long Chesapeake Bay Bridge (above) or the enormous renovation known as the Big Dig in Boston (above right), engineers must ensure that roadways and all supporting structures will withstand the unceasing flow of cars and trucks that travel the nation's highways (below).

ultimately set at 44,000 miles. Four 12-foot lanes were the stipulated minimum, and many sections would have more, along with 10-foot shoulders. The system would include 16,000 entrances and exits, dozens of tunnels, and more than 50,000 overpasses and bridges. To ensure that the roads would hold up under the anticipated truck traffic, hundreds of different pavement variations were tested for 2 years at a site in Illinois.

The price tag for the interstate highway system turned out to be five times greater than anticipated, and work went on for 4 decades—not without controversy and contention, especially on the urban front. Several cities, refusing to sacrifice cherished vistas or neighborhoods to an expressway, chose to do without. By the mid-1970s cities were permitted to apply some of their highway funds to mass transit projects such as subways or rail lines. But for the most part the great project moved inexorably forward, and by the 1990s cars, trucks, and buses were traveling half a trillion miles a year on the interstates—a good deal more safely than on other U.S. roads.

For the trucking industry the system was a boon. Trucks had been siphoning business from the railroads for decades, and the interstates contributed to a further withering of the nation's rail network by enabling trucks to travel hundreds of miles overnight to make a delivery. By the end of the 20th century, more and more American manufacturers had adopted a Japanese production system that dispenses with big stockpiles of materials. Instead, parts and supplies are delivered to a factory—generally by truck—at the exact moment when they are needed. This so-called just-in-time approach, which yields big savings in inventory expenses, turned the nation's highways into a kind of virtual warehouse. Sometimes trucking firms partner with railroads by piggybacking trailers on flatcars for long-distance legs of their journeys, but America's highways have the upper hand in freight hauling, as they do in the movement of people—far more so than in most other developed countries. Today, about 70 percent of all freight deliveries in the United States are made by trucks.

Highways continue to engender more highways by their very success. As traffic grows, engineers are working to improve pavements, markings, crash barriers, and other design elements, and they wage an unending war against congestion, sometimes by tactics as simple as adding lanes or straightening curves, sometimes with megaprojects such as the digging of a 3.5-mile, eight-lane tunnel beneath downtown Boston. It's a journey with no end in sight; Americans crave mobility, and wheels will always need roads.

Perspective

Stephen D. Bechtel, Jr.

Chairman Emeritus and Director
Bechtel Group, Inc.

I come from a long line of engineers. My grandfather, Warren A. Bechtel, began by grading railroad beds in Oklahoma Territory in 1898 and started our company in the early years of the 20th century. His son, my father Steve Bechtel, was raised around heavy construction jobs—and so was I. The first 2 years of my life were spent living in an "outfit car" with my mother and dad at a railroad construction camp as our company built railroads in the western United States.

In addition to railroads, our company has been involved in highway construction and building bridges and tunnels. In the mid-1950s my father served on President Eisenhower's Advisory Committee on a National Highway Program at the request of General Lucius Clay, the committee's chairman and Dad's good friend. "If we were going to build highways, I wanted people who knew something about it," General Clay once told an interviewer. "Steve Bechtel had more experience in the construction field than anyone in America. He wasn't involved in road building, but had a comprehensive knowledge of the construction industry." I know my dad got a great deal of satisfaction out of helping develop the concept and general plans for what became the interstate highway system—one of the great infrastructure feats of the century.

At Purdue University I majored in civil engineering, an integral part and perhaps the dominant engineering discipline in most transportation systems. In the early 1960s, shortly after I became president of Bechtel, our company partnered with Parsons Brinckerhoff and Tudor to do the feasibility study for, and then served as project managers on, the Bay Area Rapid Transit System. We also had a major role in building the Washington Rapid Transit. These projects were the pioneers of modern-day rapid transit. During my years as president of Bechtel, we were involved in a major 229-kilometer highway program in Turkey, and we are currently helping build the 200-kilometer Croatian Motorway. We are also involved in the Big Dig in Boston, where our people, in partnership with Parsons Brinckerhoff, are nearing completion of the largest road project ever in the United States.

I have been privileged in my career to have gained personal experience in a wide range of transportation projects, including highways, railroads, airports, marine terminals, and pipelines—all of which offered fascinating engineering challenges and very rewarding activities. Although each project is unique, every one demands a high level of integration. First, you must determine the need, the scope of the project, and what it will cost. The potential impact on the environment and the local community also must be considered. The engineering team then uses all the information to create the design. Equipment and materials are procured, and a skilled workforce is brought in to finish the job on time and within budget. It's a straightforward process, but on a big project it takes extraordinary organizational ability to do it well.

For those of us who are fortunate to have been trained and to serve as engineers, there is great satisfaction in working on historic and important infrastructure projects. They improve the quality of life, in both safety and convenience, and facilitate improved commerce and economic growth around the world. From a personal view, the engineers involved get the great feeling of accomplishment in participating and helping to bring these benefits to the people of the world.

HIGHWAYS TIMELINE

A reliable network of roads, bridges, and tunnels is so fundamental to any country's economic and social well-being that some of the greatest engineering feats of all time have involved solving the problem of how to get from Point A to Point B—or rather, how to get over, under, around, or through some natural obstacle in order to go from Point A to Point B. Before the advent of the automobile, roads in the United States amounted to little more than a collection of dusty two-lane trails and occasional short bridges. Today, thanks to 20th-century civil engineers, the driving public can travel coast to coast on a world-class interstate highway system that includes bridges and tunnels of phenomenal strength and beauty.

1905 The Office of Public Roads (OPR) is established, successor to the Office of Road Inquiry established in 1893. OPR's director, Logan Waller Page, who would serve until 1919, helps found the American Association of State Highway Officials and lobbies Congress to secure the Federal Aid Highway Program in 1916, giving states matching funds for highways.

1910 Gulf Oil, Texas Refining, and Sun Oil introduce asphalt manufactured from byproducts of the oil-refining process. Suitable for road paving, it is less expensive than natural asphalt mined in and imported from Venezuela. The new asphalt serves a growing need for paved roads as the number of motor vehicles in the United States soars from 55,000 in 1904 to 470,000 in 1910 to about 10 million in 1922.

Garrett Morgan, an inventor with a fifth-grade education and the first African-American in Cleveland to own a car, invents the electric, automatic traffic light.

1913 The first highway paved with portland cement, or concrete, is built near Pine Bluff, Arkansas, 22 years after Bellefontaine, Ohio, first paved its Main Street with concrete. Invented in 1824 by British stone mason Joseph Aspdin from a mix of calcium, silicon, aluminum, and iron minerals, portland cement is so-named because of its similarity to the stone quarried on the Isle of Portland off the English coast.

1917 Wisconsin is the first state to adopt a numbering system as the network of roads increases. The idea gradually spreads across the country and replaces formerly named trails and highways.

1919 Thomas MacDonald is appointed to head the federal Bureau of Public Roads (BPR), successor to OPR. During his 34-year tenure he helps create the Advisory Board on Highway Research, which becomes the Highway Research Board in 1924 and the Transportation Research Board in 1974. Among other things, BPR operates an experimental farm in Arlington, Virginia, to test road surfaces.

1920 William Potts, a Detroit police officer, refines Garrett Morgan's invention by adding the yellow light. Red and green traffic signals in some form have been in use since 1868, but the increase in automobile traffic requires the addition of a warning signal.

1923 State highway engineers across the country adopt a uniform system of signage based on shapes that include the octagonal stop sign.

1925 BPR and state highway representatives create a numbering system for interstate highways. East-west routes are designated with even numbers, north-south routes with odd numbers. Three-digit route numbers are given to shorter highway sections, and alternate routes are assigned the number of the principal line of traffic preceded by a one.

1927 Completion of the Holland Tunnel beneath the Hudson River links New York City and Jersey City, New Jersey. It is named for engineer Clifford Holland, who solves the problem of venting the build-up of deadly car exhaust by installing 84 electric fans, each 80 feet in diameter.

1932 The opening of a 20-mile section of Germany's fledgling autobahn, regarded as the world's first superhighway, links Cologne and Bonn. By the end of the decade the autobahn measures 3,000 kilometers and inspires U.S. civil engineers contemplating a similar network. Today the autobahn covers more than 11,000 kilometers.

1937 The Golden Gate Bridge opens and connects San Francisco with Marin County. To construct a suspension bridge in a region prone to earthquakes, engineer Joseph Strauss uses a million tons of concrete to hold the anchorages in place. Its two main towers each rise 746 feet above the water and are strung with 80,000 miles of cable.

The paving of Route 66 linking Chicago and Santa Monica, California, is complete. Stretching across eight states and three time zones, the 2,448-mile-long road is also known as "The Mother Road" and "The Main Street of America." For the next half-century it is the country's main thoroughfare, bringing farm workers from the Midwest to California during the Dust Bowl and contributing to California's post–World War II population growth. Officially decommissioned in 1985, the route has been replaced by sections of Interstate-55, I-44, I-40, I-15, and I-10.

Late 1930s Air-entrained concrete, one of the greatest advancements in concrete technology, is introduced. The addition of tiny air bubbles in the concrete provides room for expansion when water freezes, thus making the concrete surface resistant to frost damage.

1940 The Pennsylvania Turnpike opens as the country's first roadway with no cross streets, no railroad crossings, and no traffic lights. Built on an abandoned railroad right of way, it includes 7 miles of tunnels through the mountains, 11 interchanges, 300 bridges and culverts, and 10 service plazas. By the mid-1950s America's first superhighway extends westward to the Ohio border, north toward Scranton, and east to Philadelphia for a total of 470 route miles.

1944 The Federal Aid Highway Act authorizes the designation of 40,000 miles of interstate highways to connect principal cities and industrial centers.

1949 The first concrete pavement constructed using slipforms is built in O'Brian and Cerro Counties, Iowa.

1952 The Chesapeake Bay Bridge, the world's largest continuous over-water steel structure, opens, linking Maryland's eastern and western shores of the bay. Spanning 4.35 miles, the bridge has a vertical clearance of 186 feet to accommodate shipping traffic. In 1973 another span of the bridge opens to ease increasing traffic. By the end of the century, more than 23 million cars and trucks cross the bridge each year.

The first "Walk/Don't Walk" signal is installed in New York City.

1956 President Dwight D. Eisenhower signs a new Federal Aid Highway Act, committing $25 billion in federal funding. Missouri is the first state to award a highway construction contract with the new funding. The act incorporates existing toll roads, bridges, and tunnels into the system and also sets uniform interstate design standards.

Lake Pontchartrain Causeway opens, connecting New Orleans with its north shore suburbs. At 24 miles it is the world's longest over-water highway bridge. Made up of two parallel bridges, the causeway is supported by 95,000 hollow concrete pilings sunk into the lakebed. It was originally designed to handle 3,000 vehicles per day but now carries that many cars and trucks in an hour.

1960s Paint chemist and professor Elbert Dysart Botts develops a reflective paint for marking highway lanes. When rainwater obscures the paint's reflective quality, Botts develops a raised marker that protrudes above water level. Widely known as Botts' Dots, the raised markers were first installed in Solano County, California, along a section of I-80. They have the added benefit of making a drumming sound when driven over, warning drivers who veer from their lanes.

1962 The AASHO (American Association of State Highway Officials) road test near Ottawa, Illinois, which subjects sections of pavements to carefully monitored traffic loads, establishes pavement standards for use on the interstate system and other highways.

1964 The Chesapeake Bay Bridge-Tunnel opens, connecting Virginia Beach and Norfolk to Virginia's Eastern Shore. Its bridges and tunnels stretch 17.6 miles shore to shore and feature a pair of mile-long tunnels that run beneath the surface to allow passage above of commercial and military ships *(below)*. In 1965 the bridge-tunnel is named one of the "Seven Engineering Wonders of the Modern World" in a competition that includes 100 major projects.

1966 The Highway Safety Act establishes the National Highway Program Safety Standards to reduce traffic accidents.

1973 Interstate 70 in Colorado opens from Denver westward. It features the 1.75-mile Eisenhower Memorial Tunnel, the longest tunnel in the interstate program.

1980s and 1990s Introduction of the open-graded friction course, allowing asphalt to drain water more efficiently and thus reducing hydroplaning and skidding, and Superpave, or Superior Performing Asphalt Pavement, which can be tailored to the climate and traffic of each job, are among refinements that improve the country's 4 million miles of roads and highways, 96 percent of which are covered in asphalt. By the end of the century, 500 million tons of asphalt will be laid every year.

1986 The Fort McHenry Tunnel in Baltimore opens and at 1.75 miles is the longest and widest underwater highway tunnel ever built by the immersed-tube method. The tunnel was constructed in sections, then floated to the site and submerged in a trench. It also includes a computer-assisted traffic control system and communications and monitoring systems.

1987 The Sunshine Skyway Bridge is completed, connecting St. Petersburg and Bradenton, Florida. At 29,040 feet long, it is the world's largest cable-stayed concrete bridge. Twenty-one steel cables support the bridge in the center with two 40-foot roadways running along either side of the cable for an unobstructed view of the water.

1990s Work begins in Boston on the Big Dig, a project to transform the section of I-93 known as the Central Artery, an elevated freeway built in the 1950s, into an underground tunnel. Scheduled for completion in 2004, it will provide a new harbor crossing to Logan Airport and replace the I-93 bridge across the Charles River.

1993 The Glenn Anderson Freeway/Transitway, part of I-105, opens in Los Angeles, featuring a light rail train that runs in the median. Sensors buried in the pavement monitor traffic flow, and closed-circuit cameras alert officials to accidents.

Officially designated the Dwight D. Eisenhower System of Interstate and Defense Highways, the interstate system is praised by the American Society of Civil Engineers as one of the "Seven Wonders of the United States" and "the backbone of the world's strongest economy."

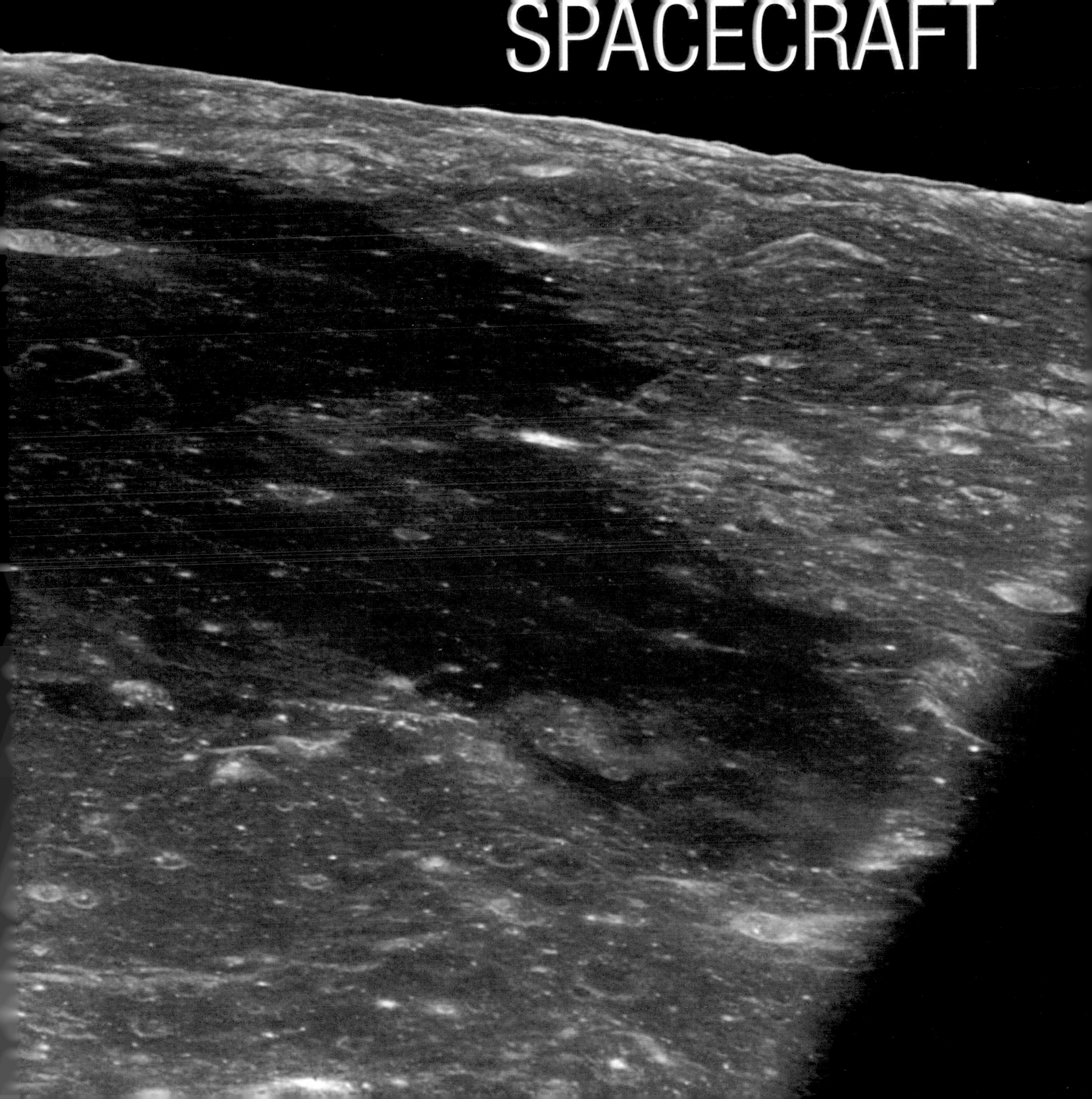
SPACECRAFT

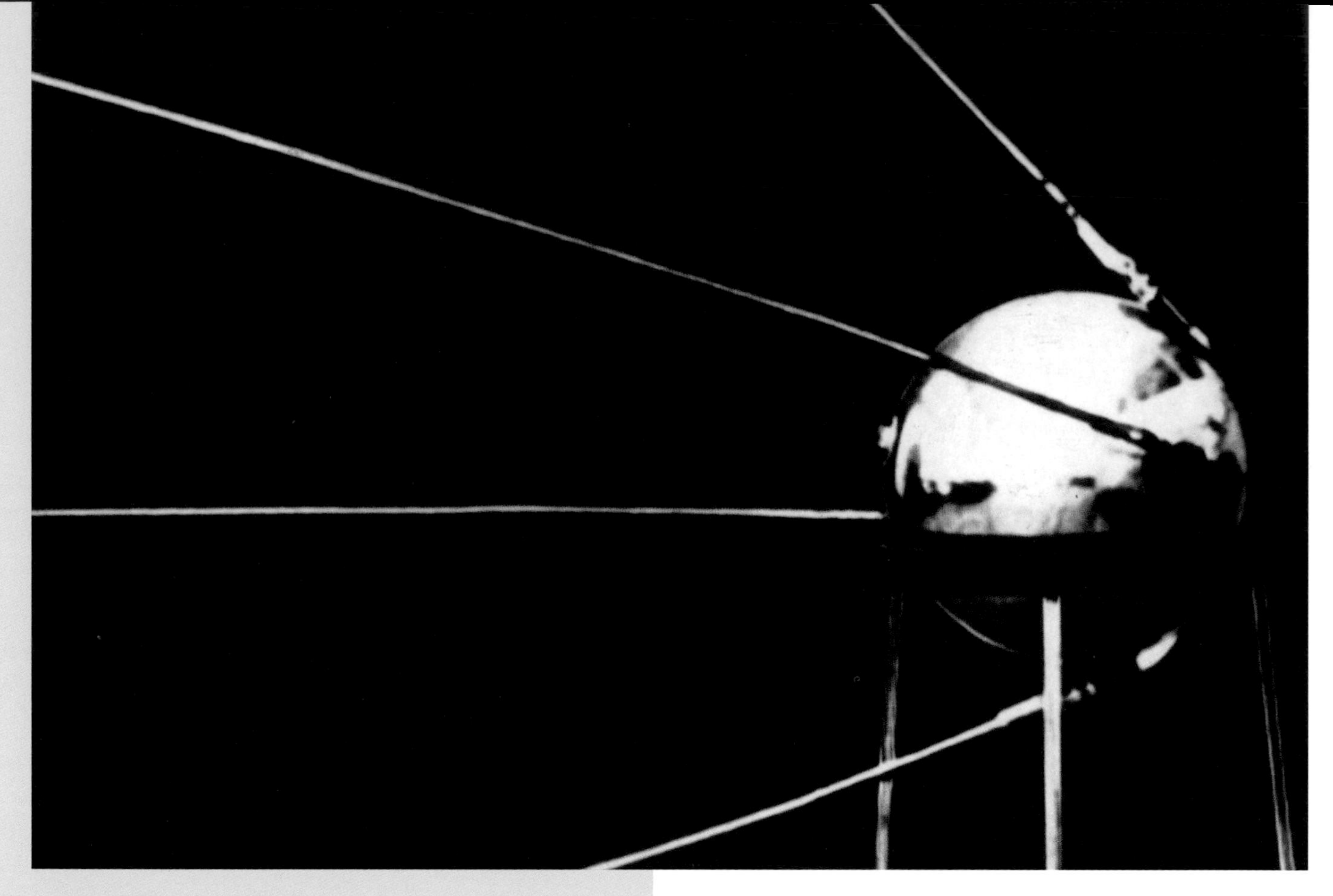

The event was so draped in secrecy that, despite its historic nature, no pictures were taken. But no one who was there—nor, for that matter, anyone else who heard of it—would ever forget the moment. With a blinding glare and a shuddering roar, the rocket lifted from its concrete pad and thundered into the early evening sky, soaring up and up and up until it was nothing more than a tiny glowing speck. On the plains of Kazakhstan, on October 4, 1957, the Soviet Union had just launched the first-ever spacecraft, its payload a 184-pound satellite called Sputnik.

In the days and weeks that followed, the whole world tracked *Sputnik*'s progress as it orbited the globe time and again. Naked-eye observers could see its pinpoint of reflected sunlight tracing across the night sky, and radios picked up the steady series of beeps from its transmitter. For Americans it was a shocking realization. Here, at the height of the Cold War, was the enemy flying right overhead. For the nascent U.S. space program, it was also a clear indication that the race into space was well and truly on—and that the United States was behind.

That race would ultimately lead to what has been called the most spectacular engineering feat of all time: landing humans on the Moon and bringing them safely back. But much more would come of it as well. Today literally thousands of satellites orbit the planet—improving global communications and weather forecasting; keeping tabs on climate change, deforestation, and the status of the ozone layer; making possible pinpoint navigation practically everywhere on Earth's surface; and, through such satellite-borne observatories as the Hubble Space Telescope, opening new eyes into the deepest reaches of the cosmos. The Space Shuttle takes astronauts, scientists, and engineers into orbit, where they perform experiments on everything from new medicines to superconductors. The Shuttle now also ferries crews to and from the International Space Station, establishing a permanent human presence in space. Venturing farther afield, robotic spacecraft have toured the whole solar system, some landing on planets and others making spectacular flybys, sending back reams of data and stunning close-up images of planets, moons, asteroids, and even comets.

In making all this possible, aerospace engineers have also propelled advances in a wide range of fields, from electronics to materials composition. Indeed, even though some critics contend that spaceflight is no more than a romantic and costly adventure, space technologies have spawned many products and services of practical use to the general public, including everything from freeze-dried foods to desktop computers and Velcro.

Humankind's venture into space began with the launch of the Soviet satellite Sputnik *(opposite) in 1957. By the end of the century we had traveled to the Moon (pages 132-133), and robotic probes were making portraits of the outer solar system, as in this composite of images returned by the* Voyager *spacecraft of the planets and four of Jupiter's moons, set against a false-color image of the Rosette Nebula, with Earth's moon in the foreground.*

Sputnik was only the beginning, but it was also the culmination of efforts to get into space that dated back to the start of the 20th century. Of the several engineering challenges that had to be addressed along the way, the first and foremost was building a rocket engine with enough thrust to overcome the pull of gravity and lift a vehicle into orbit. Rockets themselves had been around for centuries, almost exclusively as weapons of war, driven by the burning of solid fuels such as gunpowder. By the 19th century it was clear to experimenters that, although solid fuel could launch missiles along shallow trajectories, it couldn't create enough thrust to send a rocket straight up for more than a few hundred feet. You just couldn't pack enough gunpowder into a rocket to blast it beyond Earth's gravity.

Three men of the 20th century can justly lay claim to solving the problem and setting the stage for spaceflight. Working independently, Konstantin Tsiolkovsky in Russia, Robert Goddard in the United States, and Hermann Oberth in Germany designed and, in Goddard's and Oberth's cases, built rocket engines propelled by liquid fuel, typically a mixture of kerosene or liquid hydrogen and liquid oxygen. Tsiolkovsky took the first step, publishing a paper in 1903 that mathematically demonstrated how to create the needed thrust with liquid fuels. Among his many insights was the notion of using multistage rockets; as each rocket expended its fuel, it would be jettisoned to reduce the overall weight of the craft and maintain a fuel-to-weight ratio high enough to keep the flight going. He also proposed guidance systems using gyroscopes and movable vanes positioned in the exhaust stream and developed formulas still in use today for adjusting a spacecraft's direction and speed to place it in an orbit of virtually any given height.

Goddard was the first to launch a liquid-fuel rocket, in 1926, and further advanced the technology with tests of guidance and stabilization systems. He also built pumps to feed fuel more efficiently to the engine and developed the mechanics for keeping engines cool by circulating the cold liquid propellants around the engine through a network of pipes. In Germany, Oberth was garnering similar successes in the late 1920s and 1930s, his gaze fixed steadily on the future. One of the first members of the German Society for Space Travel, he postulated that rockets would someday carry people to the Moon and other planets.

One of Oberth's protégés was responsible for rocketry's next advance. Wernher von Braun was a rocket enthusiast from an early age, and when he was barely out of his teens, the German army tapped him to develop a ballistic missile. The 20-year-old von Braun saw the work as an opportunity to further his own interests in spaceflight, but in the short term his efforts led to the V-2, or Vengeance Weapon 2, used to deadly effect against

American engineer Robert Goddard, shown here in the mid-1930s in front of his workshop with one of his hand-built rockets, succeeded in launching a liquid-fueled rocket in 1926.

EXTRA The Huntsville Times

Jupiter-C Puts Up Moon

Following the successful launch in 1958 of the Explorer 1 *satellite, carried into orbit by a Jupiter-C rocket (above), program director Wernher von Braun was presented with a commemorative front page from the* Huntsville *(Alabama)* Times.

London in 1944. (His rocket design worked perfectly, von Braun told a friend, "except for landing on the wrong planet.") After the war, von Braun and more than a hundred of his rocket engineers were brought to the United States, where he became the leading figure in the nation's space program from its earliest days in the 1950s to its grand achievements of the 1960s.

With the Soviet *Sputnik* success, U.S. space engineers were under pressure not just to catch up but to take the lead. Less than 5 months later, von Braun and his team successfully launched America's first spacecraft, the satellite *Explorer 1*, on January 31, 1958. Several months after that, Congress authorized the formation of an agency devoted to spaceflight. With the birth of the National Aeronautics and Space Administration (NASA), the U.S. space program had the dedicated resources it needed for the next great achievement: getting a human being into space.

Again the Soviet Union beat the Americans to the punch. In April 1961, Yuri Gagarin became the first man in space, followed only a few weeks later by the American Alan Shepard. Gagarin's capsule managed one Earth orbit along a preset course over which he had virtually no control, except the timing of when retro-rockets were fired to begin the descent. Shepard simply went up and came back down on a suborbital flight, although he did experiment with some astronaut-controlled maneuvers during the flight, firing small rockets positioned around the capsule to change its orientation. Both were grand accomplishments, and both successes depended on key engineering advances. For example, Shepard's capsule, *Freedom 7*, was bell shaped, a design developed by NASA engineer Maxime Faget. The wide end would help slow the capsule during reentry as it deflected the heat of atmospheric friction. Other engineers developed heat-resistant materials to further protect the astronaut's capsule during reentry, and advances in computer technology helped control both flights from start to finish. But the United States was still clearly behind in the space race.

Then, barely 6 weeks after Shepard's flight and months before John Glenn became the first American to orbit Earth, President John F. Kennedy threw down the gauntlet in what was to become a major battle in the Cold War. "I believe," said Kennedy, "that this nation should commit itself to achieving the goal, before this decade is out, of landing a man on the Moon and returning him safely to the Earth."

NASA's effort to meet Kennedy's challenge was divided into three distinct programs, dubbed Mercury, Gemini, and Apollo, each of which had its own but related agenda. The Mercury program focused on the basics of getting the astronaut up and returning him safely. Gemini, named for the twins of Greek mythology, fulfilled its name in two ways. First, each Gemini flight included two astronauts, whose main task was to learn to maneuver their craft in space. Second, the overall goal of the program was to have two spacecraft rendezvous and link together, a capability deemed essential for the

Soviet cosmonaut Yuri Gagarin (below left) became the first human being in space in April 1961. A few weeks later the first American in space was Alan Shepard, (below right) being winched into a helicopter after splashing down in his Mercury capsule, which bobs on the ocean surface.

final Moon missions. Engineers had at least three different ideas about how to accomplish rendezvous. Gemini astronaut Buzz Aldrin, whose doctoral thesis had been on just this subject, advocated a method founded on the basic principle of orbital mechanics that a craft in a lower orbit travels faster than one in a higher orbit (to offset the greater pull of gravity at a lower altitude). Aldrin argued that a spacecraft in a lower orbit should chase one in a higher orbit and, as it approached, fire thrusters to raise it into the same orbit as its target. The system was adopted, and on March 16, 1966, *Gemini 8*, with Neil Armstrong and David Scott aboard, achieved the first docking in space, physically linking up with a previously launched, unmanned Agena rocket.

Armstrong and Aldrin would, of course, go on to greater fame with the Apollo program—the series of missions that would finally take humans to the surface of the Moon. Apollo had the most complex set of objectives. Engineers had to design and build three separate spacecraft components that together made up the Apollo spacecraft. The service module contained life-support systems, power sources, and fuel for in-flight maneuvering. The conical command module would be the only part of the craft to return to Earth. The lunar module would ferry two members of the three-man crew to the lunar surface and then return them to dock with the combined service and command modules. Another major task was to develop new tough but lightweight materials for the lunar module and for spacesuits that would protect the astronauts from extremes of heat and cold. And then there was what has often seemed the most impossible challenge of all. Flight engineers had to perfect a guidance system that would not only take the spacecraft across a quarter of a million miles to the Moon but also bring it back to reenter Earth's atmosphere at an extremely precise angle that left very little room for error (roughly six and half degrees, give or take half a degree). If the angle was too steep, the capsule would burn up in the atmosphere, too shallow and it would glance off the atmosphere like a stone skimming the surface of a pond and hurtle into space with no possibility of a second chance.

Launching all that hardware—40 tons of payload—called for a rocket of unprecedented thrust. Von Braun and his rocket team again rose to the challenge, building the massive Saturn V, the largest rocket ever created. More than 360 feet long and weighing some 3,000 tons, it generated 7.5 million pounds of thrust and propelled all the Apollo craft on their way without a hitch. On July 16, 1969, a Saturn V launched *Apollo 11* into space. Four days later, on July 20, Neil Armstrong and Buzz Aldrin became the first humans to set foot on the Moon, thus

On the morning of July 16, 1969, a mighty Saturn V rocket lifted off, carrying the Apollo 11 *crew of Neil Armstrong, Buzz Aldrin, and Michael Collins skyward, en route to the Moon. Four days later, Armstrong (above left) and Aldrin planted the Stars and Stripes on the lunar surface, a moment captured by a camera mounted on the lunar module.*

THE GLOBAL POSITIONING SYSTEM

PINPOINT NAVIGATION AROUND THE WORLD

Among the thousands of satellites orbiting Earth are 24 that work together as the key elements of the Global Positioning System (GPS), a navigational tool of unprecedented precision. Circling the globe once every 12 hours at an altitude of more than 12,000 miles, the satellites maintain positions relative to one another that enable handheld or vehicle-mounted units to receive signals from as many as six of them at once, from virtually any point on the planet's surface at any given time. Using signals from at least four satellites, the receivers can calculate their own location—latitude, longitude, and elevation—to within 100 feet; the more satellites the receiver has within its line of sight, the more precisely location can be determined, in some cases within 20 feet.

Each satellite continuously transmits three vital pieces of information: an identification code, data on its own location in space, and a time signal. Four atomic clocks on each satellite keep track of time to within 3 billionths of a second, a level of accuracy that enables a receiver's microprocessor to precisely measure the distance to each satellite from which it receives signals. The microprocessor then applies a sophisticated form of triangulation to determine its own position.

Known officially as NAVSTAR (Navigation Satellite Timing and Ranging), GPS was envisioned as early as the 1950s by Ivan Getting, who subsequently championed its construction. Brad Parkinson, then at the Department of Defense, directed its development for its original military functions. GPS played an important role in Operation Desert Storm, enabling U.S. forces to maneuver during sandstorms and at night; aircraft, ships, tanks, and individual troops made use of more than 9,000 receivers throughout the Gulf region.

Eventually released for civilian use, GPS began finding an increasing number of applications in the 1990s. For example, engineers building the English Channel tunnel used GPS-provided measurements to ensure that the French and British teams digging from opposite ends would meet precisely in the middle. Transportation companies employ GPS receivers to keep track of their fleets, and police cars, fire engines, and ambulances use them to speed their response to emergencies. As receivers become less expensive, GPS is becoming more and more a part of everyday life, helping us all to know, quite literally, our place in the world.

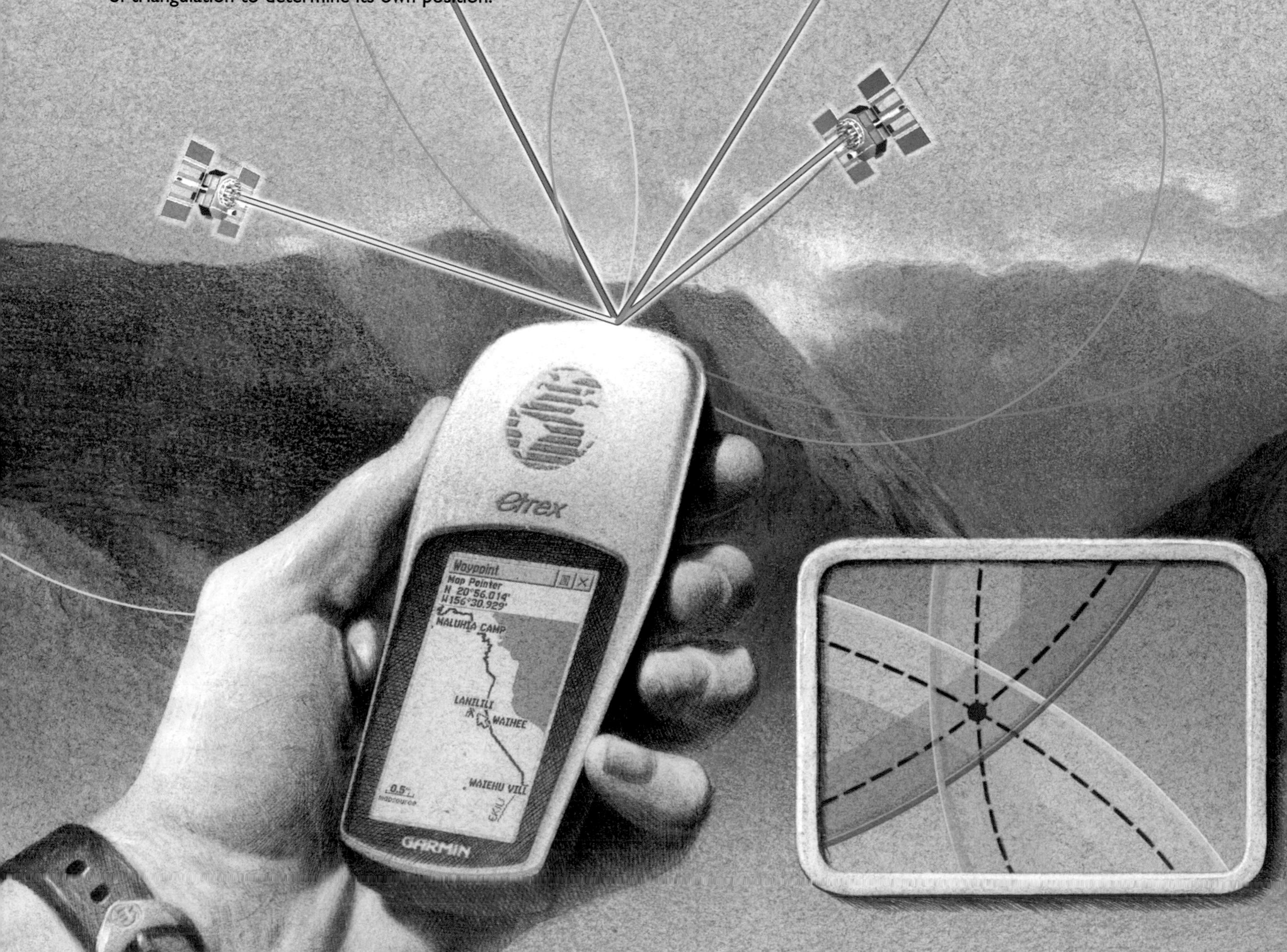

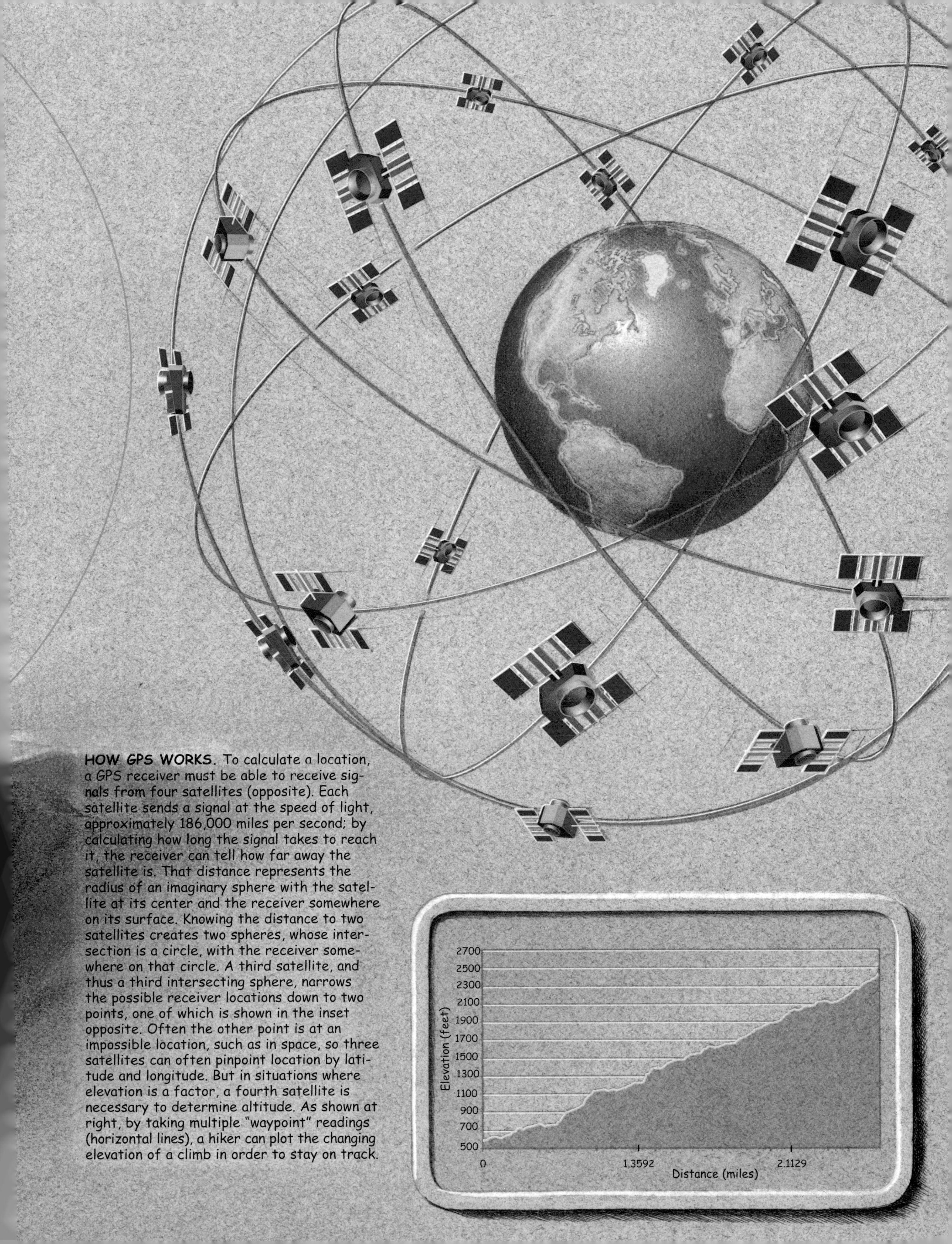

HOW GPS WORKS. To calculate a location, a GPS receiver must be able to receive signals from four satellites (opposite). Each satellite sends a signal at the speed of light, approximately 186,000 miles per second; by calculating how long the signal takes to reach it, the receiver can tell how far away the satellite is. That distance represents the radius of an imaginary sphere with the satellite at its center and the receiver somewhere on its surface. Knowing the distance to two satellites creates two spheres, whose intersection is a circle, with the receiver somewhere on that circle. A third satellite, and thus a third intersecting sphere, narrows the possible receiver locations down to two points, one of which is shown in the inset opposite. Often the other point is at an impossible location, such as in space, so three satellites can often pinpoint location by latitude and longitude. But in situations where elevation is a factor, a fourth satellite is necessary to determine altitude. As shown at right, by taking multiple "waypoint" readings (horizontal lines), a hiker can plot the changing elevation of a climb in order to stay on track.

Since the Moon landings in the 1970s, rotating crews of human astronauts have stayed in Earth orbit, constructing the International Space Station (above) while robotic probes have touched down on other worlds. In 1976 the Viking Lander 2 *traveled to the Red Planet and sent back views of the boulder-strewn Utopian plain and the pinkish Martian sky (below).*

meeting Kennedy's challenge and winning the space race. After the tragic loss of astronauts Virgil I. (Gus) Grissom, Edward H. White, and Roger B. Chaffee during a launchpad test for *Apollo 1*, the rest of the Apollo program was a spectacular success. Even the aborted *Apollo 13* mission proved how resourceful both the astronauts in space and the engineers on the ground could be in dealing with a potentially deadly catastrophe—an explosion aboard the service module. But with the space race won and with increasing cost concerns as well as a desire to develop other space programs, the Moon missions came to an end in 1972.

NASA turned its attention to a series of robotic, relatively low-cost science and discovery missions. These included the *Pioneer* probes to Jupiter and Saturn; the twin Viking craft that landed on Mars; and *Voyagers 1* and *2*, which explored the outer planets and continue to this day flying beyond the Solar System into interstellar space. Both the Soviet Union and the United States also built space stations in the 1970s. Then, in 1981 the United States ramped up its human spaceflight program again with the first of what would be, at last count, scores of Space Shuttle missions. An expensive breakthrough design, the Shuttle rises into space like any other spacecraft, powered by both solid- and liquid-fuel rockets. But upon reentering the atmosphere, the craft becomes a glider, complete with wings, rudder, and landing gear—but no power. Shuttle pilots put their craft through a series of S turns to control its rate of descent and they get only one pass at landing. Among the many roles the Shuttle fleet has played, the most significant may be as a convenient though costly space-based launchpad for new generations of satellites that have turned the world itself into a vast arena of instant communications (see *Telephony*).

As with all of the greatest engineering achievements, satellites and other spacecraft bring benefits now so commonplace that we take them utterly for granted. We prepare for an impending snowstorm or hurricane and tune in to our favorite news source for updates, but few of us think of the satellites that spied the storm brewing and relayed the warning to Earth. Directly and indirectly, spacecraft and the knowledge they have helped us gain not only contribute in myriad ways to our daily well-being but have also transformed the way we look at our own blue planet and the wider cosmos around us.

Perspective

William A. Anders

Retired Chairman
General Dynamics Corporation

As a youth I was fascinated by science and engineering. I also loved to explore and find out what was on the other side of the mountain. These traits came together beyond my wildest boyhood dreams when I was selected by NASA to be an Apollo astronaut. The Apollo program, with its preeminent role in the "space race" with the USSR, allowed me to serve my country during a critical period of the Cold War.

As astronauts we spent many hours in training, learning how to operate spacecraft systems in our effort to get to the Moon. Those of us with engineering backgrounds (I was a nuclear engineer) were assigned tasks in design and testing, such as determining methods for measuring radiation dosage and shielding. We also studied geology—Earth's surface up close with rock hammer and magnifying glass and the lunar surface from afar with telescopes—so that we would be prepared to describe the lunar surface features and materials when we got there.

I was lucky enough to be chosen for the *Apollo 8* crew. This would be man's first flight on the giant Saturn V rocket, with a mission to blaze a trail to the Moon and orbit it 10 times. The mission was set to occur over Christmas 1968. After a successful launch and orbital check of the spacecraft's systems, we reignited our rocket engines and boosted our velocity to some 35,000 feet per second. This was easily a new world speed record and enough for us to be the first humans to escape the gravity of our home planet and venture to another celestial body.

After two and a half days of first "coasting" away from Earth and later of "falling" toward the Moon, we retrofired our spacecraft's rocket to slow us down enough to be captured by the Moon's gravity. We were in lunar orbit!

As the amateur geologist and photographer of the crew (as well as spacecraft systems engineer and pilot), I was especially eager to view the lunar surface. But after several 2-hour "heads down" orbits observing the Moon and photographing lunar features, I was, frankly, getting a bit bored at the Moon's sameness. The surface was a monotonous gray that looked sandblasted. And it was repetitive—crater upon crater upon crater created by countless meteors, large and small through the eons.

Then all of a sudden we saw Earth rising majestically above the Moon's stark horizon. We might have been engineers and "right stuff" test pilots, but the beauty of this sight took our breath away. I grabbed the color camera and snapped the now famous first "Earthrise" photo.

Earth appeared quite small to us—about the size of my fist at arm's length. It was the only color in the dull black "sky"—a fragile Christmas tree ornament to be handled with utmost care, not just a chunk of rock whose inhabitants treated it carelessly. Big or small, we realized that it was mankind's only home and the place where we evolved, the center of our emotional and spiritual universe.

Apollo was designed and operated to go to the Moon and learn about its properties, but its main contribution may well have been the new human perspective it created about the fragility and finiteness of our home planet. And this was done by a bunch of engineers.

SPACECRAFT TIMELINE

From early theories on how to break Earth's gravitational pull, humankind has traveled into Earth orbit, to the Moon, and even—via robotic proxies—to the outer solar system. Even as the Hubble Space Telescope has made it possible for us to study distant galaxies and stars, other satellites give us up-to-date weather information, enhanced telecommunications, and navigation systems that allow us to pinpoint where we are. These advancements and more have all been made possible through the practical problem-solving that first launched us into space.

1903 Konstantin Tsiolkovsky publishes a paper in Russia that mathematically demonstrates how to achieve liftoff with liquid fuels. He also proposes using multi-stage rockets, which would be jettisoned as they spent their fuel, and guidance systems using gyroscopes and movable vanes positioned in the exhaust stream. His formulas for adjusting a spacecraft's direction and speed to place it in any given orbit are still in use today.

1915 Robert Goddard experiments with reaction propulsion in a vacuum and establishes that it is possible to send a rocket to the Moon. Eleven years later, in 1926, Goddard launches the first liquid-fuel rocket.

1942 Ten years after his first successful rocket launch, German ballistic missile technical director Wernher von Braun achieves the successful launch of a V-2 rocket. Thousands of V-2s are deployed during World War II, but the guidance system for these missiles is imperfect and many do not reach their targets. The later capture of V-2 rocket components gives American scientists an early opportunity to develop rocket research techniques. In 1949, for example, a V-2 mated to a smaller U.S. Army WAC Corporal second-stage rocket reaches an altitude of 244 miles and is used to obtain data on both high altitudes and the principles of two-stage rockets.

1957 On October 4 the Soviet Union launches *Sputnik I* using a liquid-fueled rocket built by Sergei Korolev. About the size of a basketball, the first artificial Earth satellite weighs 184 pounds and takes about 98 minutes to complete one orbit. On November 3 the Soviets launch *Sputnik II*, carrying a much heavier payload that includes a passenger, a dog named Laika.

1958 The United States launches its first satellite, the 30.8-pound *Explorer I*. During this mission, *Explorer I* carries an experiment designed by James A. Van Allen, a physicist at the University of Iowa, which documents the existence of radiation zones encircling Earth within the planet's magnetic field. The Van Allen Radiation Belt, as it comes to be called, partially dictates the electrical charges in the atmosphere and the solar radiation that reaches Earth. Later that year the U.S. Congress authorizes formation of the National Aeronautics and Space Administration (NASA).

1959 The Soviet Union's *Luna 3* probe flies past the Moon and takes the first pictures of its far side. This satellite carries an automated film developing unit and then relays the pictures back to Earth via video camera.

1960 Weather satellite *TIROS I* is launched to test experimental television techniques for a worldwide meteorological satellite information system. Weighing 270 pounds, the aluminum alloy and stainless steel spacecraft is 42 inches in diameter and 19 inches high and is covered by 9,200 solar cells, which serve to charge the onboard batteries. Magnetic tape recorders, one for each of two television cameras, store photographs while the satellite is out of range of the ground station network. Although it is operational for only 78 days, *TIROS I* proves that a satellite can be a useful tool for surveying global weather conditions from space.

1961 On April 12, cosmonaut Yuri Gagarin, in *Vostok I*, becomes the first human in space. Launching from Baikonur Cosmodrome, he completes one orbit of Earth in a cabin that contains radios, instrumentation, life-support equipment, and an ejection seat. Three small portholes give him a view of space. At the end of his 108-minute ride, during which all flight controls are operated by ground crews, he parachutes to safety in Kazakhstan.

On May 5 astronaut Alan B. Shepard, Jr., in *Freedom 7*, becomes the second human in space. Launched from Cape Canaveral by a Mercury-Redstone rocket, *Freedom 7*—the first piloted Mercury spacecraft—reaches an altitude of 115 nautical miles and a speed of 5,100 miles per hour before splashing down in the Atlantic Ocean. During his 15-minute suborbital flight, Shepard demonstrates that individuals can control a vehicle during weightlessness and high G stresses, supplying researchers on the ground with significant biomedical data.

1962 John Glenn becomes the first American to circle Earth, making three orbits in his *Friendship 7* Mercury spacecraft. Glenn flies parts of the last two orbits manually because of an autopilot failure and during reentry leaves the normally jettisoned retro-rocket pack attached to his capsule because of a loose heat shield. Nonetheless, the flight is enormously successful. The public, more than celebrating the technological success, embraces Glenn as the personification of heroism and dignity.

1963 On February 14 NASA launches the first of a series of Syncom communications satellites into near-geosynchronous orbit, following procedures developed by Harold Rosen of Hughes Aircraft. In July, *Syncom 2* is placed over the Atlantic Ocean and Brazil at 55 degrees longitude to demonstrate the feasibility of geosynchronous satellite communications. It successfully transmits voice, teletype, facsimile, and data between a ground station in Lakehurst, New Jersey, and the USNS *Kingsport* while the ship is off the coast of Africa. It also relays television transmissions from Lakehurst to a ground station in Andover, Maine. Forerunners of the Intelsat series of satellites, the Syncom satellites are cylinders covered with silicon solar cells that provide 29 watts of direct power when the craft is in sunlight (99 percent of the time). Nickel-cadmium rechargeable batteries provide power when the spacecraft is in Earth's shadow.

1965 The second piloted Gemini mission, *Gemini IV*, stays aloft for four days, (June 3-7), and astronaut Edward H. White, Jr. performs the first extravehicular activity (EVA)—or spacewalk—by an American. This critical task will have to be mastered before a landing on the Moon.

1968 Humans first escape Earth's gravity on the *Apollo 8* flight to the Moon and view Earth from lunar orbit. *Apollo 8* takes off from the Kennedy Space Center on December 21 with three astronauts aboard—Frank Borman, James A. Lovell,

Jr., and William A. Anders. As their ship travels outward, the crew focuses a portable television camera on Earth and for the first time humanity sees its home from afar, a tiny "blue marble" hanging in the blackness of space. When they arrive at the Moon on Christmas Eve, the crew sends back more images of the planet along with Christmas greetings to humanity. The next day they fire the boosters for a return flight and splash down in the Pacific Ocean on December 27.

1969 Neil Armstrong becomes the first person to walk on the Moon. The first lunar landing mission, *Apollo 11* lifts off on July 16 to begin the 3-day trip. At 4:18 p.m. EST on July 20, the lunar module—with astronauts Neil Armstrong and Edwin E. (Buzz) Aldrin—lands on the Moon's surface while Michael Collins orbits overhead in the command module. After more than 21 hours on the lunar surface, they return to the command module with 20.87 kilograms of lunar samples, leaving behind scientific instruments, an American flag, and other mementos, including a plaque bearing the inscription: "Here Men From Planet Earth First Set Foot Upon the Moon. July 1969 A.D. We came in Peace For All Mankind."

1971 The Soviet Union launches the world's first space station, *Salyut 1*, in 1971. Two years later the United States sends its first space station, *Skylab*, into orbit, where it hosts three crews before being abandoned in 1974. Russia continues to focus on long-duration space missions, launching the first modules of the *Mir* space station in 1986.

1972 *Pioneer 10*, the first mission to be sent to the outer solar system, is launched on March 2 by an Atlas-Centaur rocket. The spacecraft makes its closest approach to Jupiter on December 3, 1973, after which it is on an escape trajectory from the Solar System. NASA launches *Pioneer 11* on April 5, 1973, and in December 1974 the spacecraft gives scientists their closest view of Jupiter, from 26,600 miles above the cloud tops. Five years later *Pioneer 11* makes its closest approach to Saturn, sending back images of the planet's rings, and then heads out of the solar system in the opposite direction from *Pioneer 10*. The last successful data acquisitions from *Pioneer 10* occur on March 3, 2002, the 30th anniversary of its launch date, and on April 27, 2002. Its signal is last detected on January 23, 2003 after an uplink is transmitted to turn off the last operational experiment.

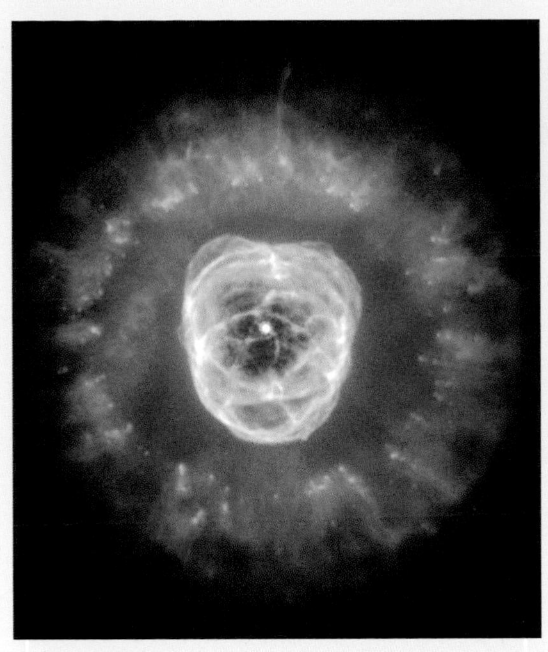

In December 1999 the Hubble Space Telescope captured a spectacular view of the Eskimo Nebula, the glowing remains of a dying star.

1975 NASA launches two Mars space probes, *Viking 1* on August 20 and *Viking 2* on November 9, each consisting of an orbiter and a lander. The first probe lands on July 20, 1976, the second one on September 3. The Viking project's primary mission ends on November 15, 11 days before Mars's superior conjunction (its passage behind the Sun), although the two spacecraft continue to operate for several more years. The last transmission reaches Earth on November 11, 1982. After repeated efforts to regain contact, controllers at NASA's Jet Propulsion Laboratory close down the overall mission on May 21, 1983.

1977 *Voyager 1* and *Voyager 2* are launched on trajectories that take them to Jupiter and Saturn. Over the next decade the Voyagers rack up a long list of achievements. They find 22 new satellites (3 at Jupiter, 3 at Saturn, 10 at Uranus, and 6 at Neptune); discover that Jupiter has rings and that Saturn's rings contain spokes and braided structures; and send back images of active volcanism on Jupiter's moon Io—the only solar body other than Earth with confirmed active volcanoes.

1981 The Space Shuttle *Columbia*, the first reusable winged spaceship, is launched on April 12 from Kennedy Space Center. Astronauts John W. Young and Robert L. Crippin fly *Columbia* on the first flight of the Space Transportation System, landing the craft at Edwards Air Force Base in Southern California on April 14. Using pressurized auxiliary tanks to improve the total vehicle weight ratio so that the craft can be inserted into its orbit, the mission is the first to use both liquid- and solid-propellant rocket engines for the launch of a spacecraft carrying humans.

1986 On the 25th shuttle flight, the Space Shuttle *Challenger* is destroyed during its launch from the Kennedy Space Center on January 28, killing astronauts Francis R. (Dick) Scobee, Michael Smith, Judith Resnik, Ronald McNair, Ellison Onizuka, Gregory Jarvis, and Sharon Christa McAuliffe. The explosion occurs 73 seconds into the flight when a leak in one of two solid rocket boosters ignites the main liquid fuel tank. People around the world see the accident on television. The shuttle program does not return to flight until the fall of 1988.

1990 The Hubble Space Telescope goes into orbit on April 25, deployed by the crew of the Space Shuttle *Discovery*. A cooperative effort by the European Space Agency and NASA, Hubble is a space-based observatory first dreamt of in the 1940s. Stabilized in all three axes and equipped with special grapple fixtures and 76 handholds, the space telescope is intended to be regularly serviced by shuttle crews over the span of its 15-year design life.

1998 The first two modules of the International Space Station are joined together in orbit on December 5 by astronauts from the Space Shuttle *Endeavour*. In a series of spacewalks, astronauts connect cables between the two modules—*Unity* from the United States and *Zarya* from Russia—affix antennae, and open the hatches between the two spacecraft.

2000 On October 31 Expedition One of the International Space Station is launched from Baikonur Cosmodrome in Kazakhstan—the same launchpad from which Yuri Gagarin became the first human in space. Prior to its return on March 21, 2001, the crew conducts scientific experiments and prepares the station for long-term occupation.

INTERNET

The conference held at the Washington Hilton in October 1972 wasn't meant to jump-start a revolution. Staged for a technological elite, its purpose was to showcase a computer-linking scheme called ARPANET, a new kind of network that had been developed under military auspices to help computer scientists share information and enable them to harness the processing power of distant machines. Traffic on the system was still very light, though, and many potential users thought it was too complex to have much of a future.

In the conference hall at the hotel was an array of terminals whose software permitted interactions with computers hundreds or thousands of miles away. The invitees were encouraged to experiment—try out an air traffic control simulator, play chess against an electronic foe, explore databases. There had been some problems in setting up the demonstrations. At one point, a file meant to go to a printer in the hall was mistakenly directed to a robotic turtle, resulting in a wild dance. But it all worked when it had to, convincing the doubters and engaging their interest so effectively that, as one of the organizers said, they were "as excited as little kids." Within a month, traffic on the network increased by two-thirds. In a few more years, ARPANET hooked up with other networks to become a connected archipelago called the Internet. By the end of the 20th century, more than 100 million people were thronging Internet pathways to exchange e-mail, chat, check the news or weather, and, often with the aid of powerful search engines to sift for useful sites, navigate the vast universe of knowledge and commerce known as the World Wide Web. Huge electronic marketplaces bloomed. Financial services, the travel industry, retailing, and many other businesses found bountiful opportunities online. Across the world, the connecting of computers via the Internet spread information and rearranged human activities with seismic force.

All this began in an obscure branch of the U.S. Department of Defense called the Advanced Research Projects Agency, or ARPA. In the 1960s a number of computer scientists at universities and research labora-

Crowded cybercafes all over the world (opposite) owe their existence to Internet pioneers (left to right) Vinton G. Cerf, Robert E. Kahn, Leonard Kleinrock, and Lawrence G. Roberts, shown here upon receiving the 2001 Charles Stark Draper Prize for development of the foundations of the Internet.

tories across the country received ARPA funding for projects that might have defense-related potential—anything from graphics to artificial intelligence. With the researchers' needs for processing power steadily growing, ARPA decided to join its scattered mainframes into a kind of cooperative, allowing the various groups to draw on one another's computational resources. Responsibility for creating the network was assigned to Lawrence Roberts, a young computer scientist who arrived at ARPA from the Massachusetts Institute of Technology in 1966.

Roberts was aware of a promising approach in the ideas of an MIT classmate, Leonard Kleinrock, and he later learned of related work by two other communications experts, Paul Baran and Donald Davies. Kleinrock had written his doctoral dissertation on the flow of messages in communications networks, exploring the complexities of moving data in small chunks. At about the same time, Baran proposed a different kind of telephone network, which would turn the analog signal of a telephone into digital bits, divide the stream into blocks, and send the blocks in several different directions across a network of high-speed switches or nodes; the node nearest the destination would put the pieces back together again. Davies proposed a similar scheme, in which he called the chunks or blocks "packets," as in packet switching, and that name stuck.

Roberts, for his part, was convinced that the telephone system's method of routing signals, called circuit switching, was poorly suited for linking computers: to connect two callers, a telephone switch opens a circuit, leaving it open until the call is finished. Computers, however, often deliver data in bursts and thus don't need full-time possession of a connection. Packet switching seemed the obvious choice for ARPA's network, not only enabling several computers to share a circuit but also countering congestion problems: when one path was in heavy use, a packet could simply take another route.

Initially, Roberts intended to have the switching done by the mainframes that ARPA wanted to connect. But small, speedy minicomputers were just then appearing, and an adviser, Wesley Clark of Washington University in St. Louis, persuaded him to assign one of them to each of the research centers as a switch. Unlike the mainframes, which came from a variety of manufacturers, these so-called interface message processors, or IMPs, could have standardized routing software, which would save on programming costs and allow easy upgrades. In early 1969 the job of building and operating the network was awarded to the consulting firm of Bolt Beranek and Newman, Inc. (BBN), in Cambridge, Massachusetts. Although modest in size, BBN employed

Paul Baran (below left) and Donald Davies (below right) independently came up with the idea of breaking data into small chunks—Davies called them packets—the underlying principle of packet-switching on the Internet. Tim Berners-Lee (below center) conceived the idea for the World Wide Web.

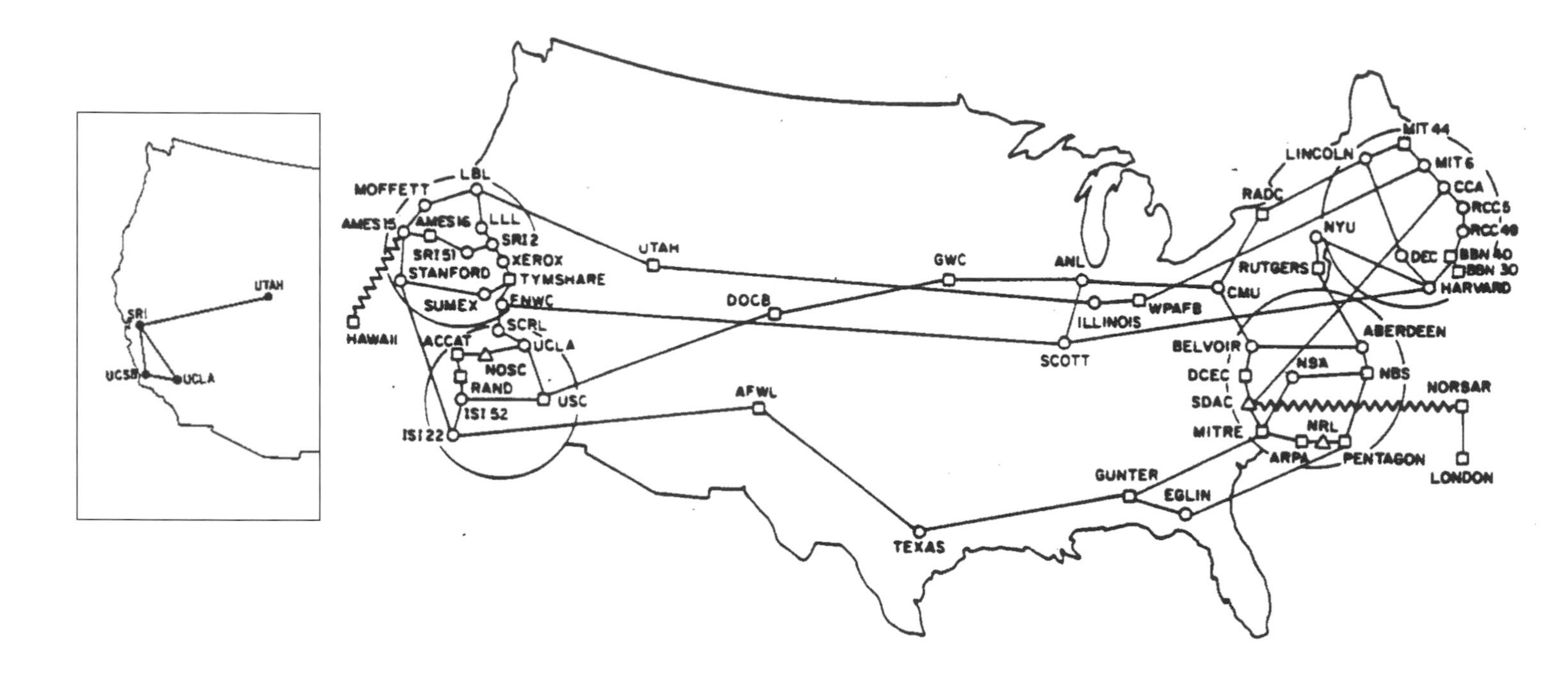

In 1969 ARPANET linked four nodes—time-sharing computers at Stanford, UCLA, UC Santa Barbara, and University of Utah—and packets began flowing on leased telephone lines. By 1977 the network was a maze of linked computers all over the country and even across the Pacific Ocean to Hawaii.

a stellar cast of engineers and scientists, drawn largely from nearby Harvard University and MIT.

Roberts had outlined what the IMPs would do. First, they would break data from a host mainframe into packets of about 1,000 bits each, attaching source and destination information to each packet, along with digits used to check for transmission errors. The IMPs would then choose optimal routes for the individual packets and reassemble the message at the other end. All the traffic would flow on leased telephone lines that could handle 50,000 bits per second. The BBN team, led by Robert Kahn of MIT, worked out the details and devised an implementation strategy. ARPANET was up and running at four sites by late 1969. At first, just four time-sharing computers were connected, but more hosts and nodes quickly followed, and the network was further expanded by reconfiguring the IMPs so they could accept data from small terminals as well as mainframes.

The nature of the traffic was not what ARPA had expected, however. As time went on, the computer scientists on the network used it primarily for personal communication rather than resource sharing. The first program for sending electronic mail from one computer to another was written in 1972—almost on a whim—by Ray Tomlinson, an engineer at BBN. He earned a kind of alphanumerical immortality in the process. For his addressing format he needed a symbol to clearly separate names from computer locations. He looked at the keyboard in front of him and made a swift choice: "The one that was most obvious was the '@' sign, because this person was @ this other computer," he later explained. "At the time, there was nobody with an '@' sign in their name that I was aware of." Trillions of e-mails would be stamped accordingly.

The @ symbol was the brainstorm of engineer Ray Tomlinson, who wrote the first e-mail program in 1972 and used it to designate an e-mail address.

Packet switching soon found favor beyond the confines of ARPANET. Roberts left ARPA in 1972 to become president of one of the first companies to offer networking services to commercial customers. Several European countries had become interested in computer networking, and the U.S. government had other packet-based projects under way. Although ARPANET was presumably destined to remain a well-guarded dominion of computer scientists, some widening of its reach by connecting with other networks seemed both desirable and inevitable.

It was clear to Robert Kahn, who had headed the BBN design team, that network-to-network linkages would require an acceptance of diversity, since ARPANET's specifications for packet sizes, delivery rates, and other features of data flow were not a standard. Commonality would instead be imposed in the form of shared rules, or protocols, for communication—some of the rules to apply to the networks themselves, others meant for gateways that would be placed between networks. The job of these gateways, called routers, would be to control traffic, nothing more. What was inside the packets wouldn't matter.

To grapple with the various issues, Kahn joined forces with Vinton Cerf, who had been involved in designing the ARPANET protocols for host computers and also had experience with time-sharing systems on the ARPANET. By mid-1974 their recommendations for an overall network-to-network architecture had been accepted. Negotiations to finalize the two sets of rules, jointly known as TCP/IP (transmission control protocol/internet protocol), took several more years, and ARPANET did not formally incorporate the new system until 1983. By then ARPA—now known as DARPA, the "D" having been added to signal a clearer focus on defense—was looking for release from its network responsibilities.

An exit presented itself in mid-decade when another U.S. government entity, the National Science Foundation (NSF), began building five supercomputing centers around the country, along with a connecting backbone of lines that were about 25 times faster than ARPANET's. At that time, research scientists of all kinds were clamoring for network access to allow the kind of easy communication and collaboration that ARPANET users had long enjoyed. NSF answered the need by helping to create a number of regional networks, then joining them together by means of the supercomputer backbone. Many foreign networks were connected. In the late 1980s ARPANET began attaching its sites to the system, and in 1990 the granddaddy of packet-switching networks was decommissioned.

Meanwhile, beyond the world of science, computer networking spread in all directions. Within corporations and institutions, small computers were being hooked together in local area networks, which typically used an extremely fast, short-range packet delivery technique called Ethernet (invented by one-time ARPANET programmer Robert Metcalfe back in 1973) and were easily attached to outside networks. On a nation-spanning scale, a number of companies built high-speed networks that could be used to process point-of-sale transactions, give corporate customers access to specialized databases, and serve various other commercial functions. Huge telecommunications carriers such as AT&T and MCI entered the business. As the 1990s proceeded, the major digital highways, including those of NSF, were linked, and on-ramps known as Internet service providers proliferated, providing customers with e-mail, chat rooms, and a variety of content via telephone lines and modems. The Internet was now a vast international community, highly fragmented and lacking a center but a miracle of connectivity.

What allowed smooth growth was the TCP/IP system of rules originally devised for attaching other networks to ARPANET. Over the years rival network-to-network protocols were espoused by various factions in the computer world, among them big telecommunications carriers and such manufacturers as IBM. But TCP/IP worked well. It was highly flexible, it allowed any number of networks to be hooked together, and it was free. The NSF adopted it, more and more private companies accepted it, and computer scientists overseas came to prefer it. In the end, TCP/IP stood triumphant as the glue for the world's preeminent network of networks.

Korean college students and teenagers, like their counterparts elsewhere in the Internet world, flock to game halls like this one to play computer games that know no geographical boundaries.

In the 1990s the World Wide Web, an application designed to ride on top of TCP/IP, accelerated expansion of the Internet to avalanche speed. Conceived by Tim Berners-Lee, a British physicist working at the CERN nuclear research facility near Geneva, it was the product, he said, of his "growing realization that there was a power in arranging ideas in an unconstrained, web-like way." He adopted a venerable computer sciences idea called hypertext—a scheme for establishing nonlinear links between pieces of information—and came up with an architectural scheme for the Internet era. His

PACKET SWITCHING
THE INTERNET'S CORE TECHNOLOGY

A profound disunity underlies the swift, smooth currents of data rushing across today's networked world: all of the content on the Internet is delivered in small chunks called packets, and the constituents of any given message may travel along a variety of paths to reach their destination. This piecemeal approach to connectivity emerged in the 1960s as an alternative to traditional communications networks like the telephone system, which used a technique known as circuit-switching—the opening of a separate, exclusive circuit for each call.

With computer-to-computer communication, such dedicated circuits are wasteful because the data tends to flow in bursts, leaving the connection idle much of the time. Splitting the data into chunks yields huge gains in efficiency. The packets of one message can be interleaved with those of another to share a path, and the flow can also be easily diverted to new routes if congestion arises or the network is damaged. In addition, the digital nature of packet-switching, as the technique is called, allows for error-free communication and lends itself to all sorts of sophisticated software processing.

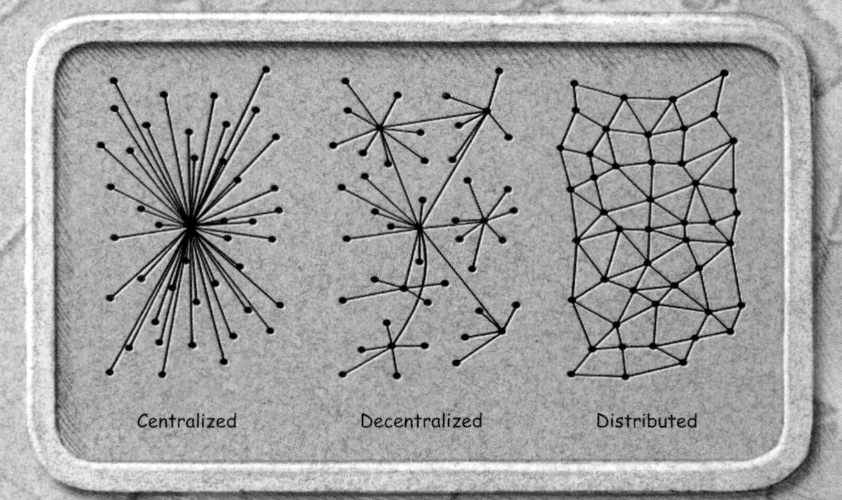

NETWORK DESIGNS. From left to right above, three designs for communications networks are progressively less vulnerable to disruption, and the meshlike pattern called "distributed" has the added virtue of allowing message chunks to take many different routes. Packet-switching pioneer Paul Baran created the diagram for a RAND Corporation report in 1964.

PACKETS. A message sent along the Internet is divided into many small packets, each one fitted with a "header" indicating the addresses of the source and destination, the packet's position in the sequence of transmission, and a so-called checksum enabling the receiver to make sure that the contents have not picked up any errors en route. All the packets move independently.

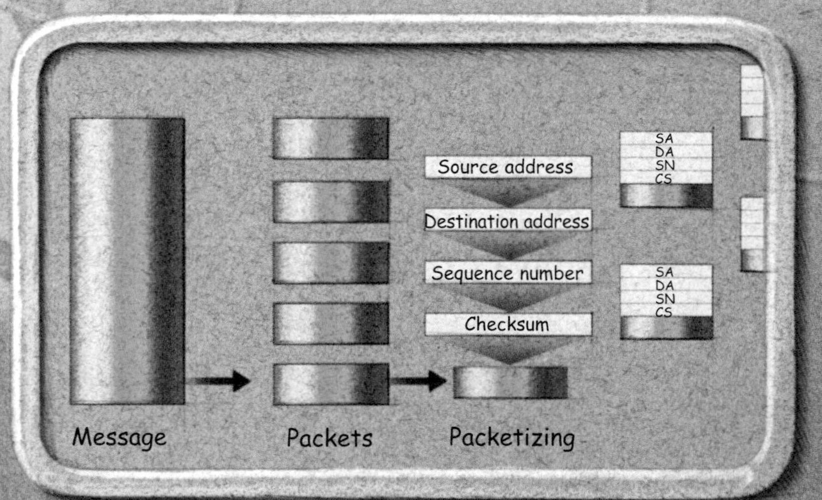

TRANSMITTING. The Internet consists of many separate networks joined by computerized traffic-control switches called routers (right), which choose links that maximize the efficiency of the flow. In the simplified diagram at right, for example, an especially heavy transmission of yellow packets is distributed among many paths so that congestion does not occur.

RECEIVING. At the receiving end of a transmission across the Internet, the packets making up a message are reassembled in their correct order. The sender keeps a copy of each packet in the numbered sequence until the receiver confirms that the original has arrived, so that any missing packet may be promptly retransmitted.

Internet-savvy children born in the late 20th century sharpen their typing skills (left) more often than practicing penmanship and communicate with far-flung family and friends more often by e-mail than by telephone. Even fast-food restaurants, like this Burger King in New York City (above), give customers a few free minutes to surf the Web or check e-mail while they eat.

Marc Andreessen (above) and Eric Bina made the Internet accessible to the general public in 1993 when they developed Mosaic, a Web browser whose commercial version, Netscape, debuted the following year.

World Wide Web allowed users to find and get text or graphics files—and later video and audio as well—that were stored on computers called servers. All the files had to be formatted in what he termed hypertext markup language (HTML), and all storage sites required a standardized address designation called a uniform resource locator (URL). Delivery of the files was guided by a set of rules known as the hypertext transfer protocol (HTTP), and the system enabled files to be given built-in links to other files, creating multiple information paths for exploration.

Although the World Wide Web rapidly found enthusiasts among skilled computer users, it didn't come into its own until appealing software for navigation emerged from a supercomputing center established at the University of Illinois by NSF. There, two young computer whizzes named Marc Andreessen and Eric Bina created a program called Mosaic, which made Web browsing so easy and graphically intuitive that a million copies of the software were downloaded across the Internet within a few months of its appearance in April 1993. The following year Andreessen helped form a company called Netscape to produce a commercial version of Mosaic. Other browsers soon followed, and staggering quantities of information moved onto servers: personal histories and governmental archives; job listings and offerings of merchandise; political tracts, artwork, and health information; financial news, electronic greeting cards, games, and uncountable other sorts of human knowledge, interest, and activity—with the whole indescribable maze changing constantly and growing exponentially.

By the end of the 20th century the Internet embraced some 300,000 networks stretching across the planet. Its fare traveled on optical fibers, cable television lines, and radio waves as well as telephone lines—and the traffic was doubling annually. Cell phones and other communication devices were joining computers in the vast weave. Some data are now being tagged in ways that allow Web sites to interact. What the future will bring is anyone's guess, but no one can fail to be amazed at the dynamism of networking. Vinton Cerf, one of the Internet's principal designers, says simply: "Revolutions like this don't come along very often."

Perspective

Robert E. Kahn

President
Corporation for National
Research Initiatives

My first experience with computers, aside from a brief episode toggling the switches on an early IBM model 650, was with the IBM 704, 709, and finally 7090 at Bell Laboratories in the early 1960s. In those days we submitted computer programs to large "batch-processing" machines via punched cards, carrying them by hand to the computer center for processing later. It usually took several tries to fully debug even relatively simple programs, which usually translated into several days of elapsed time from start to finish.

The introduction of time-sharing procedures, which allowed several users to take turns using the same computer via teletype machines, made it possible to debug programs interactively, thus shortening the time to create working programs. Eventually we were doing this over telephone lines, at initial speeds of 300 bits per second. However, errors on the phone lines, along with the high cost of long-distance telephony and error correction, made interactive computer communications impractical over long distances.

The ARPANET, which leased 50-kilobit-per-second lines from the telephone company, was a significant improvement, allowing messages to be communicated between computers in the United States in a fraction of a second using a technique called packet switching. During the 1970s, more than 100 time-sharing machines were connected to the ARPANET and many thousands of researchers could access remote machines, do computations involving multiple machines, and even make use of specialized facilities such as the ILLIAC IV, the world's first supercomputer.

When I began working at DARPA (Defense Advanced Research Projects Agency) in the early 1970s, I recognized a need for multiple networks to communicate. I invited a colleague, Vint Cerf, to work with me in creating a logical architecture for connecting multiple networks and the computers attached to them. The key ingredients of this architecture were a protocol, now known as TCP/IP (transmission control protocol/internet protocol), with Internet addresses (i.e., IP addresses) to identify the individual machines and gateways (now known as routers) to provide linkages among networks. The networks that constituted the initial Internet were the ARPANET and two wireless networks (one terrestrial, the other satellite), each with characteristics that differed from the ARPANET. Before long the most prevalent networks in the system were local area networks called Ethernets.

These networks and computers and, of course, packet-switching technology all formed the underpinnings of the Internet. However, the Internet would not have become a worldwide phenomenon had it not been for three critical developments along the way. First, the National Science Foundation (NSF) in the United States embraced the DARPA technology and expanded the nascent Internet so as to allow the entire research and educational community in the United States and, in due course, much of the international community to participate. Second, a competitive telecommunications industry in the United States offered several cost-effective industry proposals for a high-speed alternative to the ARPANET. Finally, the U.S. Congress passed legislation in early 1993 that permitted NSF to open the NSFNET for commercial use as well as research and education purposes.

The Internet has clearly changed the lives of us all. Worldwide communication is now cost effective and continuously available from home and office, with access to information of all kinds readily at hand. However, a number of nontechnological issues remain. Today we are concerned with matters of privacy and security as well as with intrusions such as viruses, spam, or other offensive material. There is also the question of how best to deal with intellectual property on the Internet. These are thorny problems. Still, I believe the greatest opportunity for innovation lies in the power of the Internet to inspire new forms of creativity and to foster collaboration between individuals and organizations on a far larger scale than most of us can yet imagine.

INTERNET TIMELINE

Beginning as a tool for a select group of engineers and scientists associated with academia or government and evolving rapidly into the World Wide Web open to anyone with a computer and a telephone connection, the Internet has transformed the way we conduct research, communicate, and make purchases ranging from groceries and airline tickets to the latest books and music or clothing and furniture. How we got from there to here on the information highway is the story of a host of individuals and breakthrough thinking.

1962 Leonard Kleinrock, a doctoral student at MIT, writes a thesis describing queuing networks and the underlying principles of what later becomes known as packet-switching technology.

J. C. R. Licklider becomes the first director of the Information Processing Techniques Office established by the Advanced Research Projects Agency (ARPA, later known as DARPA) of the U.S. Department of Defense (DOD). Licklider articulates the vision of a "galactic" computer network—a globally interconnected set of processing nodes through which anyone anywhere can access data and programs.

1964 The RAND Corporation publishes a report, principally authored by Paul Baran, for the Pentagon called *On Distributed Communications Networks*. It describes a distributed radio communications network that could survive a nuclear first strike, in part by dividing messages into segments that would travel independently.

1966 Larry Roberts of MIT's Lincoln Lab is hired to manage the ARPANET project. He works with the research community to develop specifications for the ARPA computer network, a packet-switched network with minicomputers acting as gateways for each node using a standard interface.

1967 Donald Davies, of the National Physical Laboratory in Middlesex, England, coins the term *packet switching* to describe the lab's experimental data transmission.

1968 Bolt Beranek and Newman, Inc. (BBN) wins a DARPA contract to develop the packet switches called interface message processors (IMPs).

1969 DARPA deploys the IMPs. Kleinrock, at the Network Measurement Center at the University of California at Los Angeles, receives the first IMP in September. BBN tests the "one-node" network. A month later the second IMP arrives at Stanford, where Doug Englebart manages the Network Information Center, providing storage for ARPANET documentation. Dave Evans and Ivan Sutherland, professors researching computer systems and graphics at the University of Utah, receive the third IMP, and the fourth goes to the University of California at Santa Barbara, where Glen Culler is conducting research on interactive computer graphics.

1970 In December the Network Working Group (NWG), formed at UCLA by Steve Crocker, deploys the initial ARPANET host-to-host protocol, called the Network Control Protocol (NCP). The primary function of the NCP is to establish connections, break connections, switch connections, and control flow over the ARPANET, which grows at the rate of one new node per month.

At Bell Labs, Dennis Ritchie and Kenneth Thompson complete the UNIX operating system, which gains a wide following among scientists.

1972 Ray Tomlinson at BBN writes the first e-mail program to send messages across the ARPANET. In sending the first message to himself to test it out, he uses the @ sign—the first time it appears in an e-mail address.

Robert Kahn at BBN, who is responsible for the ARPANET's system design, organizes the first public demonstration of the new network technology at the International Conference on Computer Communications in Washington, D.C., linking 40 machines and a Terminal Interface Processor to the ARPANET.

1973 In September, Kahn and Vinton Cerf, an electrical engineer and head of the International Network Working Group, present a paper at the University of Sussex in England describing the basic design of the Internet and an open-architecture network, later known as TCP (transmission control protocol), that will allow networks to communicate with each other. The paper is published as "A Protocol for Packet Network Interconnection" in *IEEE Transactions on Communications*.

1975 Initial testing of packet radio networks takes place in the San Francisco area. The SATNET program is initiated in September with one Intelsat ground station in Etam, West Virginia, and another in Goonhilly Downs, England.

1976 At DARPA's request, Bill Joy incorporates TCP/IP (internet protocol) in distributions of Berkeley Unix, initiating broad diffusion in the academic scientific research community.

1977 Cerf and Kahn organize a demonstration of the ability of three independent networks to communicate with each other using TCP protocol. Packets are communicated from the University of Southern California across the ARPANET, the San Francisco Bay Packet Radio Net, and Atlantic SATNET to London and back.

Larry Landweber, of the University of Wisconsin, creates Theorynet, to link researchers for e-mail via commercial packet-switched networks like Telenet.

1979 DARPA establishes the Internet Configuration Control Board (ICCB) to help manage the DARPA Internet program. The ICCB acts as a sounding board for DARPA's plans and ideas. Landweber convenes a meeting of computer researchers from universities, the National Science Foundation (NSF), and DARPA to explore creation of a "computer science research network" called CSNET.

USENET, a "poor man's ARPANET," is created by Tom Truscott, Jim Ellis, and Steve Belovin to share information via e-mail and message boards between Duke University and the University of North Carolina, using dial-up telephone lines and the UUCP protocols in the Berkeley UNIX distributions.

1980 U.S. Department of Defense adopts the TCP/IP (transmission control protocol/internet protocol) suite as a standard.

1981 NSF and DARPA agree to establish ARPANET nodes at the University of

Wisconsin at Madison, Purdue University, the University of Delaware, BBN, and RAND Corporation to connect ARPANET to CSNET sites on a commercial network called Telenet using TCP/IP.

1982 All hosts connected to ARPANET are required to convert to the new TCP/IP protocols by January 1, 1983. The interconnected TCP/IP networks are generally known as the Internet.

1983 Sun Microsystems introduces its UNIX scientific workstation. TCP/IP, now known as the Internet protocol suite, is included, initiating broad diffusion of the Internet into the scientific and engineering research communities.

ARPANET, and all networks attached to it, officially adopts the TCP/IP networking protocol. From now on, all networks that use TCP/IP are collectively known as the Internet. The number of Internet sites and users grow exponentially.

The Internet Activities Advisory Board (later the Internet Activities Board, or IAB) replaces the ICCB. It organizes the research community into task forces on gateway algorithms, new end-to-end service, applications architecture and requirements, privacy, security, interoperability, robustness and survivability, autonomous systems, tactical interneting, and testing and evaluation. One of the task forces, soon known as "Internet Engineering," deals with the Internet's operational needs.

1984 The advent of Domain Name Service, developed by Paul Mockapetris and Craig Partridge, eases the identification and location of computers connected to the ARPANET by linking unique IP numerical addresses to names with suffixes such as .mil, .com, .org, and .edu.

1985 NSF links scientific researchers to five supercomputer centers across the country at Cornell University, University of California at San Diego, University of Illinois at Urbana-Champaign, Pittsburgh Supercomputing Center, and Princeton University. Like CSNET, NSFNET employs TCP/IP in a 56-kilobits-per-second backbone to connect them.

1986 Senator Albert Gore, of Tennessee, proposes legislation calling for the interconnection of the supercomputers centers using fiber-optic technology.

The Internet Engineering Task Force (IETF) expands to reflect the growing importance of operations and the development of commercial TCP/IP products. It is an open informal international community of network designers, operators, vendors, and researchers interested in the evolution of the Internet architecture and its smooth operation.

1987 As the NSFNET backbone becomes saturated, NSF plans to increase capacity, supports the creation of regional networks, and initiates a program to connect academic institutions, which invest heavily in campus area networks. The Internet of administratively independent connected TCP/IP networks emerges.

NSF convenes the networking community in response to a request by Senator Gore to examine prospects for a high-speed national research network. Gordon Bell at NSF reports to the White House Office of Science and Technology Policy (OSTP) on a plan for the National Research and Education Network. Presidential Science Advisor Allan Bromley champions the high-performance computing and communications initiatives that eventually implement the networking plans.

UUNET is formed by Rick Adams and PSINET is formed by Bill Schrader to provide commercial Internet access. At DARPA's request, Dan Lynch organizes the first Interop conference for information purposes and to bring vendors together to test product interoperability.

1988 An NSFNET contract is awarded to the team of IBM and MCI, led by Merit Network, Inc. The initial 1.5-megabits-per-second NSFNET is placed in operation.

1989 The Federal Networking Council (FNC), program officers from cooperating agencies, gives formal approval for interconnection of commercial and federal networks. The following year ARPANET is decommissioned.

1991 CERN releases the World Wide Web software developed earlier by Tim Berners-Lee. Specifications for HTML (hypertext markup language), URL (uniform resource locator), and HTTP (hypertext transfer protocol) launch a new era for content distribution.

At the University of Minnesota, a team of programmers led by Mark McCahill releases a point-and-click navigation tool, the "Gopher" document retrieval system, simplifying access to files over the Internet.

1992 The nonprofit Internet Society is formed to give the public information about the Internet and to support Internet standards, engineering, and management. The society later becomes home to a number of groups, including the IAB and IETF, and holds meetings around the world to promote diffusion of the Internet.

1993 The NSF solicits proposals to manage domain names for nonmilitary registrations and awards a 5-year agreement to Network Solutions, Inc.

Marc Andreessen and Eric Bina, of the National Center for Supercomputing Applications (NCSA) at the University of Illinois at Urbana-Champaign, develop an easy-to-use graphical interface for the World Wide Web. Distribution of the "browser," NCSA Mosaic, accelerates adoption of the Web. The technology is eventually licensed to Microsoft as the basis for its initial Internet Explorer browser. In 1994 the team rewrites the browser, changing its name to Netscape. Later "browser wars" focus public attention on the emerging commercial Internet.

1995 NSF decommissions the NSFNET.

1996 President Clinton signs the Telecommunications Act of 1996. Among its provisions, it gives schools and libraries access to state-of-the-art services and technologies at discounted rates.

1998 The Internet Corporation for Assigned Names and Numbers is chartered by the U.S. Department of Commerce to transition from the federal government to the private sector the coordination and assignment of Internet domain names, IP address numbers, and various protocol parameters.

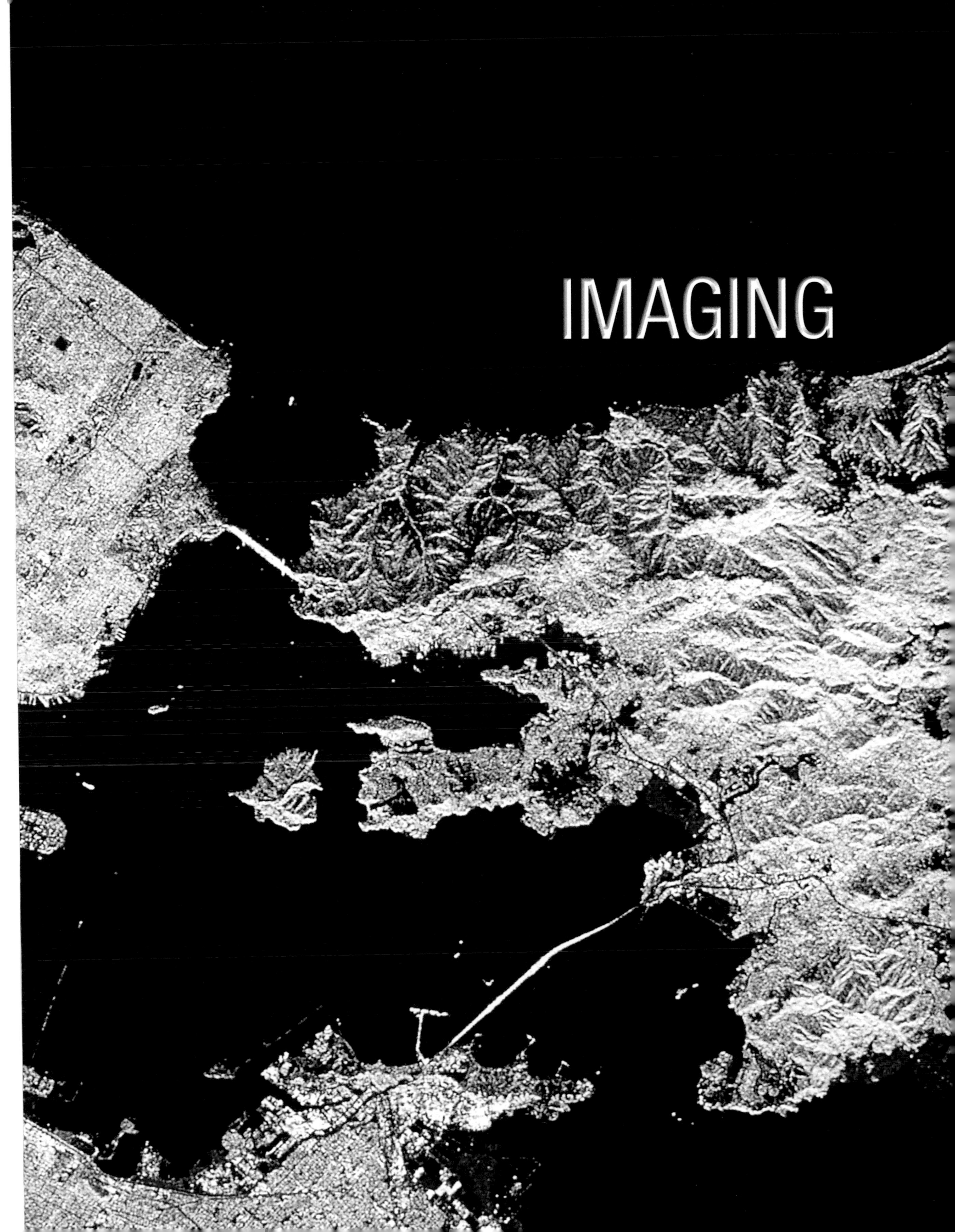

IMAGING

engineer named Hal Anger developed a camera that could record gamma rays—electromagnetic waves of even higher frequency than X rays—emitted by radioactive isotopes. By injecting small amounts of these isotopes into the body, radiologists were able to locate areas in the body where the isotopes were taken up. Known as the scintillation camera, or sometimes simply the Anger camera, the device evolved into use in several of modern medicine's most valuable imaging tools, including positron emission tomography (PET). X-ray imaging continued to evolve as well. In the 1970s medical engineers added computers to the equation, developing the technique known as computerized axial tomography (CAT), in which multiple cross-sectional x-ray views are combined by a computer to create three-dimensional images of the body's internal structures.

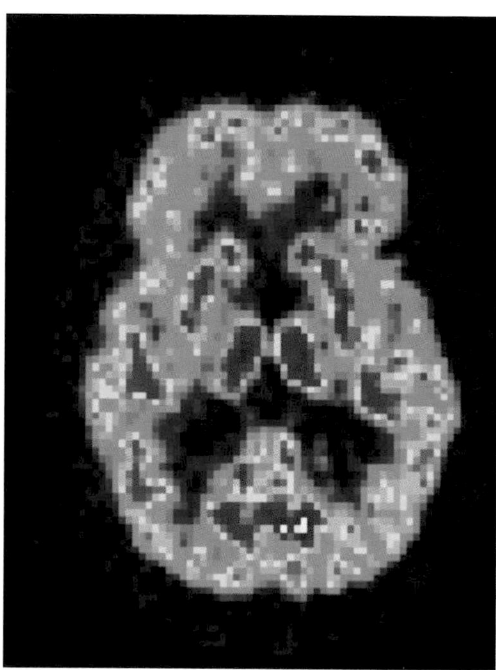

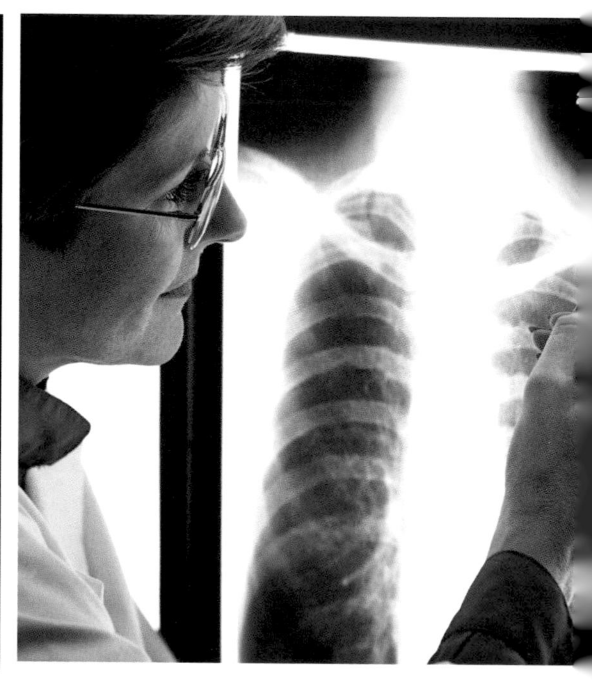

The positron scintillation camera, invented by electrical engineer Hal Anger (above left), evolved into many of modern medicine's most valuable tools, including PET (positron emission tomography) scans (above center), even as X rays continue to be the workhorse of medical diagnosis (above right).

One major drawback to these approaches was that they carried the risks associated with exposure to ionizing radiation, which even in relatively small amounts can cause irreparable damage to cells and tissues. But about the same time that CAT scanning was becoming a practical tool, a less harmful—and indeed more revealing—imaging technology appeared. Magnetic resonance imaging (MRI) relies not on X rays or gamma rays but on the interaction of harmless radio waves with hydrogen atoms in cells subjected to a magnetic field. It took a great deal of work by radiologists, scientists, and engineers to iron out all the wrinkles in MRI, but by the 1980s it too was proving to be an indispensable diagnostic tool. Not only was MRI completely noninvasive and free of ill effects, it also could create images of nearly any soft tissue. In its most developed form it can even chart blood flow and chemical activity in areas of the brain and heart, revealing details of functioning that had never been seen before.

Completing the gamut of medical imaging techniques is ultrasound, in at least one way a unique member of the family. As its name implies, rather than using electromagnetic radiation, ultrasound imaging relies on sound waves at high frequencies. The history of seeing with sound traces back to the early years of the century, when several different European engineers discovered that high-frequency sound waves bounced off metallic, underwater objects. Timing how long an echo takes to return to a transmitter/detector made it possible to determine the object's distance; other refinements eventually gave more detailed views of size and shape. Although sound navigation and ranging (sonar), a later term, was employed to some extent in World War I to detect submarines and underwater mines, it didn't really become refined enough for practical benefit until World War II. Meanwhile, as with X rays, sonar was also being used to detect flaws in metals and welded joints.

Not until the 1940s and 1950s did researchers seriously apply the principles and technology of sonar to the medical realm. One of the many pioneers was a Scottish gynecologist named Ian Donald, known by some of his colleagues as Mad Donald for his seemingly eccentric interest in all sorts of machines and inventions. Experimenting with tissue samples and an ultrasonic metal flaw detector owned by one of his patients, Donald realized that the detector could be used to create images of dense masses and growths within less dense tissue. Some of his early clinical efforts proved disappointing, but one noted success, when he correctly diagnosed an easily removed ovarian cyst that had been misread as inoperable stomach cancer, changed everything. As Donald himself said, "There could be no turning back." In 1959 he went on to discover that particularly clear echoes were returned from the heads of fetuses, and within a decade the use of ultrasound to chart fetal development throughout pregnancy was becoming more commonplace. Today, it is considered one of the safest methods of imaging in medicine and is a routine procedure in many doctors' offices and hospitals in the developed world.

Following on the heels of sonar was another imaging technology that has found an extraordinarily wide range of applications—almost everywhere, it seems, except in medicine. By 1930 researchers at the U.S. Naval Research Laboratory in Washington, D.C., had developed crude equipment that used long waves at the radio end of the spectrum to locate aircraft. Then in 1935, British physicist Robert Watson-Watt designed a more practical radio-wave detector that could determine not only range but also altitude. By 1938 dozens of Watson-Watt's devices—called radar, for radio detection and ranging—were linked to form a network of aircraft detectors along Britain's south and east coasts that proved extremely effective against attacking German planes throughout World War II. It is a little-known fact that an hour before the attack on Pearl Harbor, radar detected the incoming planes, though nothing was done with the information. By the end of the war, all the armed powers of the day employed radar in one form or another.

Radar has become one of the most ubiquitous of imaging technologies. Radar detectors create images of weather patterns and support the entire air traffic control system of the United States and other countries. Satellite-borne radar systems have mapped Earth's surface in exquisite detail, independent of weather or cloud cover. Radar aboard spacecraft venturing farther afield have returned images of other planets' surfaces, including stunningly detailed three-dimensional views of Venus obtained right through its otherwise impenetrable blanket of clouds. And, of course, radar is used today in traffic control worldwide, a byproduct of American traffic engineer John Barker's 1947 adaptation of radar to determine an automobile's speed.

Of all these diverse imaging accomplishments stretching across the century, perhaps the greatest revolution has been in telescopes. A telescope's light-gathering power is determined by the size of its aperture: the wider its diameter, the more light a telescope can gather and, therefore, the dimmer the celestial objects it can detect. Before 1900 the best telescopes were refractors, gathering and focusing light through a series of lenses arranged in a long tube. But refractors can get only so big before the weight of their lenses, which must be supported just at their edges, becomes too great; the practical limit turned out to be 40 inches. Reflecting telescopes, on the other hand, use a mirror to gather

The 100-inch telescope (below) built in 1918 atop Mount Palomar in California, was one in a series of ever larger instruments that astronomer George Ellery Hale trained on the heavens.

and focus light, and the mirror can be supported under its entire area. Improvements in mirror-making techniques after the turn of the century opened the door for the telescope revolution. Under the direction of American astronomer George Ellery Hale, engineers built a series of increasingly larger reflecting telescopes: a 60-inch version in 1908, followed by a 100-inch giant in 1918, and then a 200-inch behemoth, named after Hale and completed in 1947, 9 years after his death. This trio stood at the pinnacle for many years and unlocked a host of cosmic secrets, including the existence of galaxies and the fact that the universe is expanding. Today, the largest reflectors, using sophisticated techniques that link the light-gathering power of multiple mirrors, have effective apertures of up to 400 inches. And dramatic advances in light-sensing equipment,

IMAGING TECHNOLOGY

FROM THE MICROSCOPIC TO THE COSMIC

Human curiosity has driven the development of imaging tools that can see beyond what once seemed to be unbreachable limits. Among the most remarkable of these instruments are the scanning electron microscope (below) and the Hubble Space Telescope (opposite), respectively capable of viewing the infinitesimally small and the unimaginably distant. Besides targeting objects at opposite ends of the scale from microscopic to cosmic, these two devices also differ in the way they view the world. Like many traditional imaging devices, Hubble employs so-called passive viewing, gathering light and other forms of electromagnetic radiation in order to capture images. Electron microscopes use active viewing, projecting a beam of radiation to "illuminate" objects; in essence, they shine an electron flashlight on targeted samples, enabling computerized devices to record the reflected radiation.

Both electron microscopes and the Hubble telescope hold distinct advantages over other imaging instruments of their type. Because its electron beam has a wavelength about 100,000 times shorter than that of visible light, an electron microscope can distinguish features that are extremely close together, achieving magnifications of up to two million times life size. Hubble sees more clearly and can capture radiation from more distant objects because of its location above the distorting effects of Earth's atmosphere. Hubble's electronic cameras can also record images of extremely faint objects by achieving exposures of as long as 45 minutes because of the telescope's ability to lock onto a target without deviating more than about the width of a human hair seen at a distance of a mile.

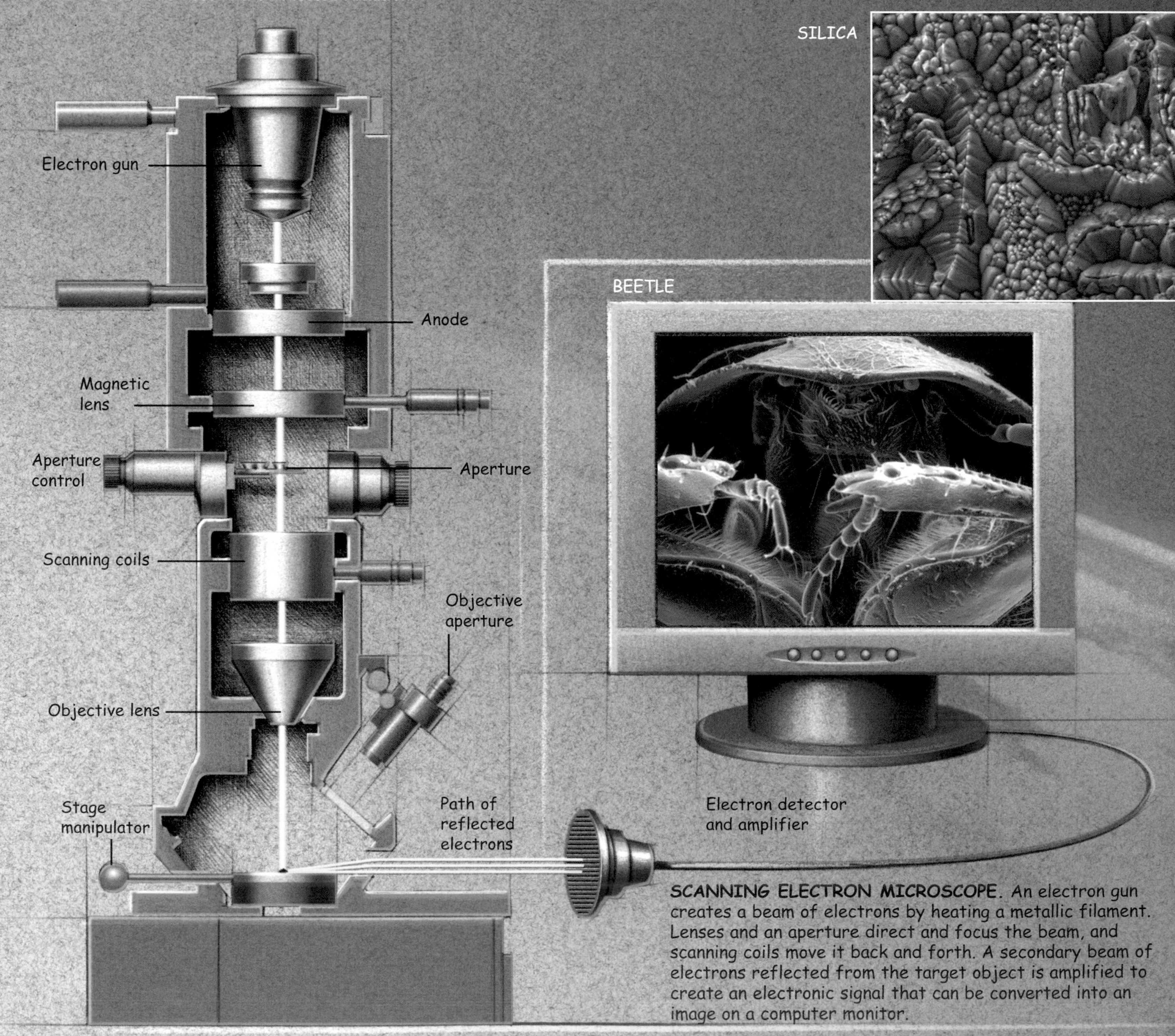

SCANNING ELECTRON MICROSCOPE. An electron gun creates a beam of electrons by heating a metallic filament. Lenses and an aperture direct and focus the beam, and scanning coils move it back and forth. A secondary beam of electrons reflected from the target object is amplified to create an electronic signal that can be converted into an image on a computer monitor.

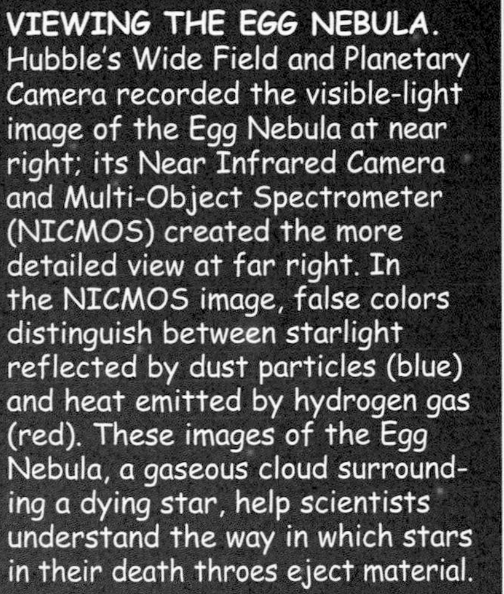

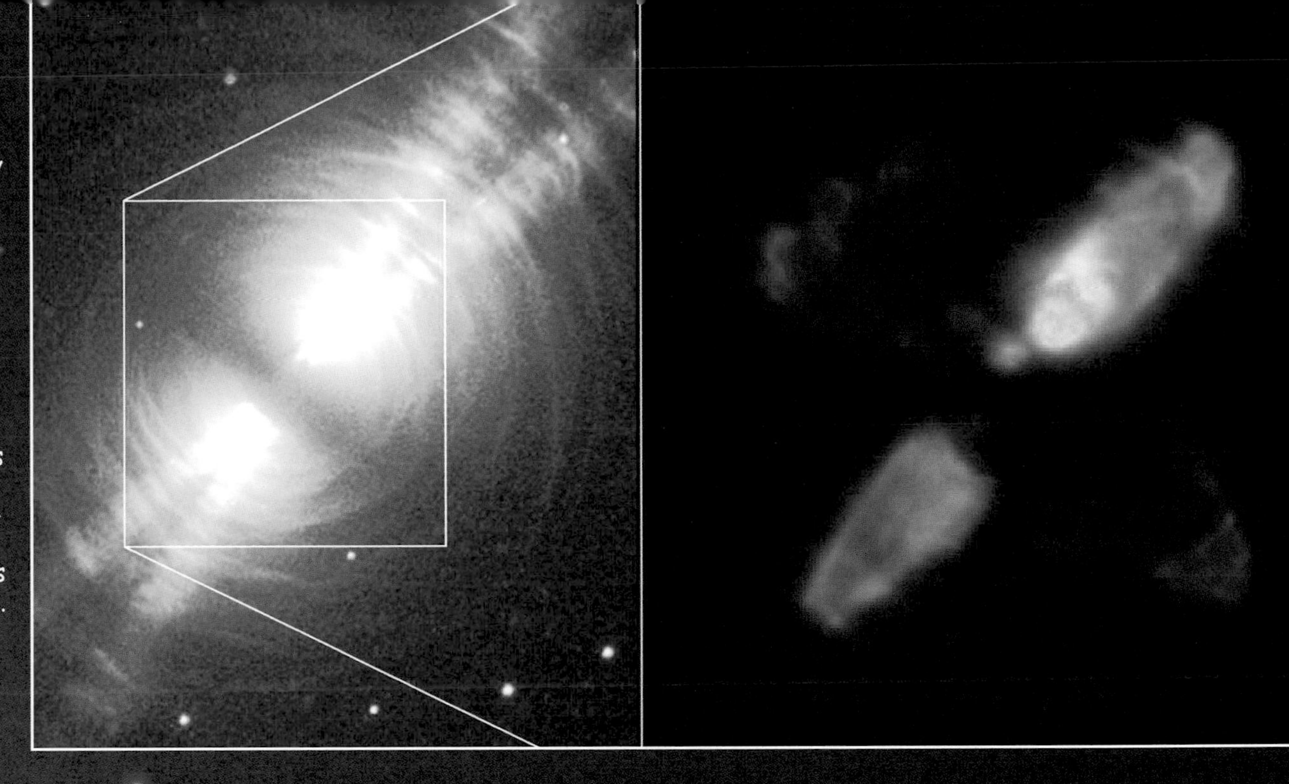

VIEWING THE EGG NEBULA. Hubble's Wide Field and Planetary Camera recorded the visible-light image of the Egg Nebula at near right; its Near Infrared Camera and Multi-Object Spectrometer (NICMOS) created the more detailed view at far right. In the NICMOS image, false colors distinguish between starlight reflected by dust particles (blue) and heat emitted by hydrogen gas (red). These images of the Egg Nebula, a gaseous cloud surrounding a dying star, help scientists understand the way in which stars in their death throes eject material.

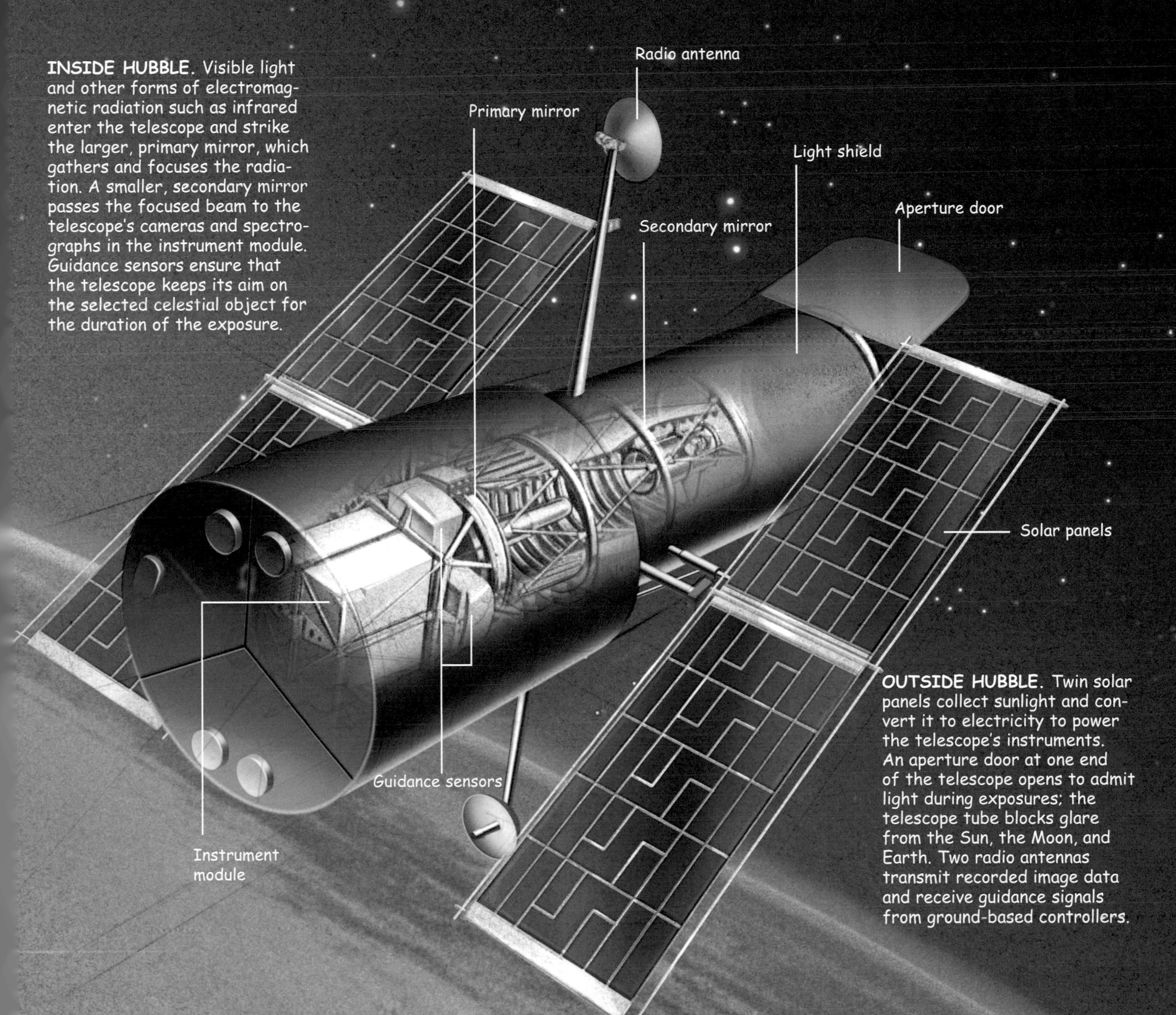

INSIDE HUBBLE. Visible light and other forms of electromagnetic radiation such as infrared enter the telescope and strike the larger, primary mirror, which gathers and focuses the radiation. A smaller, secondary mirror passes the focused beam to the telescope's cameras and spectrographs in the instrument module. Guidance sensors ensure that the telescope keeps its aim on the selected celestial object for the duration of the exposure.

OUTSIDE HUBBLE. Twin solar panels collect sunlight and convert it to electricity to power the telescope's instruments. An aperture door at one end of the telescope opens to admit light during exposures; the telescope tube blocks glare from the Sun, the Moon, and Earth. Two radio antennas transmit recorded image data and receive guidance signals from ground-based controllers.

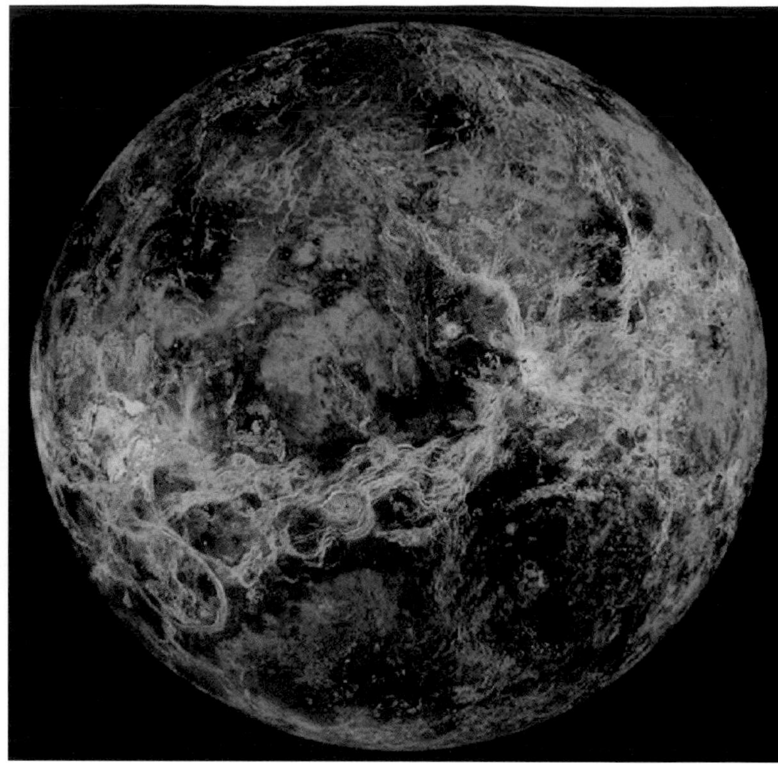

A view of Venus (above right) is the product of more than a decade of radar imaging by the Magellan *spacecraft, with gaps filled in by the planetary radar transmitter housed inside a dome suspended above the vast reflecting surface of Earth-based Arecibo Observatory (above left). The transmitter directs radar waves to objects in the solar system; return echoes are analyzed to gain information about surface properties. Nothing so high-tech or complicated is required, however, when someone on Earth wants to record a special moment for posterity. All that's needed is a point and shoot camera (left).*

including photodiodes that can detect a single photon, have added to the wonders revealed.

The year after the Hale telescope was completed, a radio engineer named Karl Jansky made a discovery that initiated yet another telescopic revolution. He determined that a constant background static being picked up by sensitive radio antennas was actually coming from space. It was ultimately identified as residual radiation from the Big Bang that gave birth to the universe. Within a few years, astronomers were training radio dishes on the heavens and learning to see with a whole new set of eyes. Radio telescopes were even easier to link together; using a technique called interferometry, engineers could create radio telescopic arrays made up of dozens of individual dishes, with a combined aperture that was not inches, but miles, across.

Telescopes now also orbit Earth on satellites, the most famous being the Hubble Space Telescope, which includes several different imaging devices—optical among others—and has produced cosmic views of astounding clarity. Crucial to that clarity are detectors called charge-coupled devices (CCDs), electronic components that convert light into electrical signals that can be interpreted and manipulated by computer. The most refined CCDs consist of hundreds of millions of individual picture elements, or pixels, each capable of distinguishing tens of thousands of shades of brightness. CCDs have become essential components not only in optical telescopes but also in digital cameras, achieving resolutions that rival the best of older photographic techniques.

Hubble is not alone out there. Its spectacular optical images are only part of what we can see of space. Orbiting x-ray observatories give us a ringside view of the most violent cosmic events, from the birth of stars to the gravitational collapses that form black holes. Gamma-ray detectors tell stories of other cataclysmic events, some still so mysterious as to defy explanation. And infrared instruments, picking up dim signals from the deepest reaches of space, reveal details about the whole history of the universe, back to its very beginning. With this new breed of imaging devices the eyepiece is long gone, but the view is still riveting.

Perspective

George M. C. Fisher

Retired Chairman and CEO
Eastman Kodak Company

Images and the process of creating them (imaging) seem to have both attracted and driven me through much of my professional life. As a schoolkid, I studied photographs of cave drawings made by Cro-Magnon people and wondered how difficult it must have been for them to communicate their stories unless someone actually came to their caves. Surely even our prehistoric brethren had stories they wanted to tell with pictures and were frustrated by their limited technological capabilities.

As a graduate student, I was awed by the high-speed photos of bullets penetrating plastics and glass that were taken in our lab at Brown University. At the same time, I was learning the marvels of developing film and printing pictures in a darkroom—the wonderful process of storing images through the magical combination of photons and photochemistry.

At Bell Labs in the 1960s I was privileged to work with some of the world's brightest scientists and engineers. Yet, once again, images and the restrictions on storing and communicating them seemed to dominate the world of so many of my colleagues. In those days a number of things limited our ability to transfer images electronically. Bandwidth was low (1.2 kilobits per second seemed good), and storing and processing images were possible only with film and central computing facilities that had such "powerful machines" as the IBM 650 or 7090 or, later, the 360.

But in the 1960s, as today, we had dreams. We talked of the "natural state of communications" and envisioned a world where wires were not necessary, memory and bandwidth were virtually free, and the human-machine interface didn't require much tactile (keyboard) input. We dreamed that the cost of memory could drop to one cent per bit some day, that modems could speed up transmission to 56 kilobits per second, and that broadband media such as fiber optics would improve enough that we could increase the distance between repeaters to more than a mile. And then there were the new silicon technologies based on insulated gate field effect transistors (IGFETs), the precursor to today's MOS (metal oxide semiconductor) devices. Perhaps eventually there would be integrated circuits with thousands of devices per chip of silicon.

We thought we were being so optimistic! Yet, today, we can only marvel at how far short our technological dreams were of the present reality. Today, multimegabit images can be captured digitally as well as on film and transmitted at multimegabit speeds. We can process images on desktop computers operating at gigacycle-per-second clock rates. We have in-camera storage on solid-state devices that cost only pennies per megabit (nearly a million times better than our dream of the early 1960s). What if we had set our expectations even higher?

Today, thanks to the dramatic gains in these underlying technologies, we are much closer to the natural state of communications in which we share and store not only voice but also images as part of our normal electronic communication. As I later came to appreciate at Eastman Kodak Company, images truly enrich our lives, whether they are for information, as primarily discussed in this chapter; for entertainment, as we see in movies and television; or for preserving precious memories, as with family photographs. Truly, imaging is an important manifestation of the symbiotic relationship between our technologies and our need to tell a story. We will be limited only by our too-modest expectations of what technology can bring to fulfilling the imaging needs of humankind.

TIMELINE

Efforts to capture visions beyond the range of the normal eye have long engaged scientists and engineers. By the mid-1880s George Eastman had improved upon celluloid and at the turn of the 20th century used it with his new camera, the Brownie. That boxy little phenomenon is still remembered by many adults today, even as digital cameras record the world around us by harnessing electrons. The discovery of X rays was only the first of many achievements leading to the development of picture-making devices that today support all manner of endeavors—in the military, medical, meteorological, computer technology, and space exploration communities. As the preceding pages make clear, images—microscopic, mundane, magnificent—affect us in all aspects of our lives.

1900 Eastman introduces the Kodak Brownie camera *(above)*. Named after popular children's book characters, it sells for $1 and uses film that sells for 15¢ a roll. For the first time, photography is inexpensive and accessible to anyone who wants to take "snapshots." In the first year 150,000 cameras are sold, and many of the first owners are children. In the course of its long production life, the Brownie has more than 175 models; the last one is marketed as late as 1980 in England.

1913 William David Coolidge invents the hot cathode x-ray tube, using a thermionic tube with a heated cathode electron emitter to replace the cold, or gas, tube. All modern x-ray tubes are of the thermionic type.

Albert Solomon, a pathologist in Berlin, uses a conventional x-ray machine to produce images of 3,000 gross anatomic mastectomy specimens, observing black spots at the centers of breast carcinomas. Mammography, the resulting imaging, has been used since 1927 as a diagnostic tool in the early detection of breast cancer.

1915 French professor and physicist Paul Langevin, working with Swiss physicist and engineer Constantin Chilowsky, develops the hydrophone, a high-frequency, ultrasonic echo-sounding device. The pioneering underwater sound technique is improved by the U.S. Navy and used during World War I in antisubmarine warfare as well as in locating icebergs. The work forms the basis for research and development into pulse-echo sonar (sound navigation and ranging), used on naval ships as well as ocean liners.

1931-1933 Ernst Ruska, a German electrical engineer working with Max Kroll, constructs and builds an electron microscope, the first instrument to provide better definition than a light microscope. Electron microscopes can view objects as small as the diameter of an atom and can magnify objects one million times. (In 1986 Ruska is awarded half of the Nobel Prize in physics. The other half is divided between Heinrich Rohrer and Gerd Binnig for their work on the scanning tunneling microscope; *see 1981*.)

1935 British scientist Sir Robert Watson-Watt patents the first practical radar (for radio detection and ranging) system for meteorological applications. During World War II radar is successfully used in Great Britain to detect incoming aircraft and provide information to intercept bombers.

1939 Henry Boot and John Randall, at the University of Birmingham in England, develop the resonant-cavity magnetron, which combines features of two devices, the magnetron and the klystron. The magnetron, capable of generating high-frequency radio pulses with large amounts of power, significantly advances radar technology and assists the Allies during World War II.

1940s MIT's Radiation Laboratory begins investigating the development of microwave radar systems, physical electronics, microwave physics, electromagnetic properties of matter, and microwave communication principles.

1943 The use of radar to detect storms begins. The U.S. Weather Radar Laboratory conducts research in the 1950s on Doppler radar, the change in frequency that occurs as a moving object nears or passes (an effect discovered for sound waves in 1842 by Austrian scientist Christian Doppler).

1946 The Civil Aviation Authority unveils an experimental radar-equipped tower for control of civil flights. Air traffic controllers soon are able to track positions of aircraft on video displays for air traffic control and ground controlled approach to airports.

1950s British chemists Max Perutz and Sir John Kendrew use x-ray crystallography to solve the structure of the oxygen-carrying proteins myoglobin and hemoglobin. They win the Nobel Prize in chemistry in 1962.

Rosalind Franklin uses x-ray crystallography to create crystal-clear x-ray photographs that reveal the basic helical structure of the DNA molecule.

Russell Morgan, a professor of radiological science at Johns Hopkins University, Edward Chamberlain, a radiologist at Temple University, and John W. Coltman, a physicist and associate director of the Westinghouse Research Laboratories, perfect a method of screen intensification that reduces radiation exposure and improves fluoroscopic vision. Their image intensifier in fluoroscopy is now universally used in medical fluoroscopy and in military applications, including night vision.

1958 Hal Anger invents a medical imaging device that enables physicians to detect tumors and make diagnoses by imaging gamma rays emitted by radioactive isotopes. Now the most common nuclear medicine imaging instrument worldwide, the camera uses photoelectron multiplier tubes closely packed behind a large scintillation crystal plate. The center of the scintillation is determined electronically by what is known as Anger logic.

1959 Ian Donald, a professor working at the University of Glasgow's Department of Midwifery, and his colleagues develop practical technology and applications for ultrasound as a diagnostic tool in obstetrics and gynecology. Ultrasound displays images on a screen of tissues or organs formed by the echoes of inaudible sound waves at high frequencies (20,000 or more vibrations per second) beamed into the body. The technique is used to look for tumors, analyze bone structure, or examine the health of an unborn baby *(below)*.

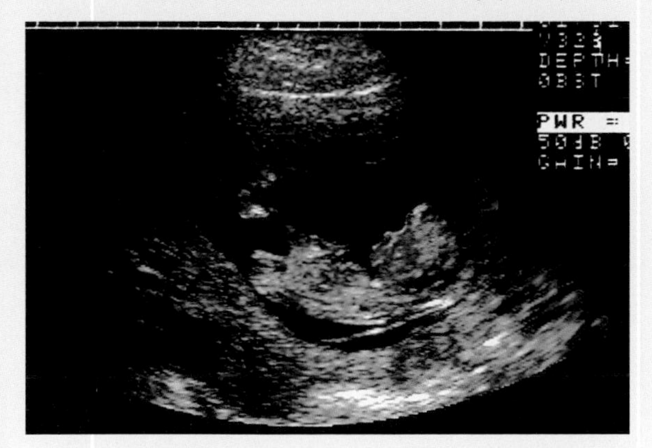

1960s Semiconductor manufacturing begins using optical lithography, an innovative technology using a highly specialized printing process that places intricate patterns onto silicon chips, or wafers. In the first stage an image containing the defining pattern is projected onto the silicon wafer, which is coated with a very thin layer of photosensitive material called "resist." The process is still used to manufacture integrated circuits and could continue to be used through the 100-nanometer generation of devices.

1960 Powell Richards and Walter Tucker, and many colleagues at the Bureau of Engineering Research at the U.S. Department of Energy's Brookhaven National Laboratory, invent a short half-life radionuclide generator that produces technetium-99m for use in diagnostic imaging procedures in nuclear medicine—a branch of medicine that uses radioisotopes for research, diagnosis, and treatment of disease. (Technetium-99m was discovered in 1939 by Emilio Segrè and Glenn Seaborg.)

1962 Sy Rankowitz and James Robertson, working at Brookhaven National Laboratory, invent the first positron emission tomography (PET) transverse section instrument, using a ring of scintillation crystals surrounding the head. (The first application of positron imaging for medical diagnosis occurred in 1953, when Gordon Brownell and William Sweet at Massachusetts General Hospital imaged patients with suspected brain tumors.) The following year David Kuhl introduces radionuclide emission tomography leading to the first computerized axial tomography, as well as to refinements in PET scanning, which is used most often to detect cancer and to examine the effects of cancer therapy. A decade later single-photon emission tomography (SPECT) methods become capable of yielding accurate information similar to PET by incorporating mathematical algorithms by Thomas Budinger and Grant Gullberg of the University of California at Berkeley.

1960s and 1970s Space-based imaging gets under way throughout the 1960s as Earth-observing satellites begin to trace the planet's topography. In 1968 astronauts on *Apollo 7*, the first piloted Apollo mission, conduct two scientific photographic sessions and transmit television pictures to the American public from inside the space capsule. In 1973 astronauts aboard *Skylab*, the first U.S. space station, conduct high-resolution photography of Earth using photographic remote-sensing systems mounted on the spacecraft as well as a Hasselblad handheld camera. Landsat satellites launched by NASA between 1972 and 1978 produce the first composite multispectral mosaic images of the 48 contiguous states. Landsat imagery provides information for monitoring agricultural productivity, water resources, urban growth, deforestation, and natural change.

1972 Engineer Godfrey Hounsfield of Britain's EMI Laboratories and South African–born American physicist Allan Cormack of Tufts University develop the computerized axial tomography scanner, or CAT scan. With the help of a computer, the device combines many x-ray images to generate cross-sectional views as well as three-dimensional images of internal organs and structures. Used to guide the placement of instruments or treatments, CAT eventually becomes the primary tool for diagnosing brain and spinal disorders. (In 1979, Hounsfield and Cormack are awarded the Nobel Prize in physiology or medicine.)

Using high-speed computers, magnetic resonance imaging (MRI) is adapted for medical purposes, offering better discrimination of soft tissue than x-ray CAT and is now widely used for noninvasive imaging throughout the body. Among the pioneers in the development of MRI are Felix Bloch and Edward Purcell (Nobel Prize winners in 1952), Paul Lauterbur, and Raymond Damadian.

1981 Gerd Binnig and Heinrich Rohrer, German physicists working at the IBM Research Laboratory in Zürich design and build the first scanning tunneling microscope (STM), with a small tungsten probe tip about one or two atoms wide. In 1986, Binnig, Cal Quate, and Christoph Gerber introduce the atomic force microscope (AFM), which is used in surface science, nanotechnology, polymer science, semiconductor materials processing, microbiology, and cellular biology. For invention of the STM Binnig and Rohrer share the 1986 Nobel Prize in physics with Ernst Ruska, who receives the award for his work on electron optics.

1987 Echo-planar imaging (EPI) is used to perform real-time movie imaging of a single cardiac cycle. (Peter Mansfield of the School of Physics and Astronomy, University of Nottingham, first developed the EPI technique in 1977.) In 1993 the advent of functional MRI opens up new applications for EPI in mapping regions of the brain responsible for thought and motor control and provides early detection of acute stroke.

1990 NASA launches the Hubble Space Telescope, a space-based observatory placed into orbit around Earth. Because it is outside the atmosphere, the telescope can provide the clearest views of the universe yet obtained in optical astronomy.

1990s–2000 NASA launches robotic spacecraft equipped with a variety of imaging instruments as part of a program of solar system exploration. Spacecraft have returned images not only from the planets but also from several of the moons of the gas giants.

This composite image of Jupiter and its four largest moons was assembled from data returned by instruments aboard the Galileo *spacecraft.*

HOUSEHOLD APPLIANCES

As a frequent purveyor of domestic dreams, Good Housekeeping magazine was on familiar ground in 1930 when it rhetorically asked its readers: "How many times have you wished you could push a button and find your meals deliciously prepared and served, and then as easily cleared away by the snap of a switch?" No such miraculous button or switch was in prospect, of course—not for cooking meals, cleaning the house, washing clothes, or any of the other home-making chores that, by enduring custom, mainly fell to women.

Seven decades later American women averaged 4 hours of housework a day, only a moderate decline since 1930, accompanying the movement of large numbers of women into the workforce. What changed—and had been changing since the beginning of the century—was the dramatic easing of drudgery by new household appliances. Effort couldn't be engineered out of existence by stoves, washing machines, vacuum cleaners, dishwashers, and other appliances, but it was radically redefined.

Consider cooking. In practically all American households by the turn of the 20th century, the work was done on cast iron stoves that burned wood or coal. A few people mourned the passing of fireplace cooking—"The open fire was the true center of home-life," wrote one wistful observer of the changeover in the middle decades of the 19th century—but the advantages of a stove were overwhelming. It used substantially less fuel than a blaze in an open hearth, didn't require constant tending, didn't blacken the walls with soot, didn't spit out dangerous sparks and embers, and, if centrally positioned, would warm a kitchen in winter much more effectively than a fireplace. It was also versatile. Heat from the perforated fire chamber was distributed to cooking holes on the top surface and to several ovens; some of it might also be directed to a compartment that kept food warm or to an apparatus that heated water. But the stove could be exasperating and exhausting, too. The fire had to be started anew each morning and fed regular helpings of fuel—an average of 50 pounds of it over the course of a day. Controlling the heat with dampers and flues was a tricky business. Touching any part of the stove's surface might produce a burn. Ashes were usually emptied twice a day. And a waxy black polish had to be applied from time to time to prevent rusting. In all, an hour or more a day was spent simply tending the stove.

As a heat source for cooking, gas began to challenge coal and wood in the closing years of the 19th century. At that time piped gas made from coke or coal was widely available in cities for illumination, but incandescent lights were clearly the coming thing. To create an alternative demand for their product, many gas companies started to make and market gas stoves, along with water heaters and furnaces. A gas stove had some powerful selling points. It could be smaller than a coal- or wood-burning stove; most of its surface remained cool; and all the labor of toting fuel, starting and tending the fire, and removing the ashes was eliminated. The development of an oven thermostat in 1915 added to its appeal, as did the increasing use of natural gas, which was cheaper and less toxic than the earlier type. By 1930 gas ranges outnumbered coal or wood burners by almost two to one.

Electric stoves were still uncommon. Although they had originated around the turn of the century, fewer than one U.S. residence in 10 was wired for electricity at the time; moreover, such power was expensive, and the first electric stoves used it gluttonously. Another deficiency was the short life of their heating elements, but in 1905 an engineer named Albert Marsh solved that problem with a patented nickel-chrome alloy that could take the heat. In the next decade electric stoves acquired an oven thermostat, matching an important feature of their gas rivals. Meanwhile America was steadily being wired. By the mid-1920s, 60 percent of residences had electricity, and it was fast falling in price. As electric stoves became more competitive, they, like gas stoves, were given a squared-off shape and a white porcelain enamel surface that was easy to clean. They continued to gain ground, receiving a major boost with the introduction in 1963 of the self-cleaning oven, which uses very high temperatures—about 900°F—to burn food residue from oven walls. Today, many households split the difference in stove types, choosing gas for the range and electricity for the oven.

The electric stove is just one of a host of household appliances based on resistance heating—the production of heat energy as current passes through an electrically resistant material. Others that appeared in the early days of electrification (especially after Albert Marsh developed the nickel-chrome resistor) included toasters, hot plates, coffee percolators, and—most welcome of all—the electric iron. The idea of a self-heated iron wasn't new; versions that burned gas, alcohol, or even gasoline were available, but for obvious reasons they were regarded warily. The usual implement for the job was a flatiron, an arm-straining mass of metal that weighed up to 15 pounds; flatirons were used several at a time, heated one after the other on the top of a stove. An electric iron, by contrast, weighed only about 3 pounds, and the ironing didn't have to be done in the vicinity of a hot stove. In short order it displaced the flatiron and became the best selling of all electric appliances. Its popularity rose still further with the introduc-

Standing over the proverbial hot stove was designated "women's work" at the turn of the 20th century (opposite). Fifty years later (below) electric "cookers" with thermostat controls had replaced balky wood burners, and by the end of the century meal preparation was no longer the province of women alone (left).

Electricity enabled a number of improvements over irons that had to be replenished with hot coals (above left), including steam irons (right), which made their debut in the late 1930s.

tion of an iron with thermostatic heat control in 1927 and the appearance of household steam irons a decade later.

Another hit was the electric toaster. The first successful version, brought out by General Electric in 1909, had no working parts, no controls, no sensors, not even an exterior casing. It consisted of a cage-like contraption with a single heating element. A slice of bread had to be turned by hand to toast both sides, and close attention was required to prevent burning. Better models soon followed—some with sliding drawers, some with mechanical ways of turning the bread—but the real breakthrough was the automatic pop-up toaster, conceived by a master mechanic named Charles Strite in 1919. It incorporated a timer that shut off the heating element and released a pop-up spring when the single slice of toast was done. After much tinkering, Strite's invention reached the consumer market in 1926, and half a million were sold within a few years. Advertisements promised that it would deliver "perfect toast every time—without watching, without turning, without burning," but that wasn't necessarily the case. When more than one slice was desired, the timer didn't allow for heat retention by the toaster, producing distinctly darker results with the second piece. The manufacturer recommended allowing time between slices for cooling—not what people breakfasting in a hurry wanted to hear. Happily, toasters were soon endowed with temperature sensors that determined doneness automatically.

GE's model D-12 toaster with warming tray required turning bread by hand.

Electricity revolutionized appliances in another way, powering small motors that could perform work formerly done by muscles. The first such household device, appearing in 1891, was a rotary fan made by the Westinghouse Electric and Manufacturing Company; its blades were driven by a motor developed chiefly by Nikola Tesla, a Serbian genius who pioneered the use of alternating current. The second was a vacuum cleaner, patented by a British civil engineer named H. Cecil Booth in 1901. He hit on his idea after observing railroad seats being cleaned by a device that blew compressed air at the fabric to force out dust. Sucking at the fabric would be better, he decided, and he designed a motor-driven reciprocating pump to do the job. Soon the power of the electric motor was applied to washing machines, sewing machines, refrigerators, dishwashers, can openers, coffee grinders, egg beaters, hair dryers, knife sharpeners, and many other devices.

At the turn of the century, only about one American family in 15 employed servants, but having such a source of muscle power was devoutly craved by many and was seen as a key indicator of status. As housework was eased by electric motors and the number of servants dropped, such views changed, but some advertising copywriters insisted on describing appliances in social terms: "Electric servants can be depended on—to do the muscle part of the washing, ironing, cleaning and sewing," said a General Electric advertisement in 1917; "Don't go to the Employment Bureau. Go to your Lighting Company or leading Electric Shop to solve your servant problem."

The electric servant brigade was rapidly improved. In 1907 an

American inventor named James Murray Spangler created a vacuum cleaner that basically consisted of an old-fashioned carpet sweeper to raise dust and a vertical shaft electric motor to power a fan and blow the dust into an external bag. Manufactured by the Hoover Company, which bought the patent in 1908, it was hugely successful, especially after Hoover in 1926 extended the fan motor's power to a rotating brush that "beats as it sweeps as it cleans." Meanwhile, the Electrolux company in Sweden grabbed a sizable share of the market with a very different design for a vacuum cleaner—a small rolling cylinder that had a long hose and a variety of nozzles to clean furniture and curtains as well as carpets.

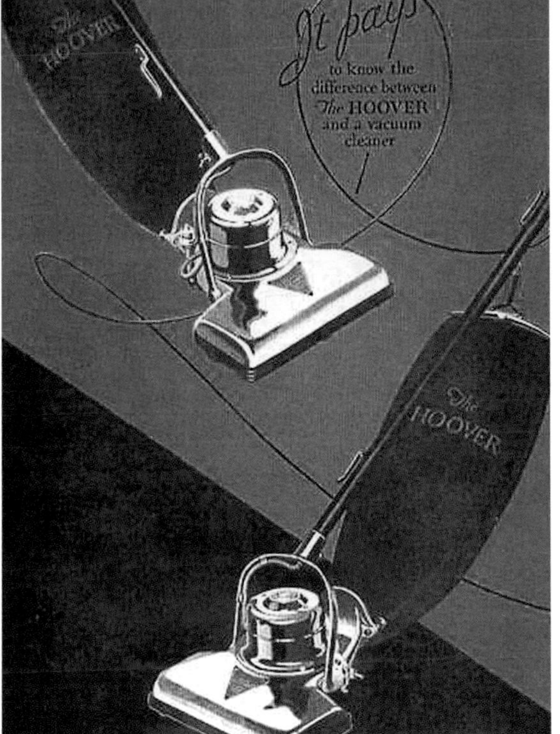

The Hoover 700 model, with its characteristic orange triangle, introduced a beater bar that reinforced the Hoover slogan "It Beats as it Sweeps as it Cleans."

No aspect of housework stood in greater need of motor power than washing clothes, a job so slow and grueling when performed manually that laundresses were by far the most sought-after domestic help. In the preelectric era, Mondays were traditionally devoted to doing the laundry. First, the clothes were rubbed against a washboard in soapy water to remove most of the dirt; next they were wrung out, perhaps by running them through a pair of hand-cranked rollers; they were then boiled briefly in a vat on top of the stove; then, after removal with a stick, they were soaped, rinsed, and wrung out again; finally they were hung on a line to dry—unless it was raining. The arrival of electricity prompted many efforts to mechanize parts of this ordeal. Some early electric washing machines worked by rocking a tub back and forth; others pounded the clothes in a tub with a plunger; still others rubbed them against a washboard. A big improvement came in 1922 when Howard Snyder of the Maytag Company designed a tub with an underwater agitator whose blade forced water through the clothes to get the dirt out.

The following decade saw the introduction of completely automatic washing machines that filled and emptied themselves. Then wringers were rendered unnecessary by perforated tubs that spun rapidly to drive the water out by centrifugal force. An automatic

Washing, wringing, and "mangling"—using heavy rollers to squeeze wash water out of clothes—characterized pre-20th century hand-cranked machines designed to ease the endless chore of laundering (below left). By the 1950s electric power wringers had come along (below right).

A PAIR OF HOUSEHOLD HELPERS

PUTTING RESISTANCE HEATING TO WORK

In most households, the breakfast rituals of making coffee and toast are assigned to a pair of small appliances that practically operate themselves. Coffeemakers and toasters are kin in one fundamental way: both do their job with heat energy created as electrons are forced through wire and collide with atoms along the way. This phenomenon, called resistance heating, was domesticated for kitchen work early in the 20th century by the creation of a nickel-chromium alloy that was ideally suited for the current-conducting role. The alloy spawned dozens of specialized cooking devices, everything from corn poppers to egg cookers, but none proved more popular than those dedicated to the age-old breakfast pleasures of coffee and toast. A modern coffeemaker has no moving parts: a heating element boils water, causing it to rise up a tube and drip through grounds to make the coffee, which is kept warm by the same heat source. In a toaster, infrared radiation from wires heats the surface of bread until its sugars and starches begin to caramelize at 310°F, taking on savor as they brown.

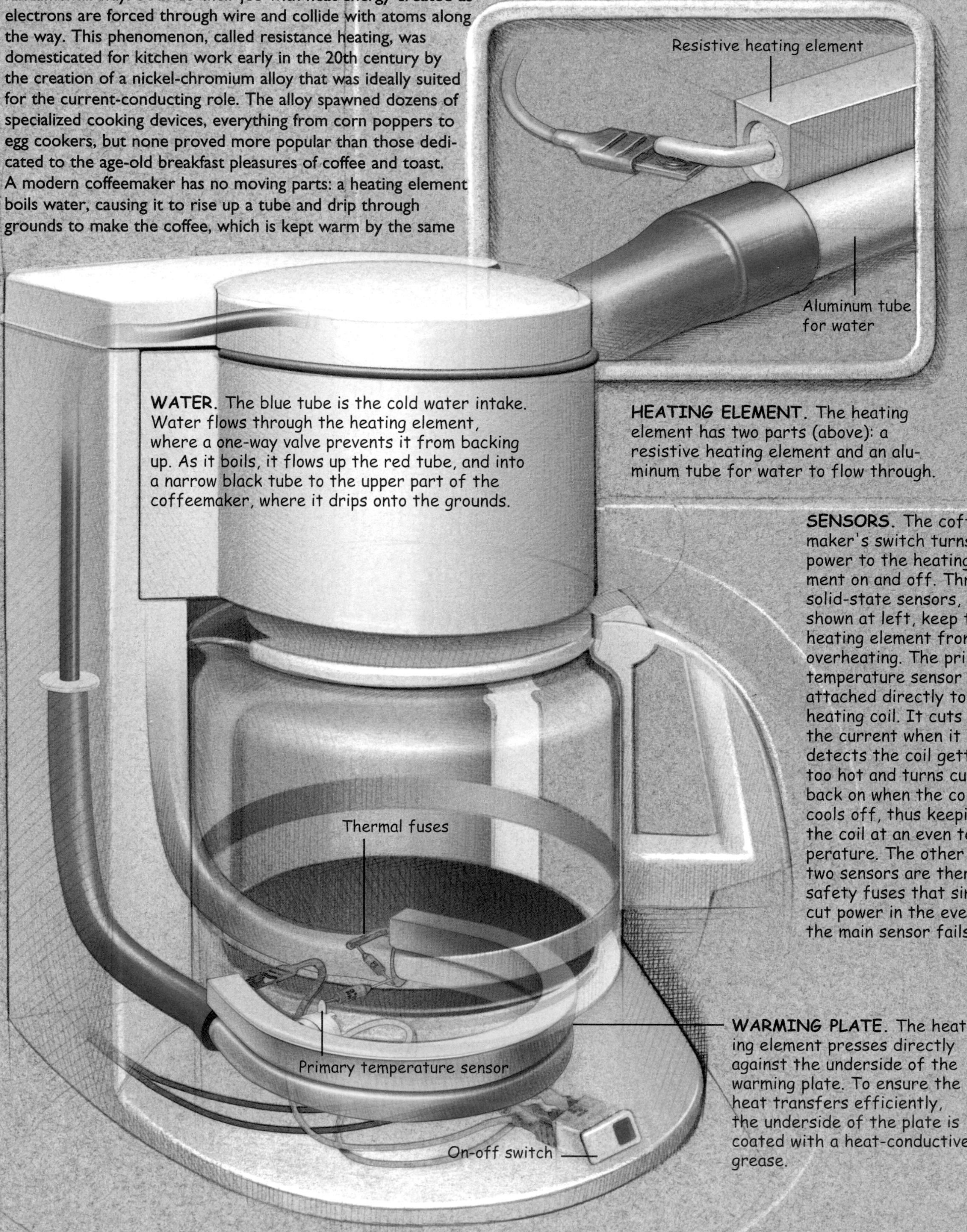

WATER. The blue tube is the cold water intake. Water flows through the heating element, where a one-way valve prevents it from backing up. As it boils, it flows up the red tube, and into a narrow black tube to the upper part of the coffeemaker, where it drips onto the grounds.

HEATING ELEMENT. The heating element has two parts (above): a resistive heating element and an aluminum tube for water to flow through.

SENSORS. The coffeemaker's switch turns power to the heating element on and off. Three solid-state sensors, as shown at left, keep the heating element from overheating. The primary temperature sensor is attached directly to the heating coil. It cuts off the current when it detects the coil getting too hot and turns current back on when the coil cools off, thus keeping the coil at an even temperature. The other two sensors are thermal safety fuses that simply cut power in the event the main sensor fails.

WARMING PLATE. The heating element presses directly against the underside of the warming plate. To ensure the heat transfers efficiently, the underside of the plate is coated with a heat-conductive grease.

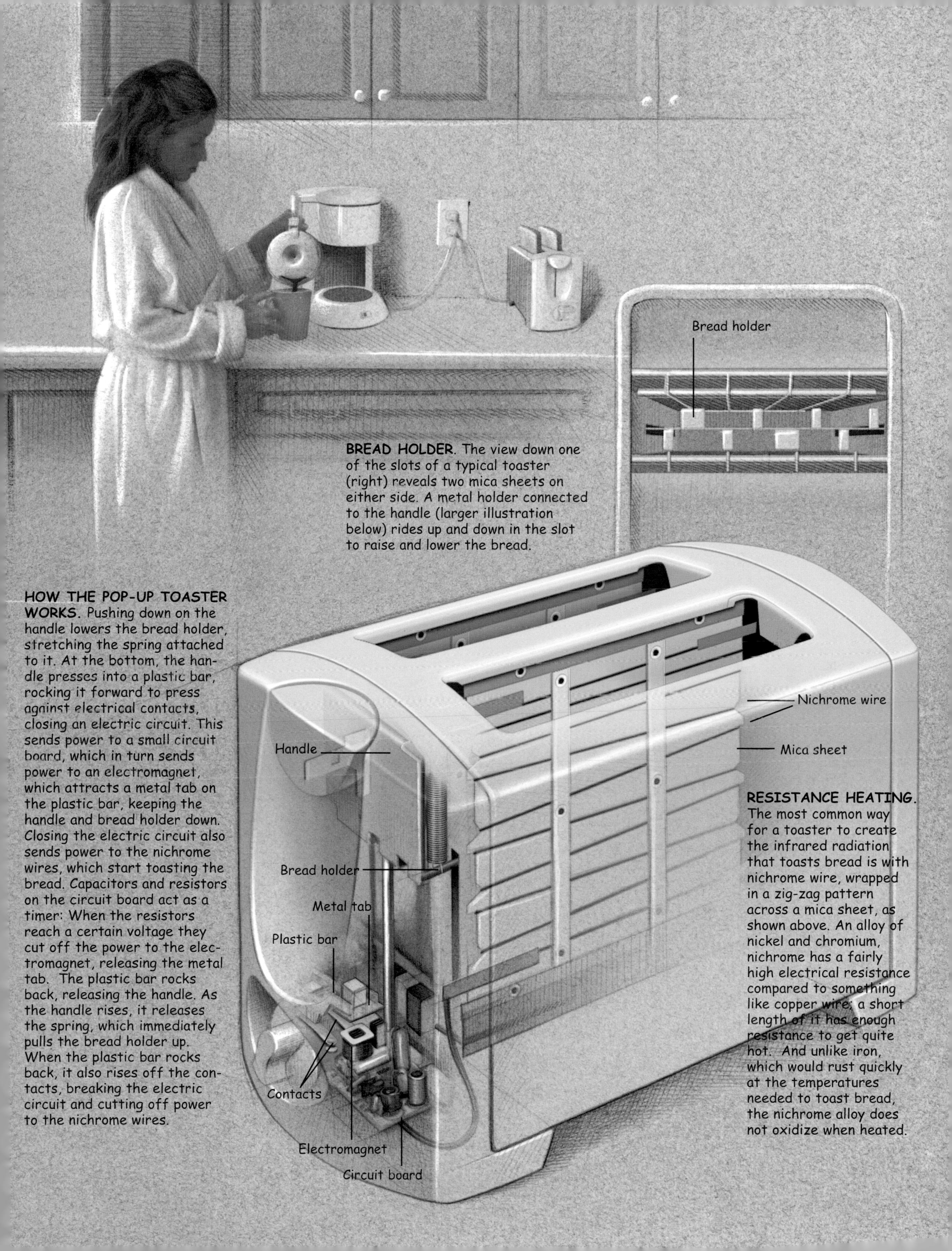

BREAD HOLDER. The view down one of the slots of a typical toaster (right) reveals two mica sheets on either side. A metal holder connected to the handle (larger illustration below) rides up and down in the slot to raise and lower the bread.

HOW THE POP-UP TOASTER WORKS. Pushing down on the handle lowers the bread holder, stretching the spring attached to it. At the bottom, the handle presses into a plastic bar, rocking it forward to press against electrical contacts, closing an electric circuit. This sends power to a small circuit board, which in turn sends power to an electromagnet, which attracts a metal tab on the plastic bar, keeping the handle and bread holder down. Closing the electric circuit also sends power to the nichrome wires, which start toasting the bread. Capacitors and resistors on the circuit board act as a timer: When the resistors reach a certain voltage they cut off the power to the electromagnet, releasing the metal tab. The plastic bar rocks back, releasing the handle. As the handle rises, it releases the spring, which immediately pulls the bread holder up. When the plastic bar rocks back, it also rises off the contacts, breaking the electric circuit and cutting off power to the nichrome wires.

RESISTANCE HEATING. The most common way for a toaster to create the infrared radiation that toasts bread is with nichrome wire, wrapped in a zig-zag pattern across a mica sheet, as shown above. An alloy of nickel and chromium, nichrome has a fairly high electrical resistance compared to something like copper wire; a short length of it has enough resistance to get quite hot. And unlike iron, which would rust quickly at the temperatures needed to toast bread, the nichrome alloy does not oxidize when heated.

Percy Spencer of Raytheon Corporation, shown above with early microwave equipment, discovered that the high-frequency radio waves could cook food from the inside out. Today we push buttons on microwave ovens (right) for cooking, reheating, and thawing food in a matter of minutes or seconds. Push-button housecleaning is still a homemaker's fantasy, but devices such as the robotic vacuum cleaner (below right), which navigates around the room with internal sensors, have begun to enter the marketplace.

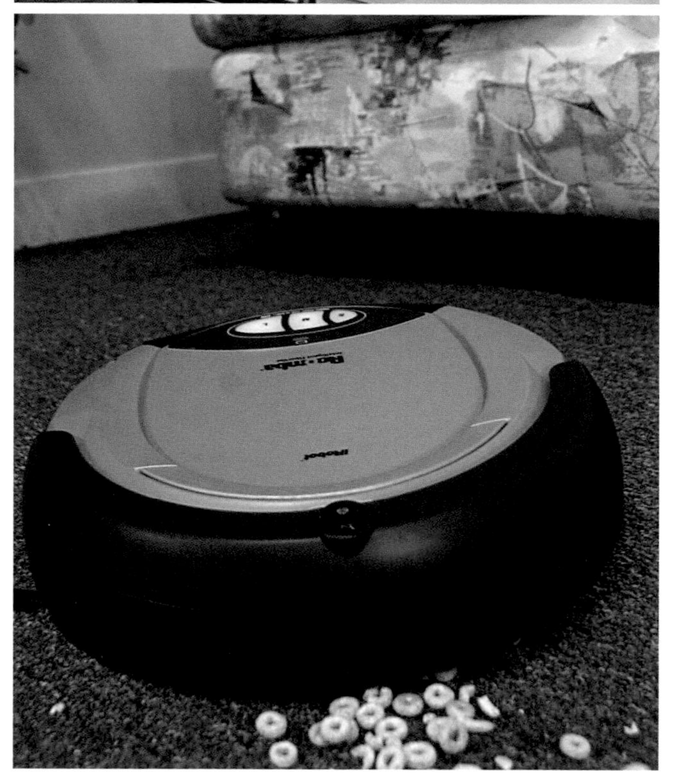

dryer arrived in 1949, and it was soon followed by models that were equipped with sensors that allowed various temperature settings for different fabrics, that measured the moisture in the clothes, and that signaled when the drying job was done.

Like the vacuum cleaner and washing machine, most modern appliances have a long lineage. One, however, seemed to appear out of the blue, serendipitously spawned by the development of radar during World War II. Much of that work focused on a top-secret British innovation called a cavity magnetron, an electronic device that could produce powerful, high-frequency radio waves—microwaves. In 1945 a radar scientist at Raytheon Corporation, Percy Spencer, felt his hand becoming warm as he stood in front of a magnetron, and he also noted that a candy bar in his pocket had softened. He put popcorn kernels close to the device and watched with satisfaction as they popped vigorously. Microwaves, it turned out, are absorbed by water, fats, and sugars, producing heat and rapidly cooking food from the inside. From Spencer's discovery came the microwave oven, first manufactured for commercial use in 1947 and ultimately a fixture in millions of kitchens, although the household versions were not produced until the mid-1960s.

The magic of electronics has now touched virtually every household appliance. Washing machines, dryers, and dishwashers offer a variety of cycles for different loads. Bread machines and coffeemakers complete their work at a time programmed in advance. Some microwave ovens hold scores of recipes in their electronic memory and can download more from the Internet. Robotic vacuum cleaners have made their debut. Where appliance technology will go from here is no more predictable than how habits of housework will be altered by it, but a century's worth of progress suggests that an eventful road lies ahead.

Perspective

Roland W. Schmitt

President Emeritus, Rensselaer Polytechnic Institute, and Retired Senior Vice President, General Electric Company

Before joining General Electric, I'd never really thought of household appliances as "high tech." The functions they perform—heating, cooling, cleaning, blowing, mixing—are as old as civilization itself. In my youth our kitchen had an icebox that was periodically supplied with 25-pound blocks of ice. We also had a wood stove in the kitchen, a fireplace in the living room, and space heaters scattered about elsewhere. My grandmother, who lived next door, had an electric-powered washing machine that sloshed clothes in a rotating drum. (She would always scrub the clothes on a washing board before putting them in the washer, not totally trusting this newfangled machine!). And she mixed the ingredients of the cakes she baked with a sturdy wooden spoon in a bowl.

Today, our kitchen has an electric range replete with electronic controls, a microwave oven similarly endowed, a toaster oven, several mixers, a dishwasher that's sometimes smarter than I am, a refrigerator-freezer, and a sturdy disposal. Our utility room has a brainy clothes washer, a smart dryer, a freezer, and a vacuum cleaner. Another refrigerator-freezer resides in the basement along with the equipment for central heating, dehumidifying, and air conditioning. But all of these fancy pieces of equipment still only heat, cool, clean, blow, and mix!

We take high tech for granted in household appliances and hardly notice it while seeing it prominently in our "electronic" appliances: televisions; audio equipment; mobile telephones; VCR, CD, and DVD recorders and players; digital cameras; pocket organizers; GPS devices; and, of course, in our Internet-connected computers. These items do things that our ancestors couldn't even dream of. So our household appliances live as sturdy, functional "wall flowers" among the active, glamorous, dancing electronic crowd.

The high tech of household appliances is a lot more than "under-the-hood" electronics. New and improved materials enable designs of convenience and efficiency. High performance plastics, especially, allow us to build style as well as functionality into our appliances. Household appliance engineers have just as rich an array of "high technologies" to feed their inventive minds as any other engineers. Innovation continues: cooking food to perfection up to eight times faster than with conventional ovens, using light. Washers and dryers that "talk" to each other, letting the dryer know what's coming, improving clothes care and saving time. The opportunity for innovation is as great as ever in this world of classical functions.

There is yet another dimension of high tech in household appliances: the way we make them. I've often thought it curious that Wall Street distinguishes between high-tech industries and manufacturing industries. Walk through any plant that makes household appliances and you're likely to see robots, lasers, intelligent conveyors, electronically controlled machine tools, computer-driven assembly stations, and smart test equipment. And, behind the scenes will be software that keeps track of everything, from incoming orders, in-process and final inventory, custom orders, shipments, and supply chain status. And when these products leave the factory into the hands of marketing and sales, they increasingly will be tracked and supervised by more and more sophisticated systems controlled by software with an array of acronyms that would make the U.S. Department of Defense envious: CRM, PLM, PDM, CIS, BPM, etc. (Customer Relationship Management, Product Lifecycle Management, Product Data Management, Customer Information System, Business Process Management. New categories and acronyms pop up faster than computer systems crash!)

The incorporation of high-tech advances into the realm of classic functions makes household appliances one of the great achievements of modern engineering. For the engineer there is something especially alluring about doing something that is functionally very, very old with ideas that are the newest of high tech.

The technologies that created the 20th century's laborsaving household devices owe a huge debt to electrification, which brought light and power into the home. Then two major engineering innovations—resistance heating and small, efficient motors—led to electric stoves and irons, vacuum cleaners, washers, dryers, and dishwashers. In the second half of the century advances in electronics yielded appliances that could be set on timers and even programmed, further reducing the domestic workload by allowing washing and cooking to go on without the presence of the human launderer or cook.

1901 British civil engineer H. Cecil Booth patents a vacuum cleaner powered by an engine and mounted on a horse-drawn cart. Teams of operators would reel the hoses into buildings to be cleaned.

1903 Earl Richardson of Ontario, California, introduces the lightweight electric iron. After complaints from customers that it overheated in the center, Richardson makes an iron with more heat in the point, useful for pressing around buttonholes and ruffles. Soon his customers are clamoring for the "iron with the hot point"—and in 1905 Richardson's trademark iron is born.

1905 Engineer Albert Marsh patents the nickel and chromium alloy nichrome, used to make electric filaments that can heat up quickly without burning out. The advent of nichrome paves the way, 4 years later, for the first electric toaster.

1907 James Spangler, a janitor at an Ohio department store who suffers from asthma, invents his "electric suction-sweeper," the first practical domestic vacuum cleaner. It employs an electric fan to generate suction, rotating brushes to loosen dirt, a pillowcase for a filter, and a broomstick for a handle. Unsuccessful with his heavy, clumsy invention, Spangler sells the rights the following year to a relative, William Hoover, whose redesign of the appliance coincides with the development of the small, high-speed universal motor, in which the same current (either AC or DC) passes through the appliance's rotor and stator. This gives the vacuum cleaner more horsepower, higher airflow and suction, better engine cooling, and more portability than was possible with the larger, heavier induction motor. And the rest, as they say, is history *(above)*.

1909 Frank Shailor of General Electric files a patent application for the D-12, the first commercially successful electric toaster (*right*). The D-12 has a single heating element and no exterior casing. It has no working parts, no controls, and no sensors; a slice of bread must be turned by hand to toast on both sides.

1913 Fred W. Wolf of Fort Wayne, Indiana, invents the first refrigerator for home use, a small unit mounted on top of an old-fashioned icebox and requiring external plumbing connections. Only in 1925 would a hermetically sealed standalone home refrigerator of the modern type, based on pre-1900 work by Marcel Audiffren of France and by self-trained machinist Christian Steenstrup of Schenectady, New York, be commercially introduced. This and other early models use toxic gases such as methyl chloride and sulfur dioxide as refrigerants. On units not hermetically sealed, leaks—and resulting explosions and poisonings—are not uncommon, but the gas danger ends in 1929 with the advent of Freon-operated compressor refrigerators for home kitchens.

The Walker brothers of Philadelphia produce the first electric dishwasher to go on the market, with full-scale commercialization by Hotpoint and others in 1930.

1915 Charles C. Abbot of General Electric develops an electrically insulating, heat conducting ceramic "Calrod" that is still used in many electrical household appliances as well as in industry.

1919 Charles Strite's first automatic pop-up toaster uses a clockwork mechanism to time the toasting process, shut off the heating element when the bread is done, and release the slice with a pop-up spring. The invention finally reaches the

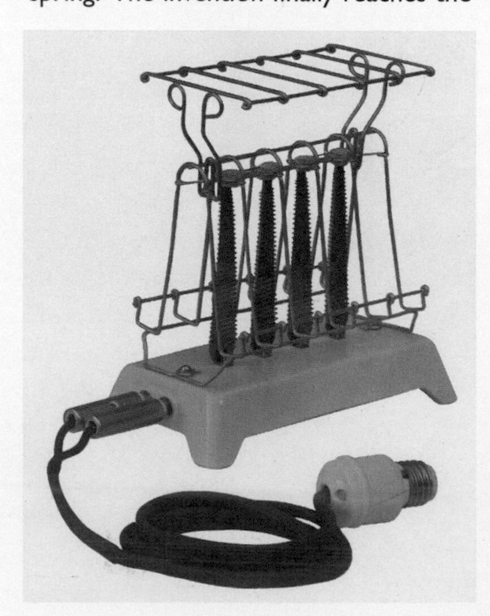

marketplace in 1926 under the name Toastmaster.

1927 The Silex Company introduces the first iron with an adjustable temperature control. The thermostat, devised by Joseph Myers, is made of pure silver.

John W. Hammes, a Racine, Wisconsin, architect, develops the first garbage disposal in his basement because he wants to make kitchen cleanup work easier for his wife. Nicknamed the "electric pig" when first introduced by the Emerson Electric Company, the appliance operates on the principle of centrifugal force to pulverize food waste against a stationary grind ring so it would easily flush down the drain.

Mid-1930s John W. Chamberlain of Bendix Corporation invents a device that enables a washing machine to wash, rinse, and extract water from clothes in a single operation (*above*). This eliminates the need for cumbersome and often dangerous powered wringer rolls atop the machine.

1935 To spare his mother having to hang wet laundry outside in the brutal North Dakota winter, J. Ross Moore builds an oil-heated drum in a shed next to his house, thereby creating the first clothes dryer. Moore's first patented dryers run on either gas or electricity, but he is forced to sell the design to the Hamilton Manufacturing Company the following year because of financial difficulties.

1945 Raytheon Corporation engineer Percy L. Spencer's realization that the vacuum tube, or magnetron, he is testing can melt candy, pop corn, and cook an egg leads to the first microwave oven. Raytheon's first model, in 1947, stands 5.5 feet tall, weighs more than 750 pounds, and sells for $5,000. It is quickly superseded by the equally gigantic but slightly less expensive Radarange; easily affordable countertop models are not marketed until 1967.

1947 The Nineteen Hundred Corporation introduces the first top-loading automatic washer, which Sears markets under the Kenmore label. Billed as a "suds saver," the round appliance sells for $239.95.

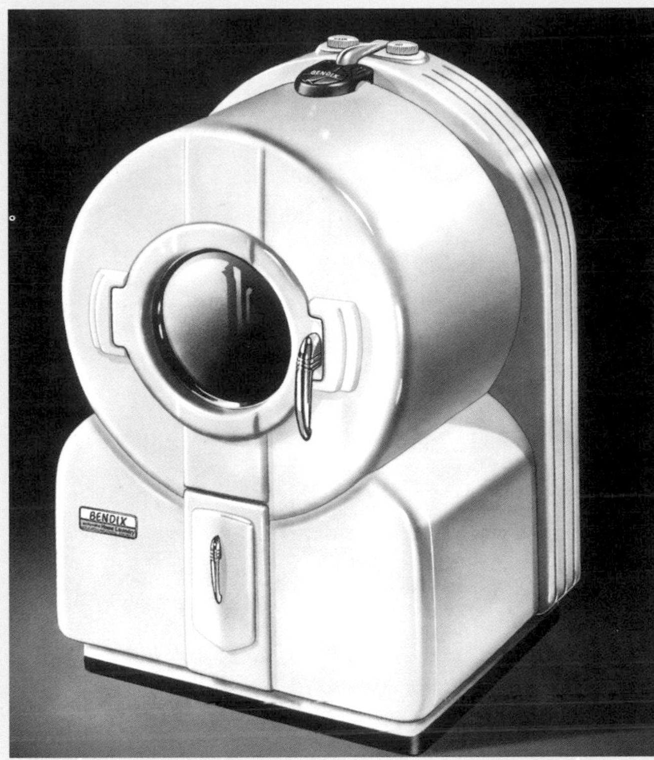

A Bendix washing machine in the mid-1940s was billed as the "automatic home laundry."

1952 Russell Hobbs invents the CP1, the first automatic coffeepot as well as the first of what would become a successful line of appliances. The percolator regulates the strength of the coffee according to taste and has a green warning light and bimetallic strip that automatically cuts out when the coffee is perked.

1962 Sunbeam ushers in a new era in iron technology by adding "spray mist" to the steam and dry functions of its S-5A model. The S-5A is itself an upgrade of the popular S-4 steam or dry iron that debuted in 1954.

1963 General Electric introduces the self-cleaning electric oven and in 1967 the first electronic oven control—beginning the revolution that would see microprocessors incorporated into household appliances of all sorts.

1972 Sunbeam develops the Mr. Coffee (*right*), the first percolator with an automatic drip process as well as an automatic cut-off control that lessens the danger of overbrewing. Mr. Coffee quickly becomes the country's leading coffeemaker.

1978 Singer introduces the Athena 2000, the world's first electronic sewing machine. A wide variety of stitches, from basic straight to complicated decorative, are available at the touch of a button. The "brain" of the system is a chip that measures less than one-quarter of an inch and contains more than 8,000 transistors.

1990s Environmentally friendly washers and dryers that save water and conserve energy are introduced. They include the horizontal-axis washer, which tumbles rather than agitates the clothes and uses a smaller amount of water, and a dryer with sensors, rather than a timer, that shuts the appliance off when the clothes are dry.

1997 Swedish appliance company Electrolux presents the first prototype of a robotic vacuum cleaner. The device, billed as "the world's first true domestic robot," sends and receives high-frequency ultrasound to negotiate its way around a room, much as bats do. In the production model, launched in Sweden a few years later, eight microphones receive and measure the returning signals to give the vacuum an accurate picture of the room. It calculates the size of a room by following around the walls for 90 seconds to 15 minutes, after which it begins a zigzag cleaning pattern and turns itself off when finished.

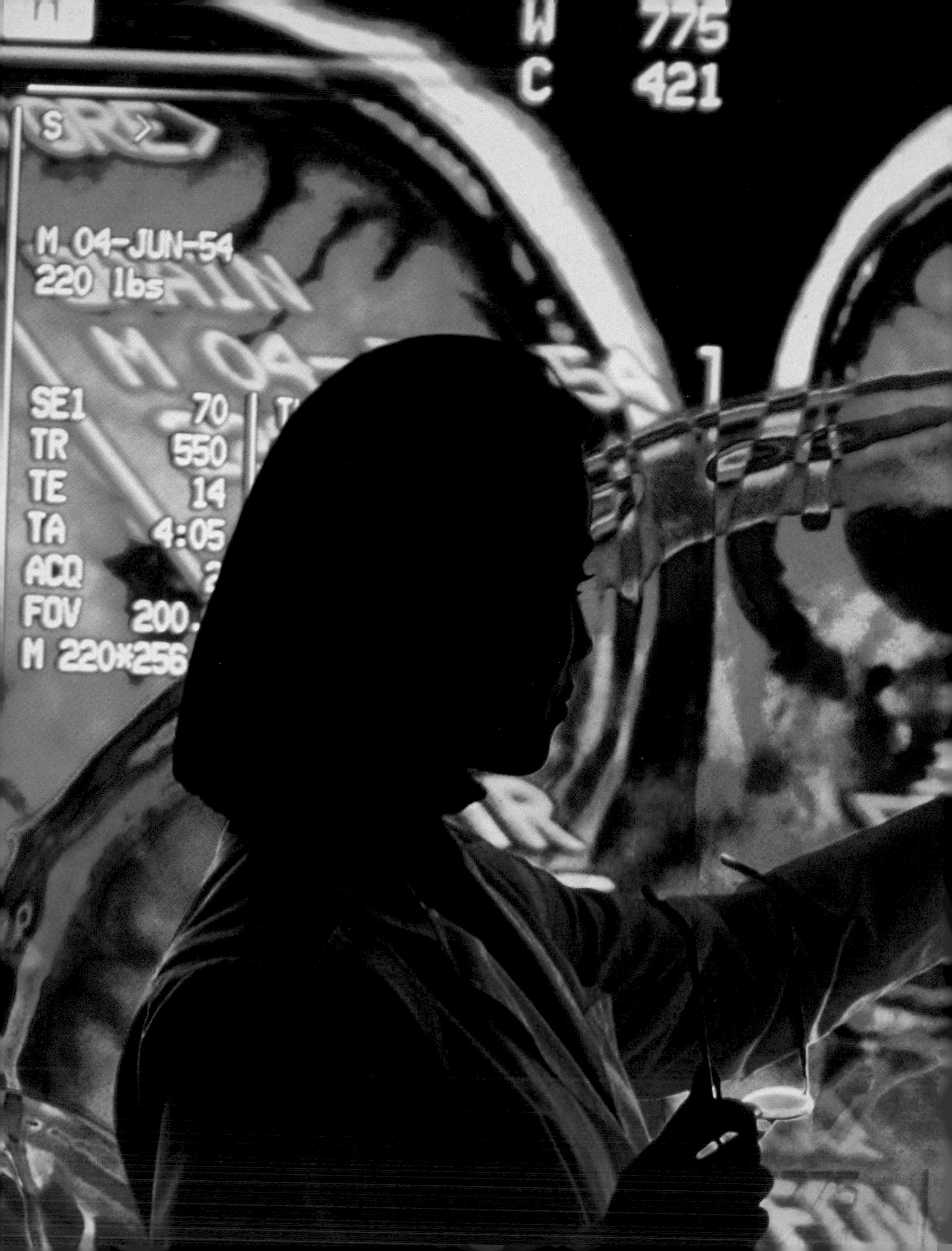

W 775
C 421
M 04-JUN-54
220 lbs
SE1 70
TR 550
TE 14
TA 4:05
FOV 200.
M 220*256

HEALTH TECHNOLOGIES

In 1900 the average life expectancy in the United States was 47 years. By 2000 it was nearing 77 years. That remarkable 30-year increase was the result of a number of factors, including the creation of a safe water supply. But no small part of the credit should go to the century's wide assortment of medical advances in diagnosis, pharmaceuticals, medical devices, and other forms of treatment.

Many of these improvements involved the combined application of engineering and biological principles to the traditional medical arts, giving physicians new perspectives on the body's workings and new solutions for its ills. From providing better diagnostic tools and surgical procedures to creating more effective replacements for the body's own tissues, engineering helped the 20th century's doctors successfully address such long-standing problems of human health as heart disease and infectious disease.

All through the century, improvements in imaging techniques wrought by the development of new systems—from x-ray machines to MRI (magnetic resonance imaging) scanners—enabled doctors to diagnose more accurately by providing a more exacting view of the body (see *Imaging*). One of the century's first such diagnostic devices created not a visual, but an electrical, image. In 1903, when Dutch physiologist Willem Einthoven developed the electrocardiograph, he paved the way for a more intensive scrutiny of the heart, spurring others to find better approaches and technologies for fixing its problems.

Working on the heart had long been considered too dangerous. In fact, in the last decade of the 19th century, famed Austrian surgeon Theodor Billroth declared: "Any surgeon who would attempt an operation of the heart should lose the respect of his colleagues." Even though doctors knew from electrocardiograph readings and other evidence that a heart might be malfunctioning or have anatomical defects, it was practically impossible to do anything about it while the heart was still beating. And stopping it seemed out of the question because blood had to circulate through the body continuously to keep tissues alive. In the first decades of the 20th century, surgeons performed some cardiac procedures on beating hearts, but with limited success.

Then in 1931, while caring for a patient with blood clots that were interfering with blood circulation to her lungs, a young surgeon named John Gibbon had a bold thought: What if oxygen-poor blood was pumped through an apparatus outside the body that would oxygenate it, and then was pumped back into the body? He began working on the problem, despite the skepticism of his fellow doctors. Teaming with his wife, laboratory technician Mary Hopkins, Gibbon fashioned a rudimentary heart-lung machine from a secondhand air pump, glass tubes, and a rotating drum that exposed blood to air and allowed it to pick up oxygen. Perfecting the device took more than two decades and countless experiments on animals. Then in 1953 Gibbon performed the first-ever successful procedure on a human using a heart-lung pump to maintain the patient's circulation while a hole in her heart was surgically closed. The era of open-heart surgery (so called because the chest cavity was opened up and the heart exposed) was born, and in the next decades surgeons would rely on what was simply called "the pump" to repair damaged hearts, replace defective heart valves with bioengineered substitutes, and perform thousands and thousands of life-extending coronary artery bypass operations to curb heart attacks.

The development of the pacemaker involved similar moments of insight and the nuts-and-bolts efforts of inspired individuals. For Wilson Greatbatch, an electronics wizard with an interest in medicine, the light flashed on in 1951 when he heard a discussion about a cardiac ailment called heart block, a flaw in the electrical signals regulating the basic heartbeat. "When they described it, I knew I could fix it," Greatbatch later recalled. Over the next few years he continued trying to create a device that could supply a regular signal for the heart. Then, while working on a device for recording heart sounds, he accidentally plugged the wrong resistor into a circuit, which began pulsing in a pattern he instantly recognized: the natural beat of a human heart.

Meanwhile other researchers had devised a pacemaker in 1952 that was about the size of a large radio; the patient had to be hooked up to an external power source. A few years later electrical engineer Earl Bakken devised a battery-powered handheld pacemaker that allowed patients in hospitals to move around. In 1958 Rune Elmqvist and Åke Senning devised the first pacemaker to be implanted in a human patient. Greatbatch's major contribution in the late 1950s was to incorporate recently available silicon transistors into an implantable pacemaker, the first of which was successfully tested in animals in 1958. By 1960 Greatbatch's pacemaker was working successfully in human hearts. He went on to improve the battery power source, ultimately devising a lithium battery that could last 10 years or more. Such pacemakers are now regulating the heartbeats of more than three million people worldwide.

Procedures such as heart bypass surgery (left)—made possible with the development of the heart-lung machine—are among the many advances in health technologies that account for the better health and longer life spans of today's populations.

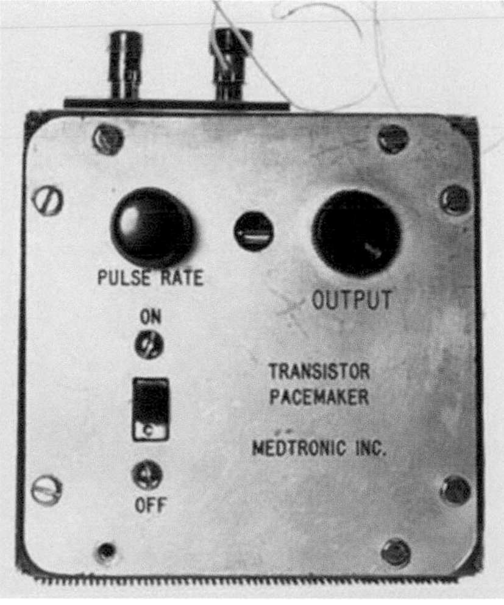

Among the medical contributions to heart health are (from left to right) the heart-lung machine, developed in 1953 by John H. Gibbon and Mary Hopkins Gibbon; the first wearable pacemaker, developed by Earl Bakken in 1957 and used on a patient in 1958; and the first pacemaker implanted in a human patient, developed in 1958 by engineer Rune Elmqvist and Åke Senning in Sweden and implanted in 43-year-old Arne Larssen, who lived 43 more years with a series of pacemakers.

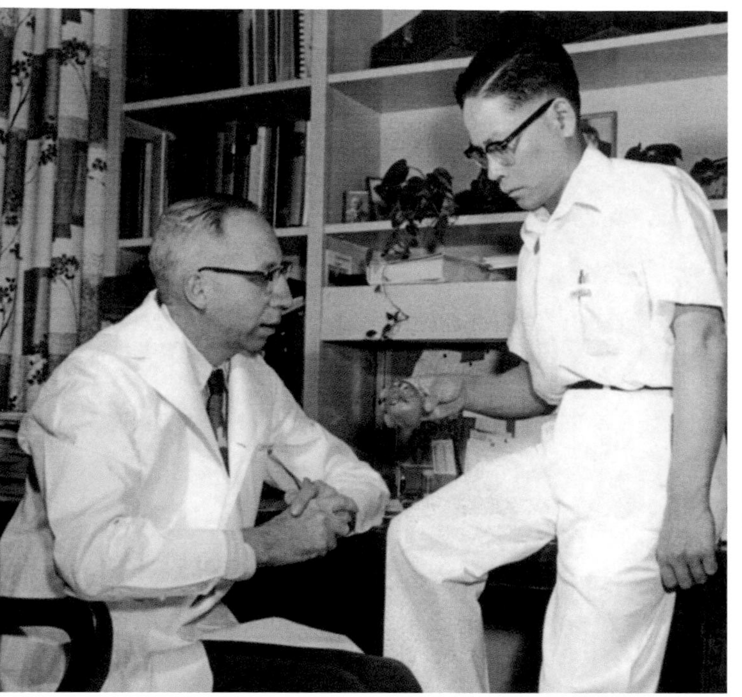
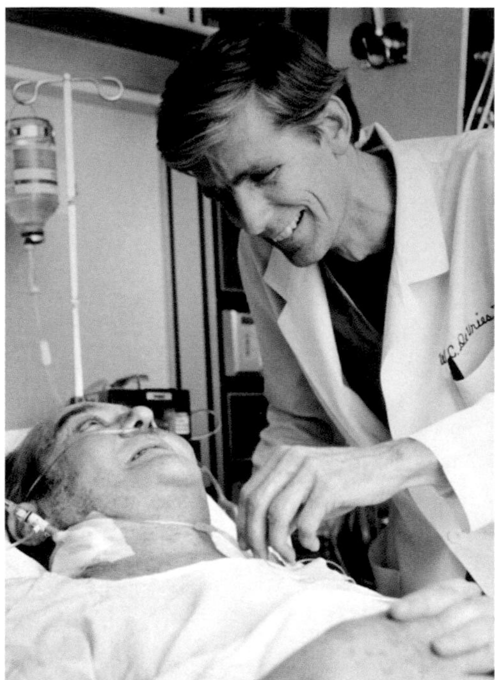
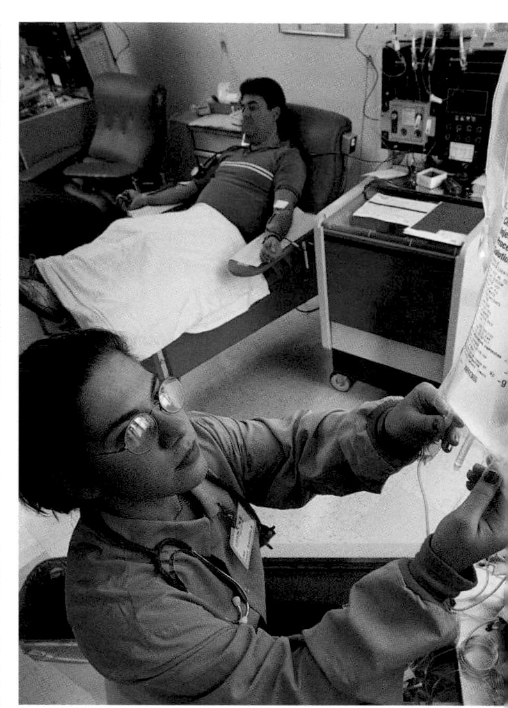

Dutch physician Willem Kolff (above left, seated) is often called the father of artificial organs for development of the artificial kidney and his work on the artificial heart. Shown with Kolff is Tetsuzo Akutsu, with whom Kolff did some of his earliest work on artificial hearts. Kolff later teamed with inventor Robert Jarvik and surgeon William DeVries, shown above (center) with Barney Clark, recipient of the first Jarvik artificial heart in 1982. The development of kidney dialysis machines (above right) has extended life for countless patients around the world.

Both the pump and the pacemaker are examples of a key application of engineering to medicine: bionic engineering, or the replacement of a natural function or body organ with an electronic or mechanical substitute. One of the foremost champions in this field was Dutch physician Willem Kolff, inventor of the kidney dialysis machine. Though severely hampered by the Nazi occupation of his country during World War II, Kolff was able to build a machine that substituted for the kidneys' role in cleansing the blood of waste products. Like Gibbon's heart-lung device, it consisted of a pump, tubing, and a rotating drum, which in this case pushed blood through a filtering layer of cellophane. Ironically, the first patient to benefit from his dialysis machine was a Nazi collaborator.

After the war Kolff moved to the United States, where he continued to work on bionic engineering problems. At the Cleveland Clinic he encouraged Tetsuzo Akutsu to design a prototype artificial heart. Together they created the first concept for a practical artificial heart. To others it seemed like an impossible challenge, but to Kolff the issue was simple: "If man can grow a heart, he can build one," he once declared. These first efforts, beginning in the late 1950s, did little more than eliminate fruitless lines of research. Later, as a professor of surgery and bioengineering at the University of Utah, Kolff formed a team that included physician-inventor Robert Jarvik and surgeon William DeVries. After 15 difficult years of invention and experimentation, DeVries implanted one of Jarvik's hearts—a silicone and rubber unit powered by compressed air from an external pump—in Barney Clark, who survived for 112 days. Negative press about Clark's condition during his final days slowed further progress for a while, but today more sophisticated versions of artificial hearts and ventricular-assist devices, including self-contained units that allow greater patient mobility, routinely serve as temporary substitutes while patients await heart transplants.

Kolff was not done. With his colleagues he helped improve the prosthetic arm—another major life-improving triumph of "spare parts" medicine—as well as contributing to the development of both an artificial eye and an artificial ear. Progress in all these efforts has depended on advancements in a number of engineering fields, including computers, electronics, and high performance materials. Computers and microelectronic components, for example, have made it possible for bioengineers to design and build prosthetic limbs that better replicate the mechanical actions of natural arms and legs. And first-generation biomaterials—polymers, metals, and acrylic fibers among others—have been used for almost everything from artificial heart valves and eye lenses to replacement hip, knee, elbow, and shoulder joints.

Engineering processes have had an even broader effect on the practice of medicine. The 20th century's string of victories over microbial diseases resulted from the discovery and creation of new drugs and vaccines, such as the polio vaccine and the whole array of antibiotics. Engineering approaches—including manufacturing techniques and systems design—played significant roles in both the development of these medications and their wide availability to the many people around the world who need them. For example, engineers are involved in designing processes for chemical synthesis of medicines and building such devices as bioreactors to "grow" vaccines. And assembly line know-how, another product of the engineering mind, is crucial to the mixing, shaping, packaging, and delivering of drugs in their myriad forms.

It may be in the operating room rather than the pharmaceutical factory, however, that engineering has had a more obvious impact. A number of systems have increased the surgeon's operating capacity, especially during the last half of the century. One of the first was the operating microscope, invented by the German company Zeiss in the early 1950s. By giving surgeons a magnified view, the operating microscope made it possible to perform all manner of intricate procedures, from delicate operations on the eye and the small bones of the inner ear to the reconnection of nerves and even the tiniest blood vessels—a skill that has enabled more effective skin grafting as well as the reattachment of severed limbs.

At about the same time as the invention of the operating microscope, a British researcher named Harold Hopkins helped perfect two devices that further revolutionized surgeons' work: the fiber-optic endoscope and the laparoscope. Both are hollow tubes containing a fiber-optic cable that allows doctors to see and work inside the body without opening it up. Endoscopes, which are flexible, can be fed into internal organs such as the stomach or intestines without an incision and are designed to look for growths and other anomalies. Laparoscopes are rigid and require a small incision, but because they are stiff, they enable the surgeon to remove or repair internal tissues by manipulating tiny blades, scissors, or other surgical tools attached to the end of the laparoscope or fed through it.

Further advances in such minimally invasive techniques began to blur the line between diagnosis and treatment. In the 1960s a radiologist named Charles Dotter erased that line altogether when he developed methods of using radiological catheters—narrow flexible tubes that can be seen with imaging devices—not just to gain views of blood vessels in and around the kidney but also to clear blocked arteries. Dotter was a tinkerer of the very best sort and was constantly inventing

Twentieth-century medical engineering enables today's surgeons to carry out a variety of procedures, including surgery on internal organs through minimal incisions with laser laparoscopes (below left) and replacing hip joints with prosthetics made of titanium (below right).

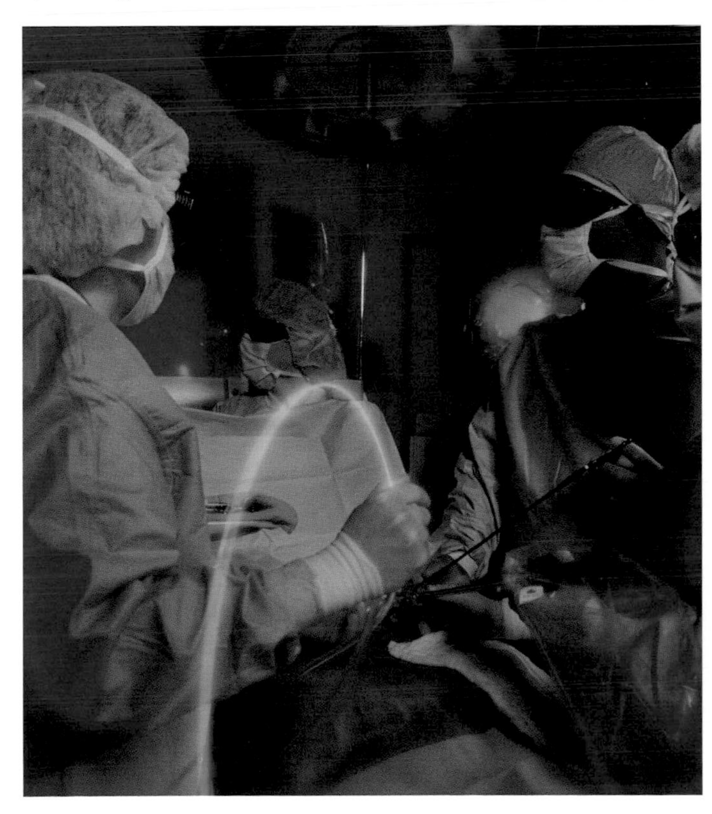

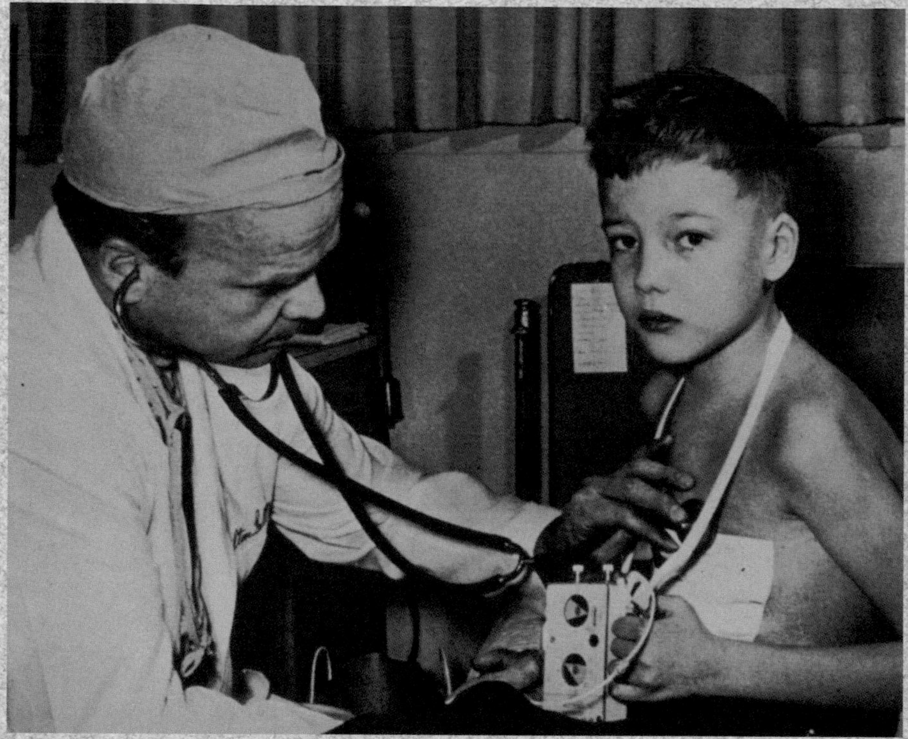

C. Walton Lillehei checks the heart rate of a young patient hooked up to one of the first portable external pacemakers, in the late 1950s. Lillehei, a pioneer in open-heart surgery at the University of Minnesota Medical School, had urged Earl Bakken to devise a pacemaker that ran on batteries after one of his young patients died during a power failure.

1966
Chardack-Greatbatch (5841)
Diameter: 70 mm
Thickness: 28mm
Weight: 180 grams

PACEMAKERS AND DEFIBRILLATORS
KEEPING THE HEART ON TIME

In 1992 10-year-old "Michael" was diagnosed with a condition in which some of the electrical impulses from the upper chambers of the heart (atria) do not reach the lower chambers (ventricles). As a result, his heart beat slowly or irregularly. In September doctors implanted a dual-chamber pacemaker and told the youngster he could play any non–contact sport he wanted. Michael tried out for baseball and eventually became one of the varsity squad's starting pitchers, as well as the second baseman. He also made the school's varsity swim team and was named most valuable player 2 years in a row, shattering four school records. By the time he headed off to college, he had a new, smaller device implanted and was looking to try out for the college baseball team.

Pacemakers and defibrillators perform different but related functions for people with heart rhythm problems. Pacemakers correct the pace of a heart that is beating too slowly with a steady pulse of electrical signals that stimulate either one or two chambers of the heart to contract in rhythm. Implanted defibrillators sense when the heart is beating too rapidly and correct the irregularity with a quick series of pulses. They can also transmit a small jolt of current if the heart goes into fibrillation, a life-threatening circumstance in which individual muscle fibers twitch rapidly in an uncoordinated fashion and can lead to sudden cardiac arrest.

Both types of devices have grown more reliable, more resourceful, and smaller, as shown by the series of pacemakers shown opposite. Most defibrillators include pacemaking functions, and rate-responsive pacemakers automatically adjust to the patient's level of activity. All contain a microprocessor, electrical leads to the targeted areas of the heart, a battery that can last up to 10 years, and in some cases monitoring and diagnostics features that enable doctors to gather data on the heart's functioning and make adjustments from a desktop computer.

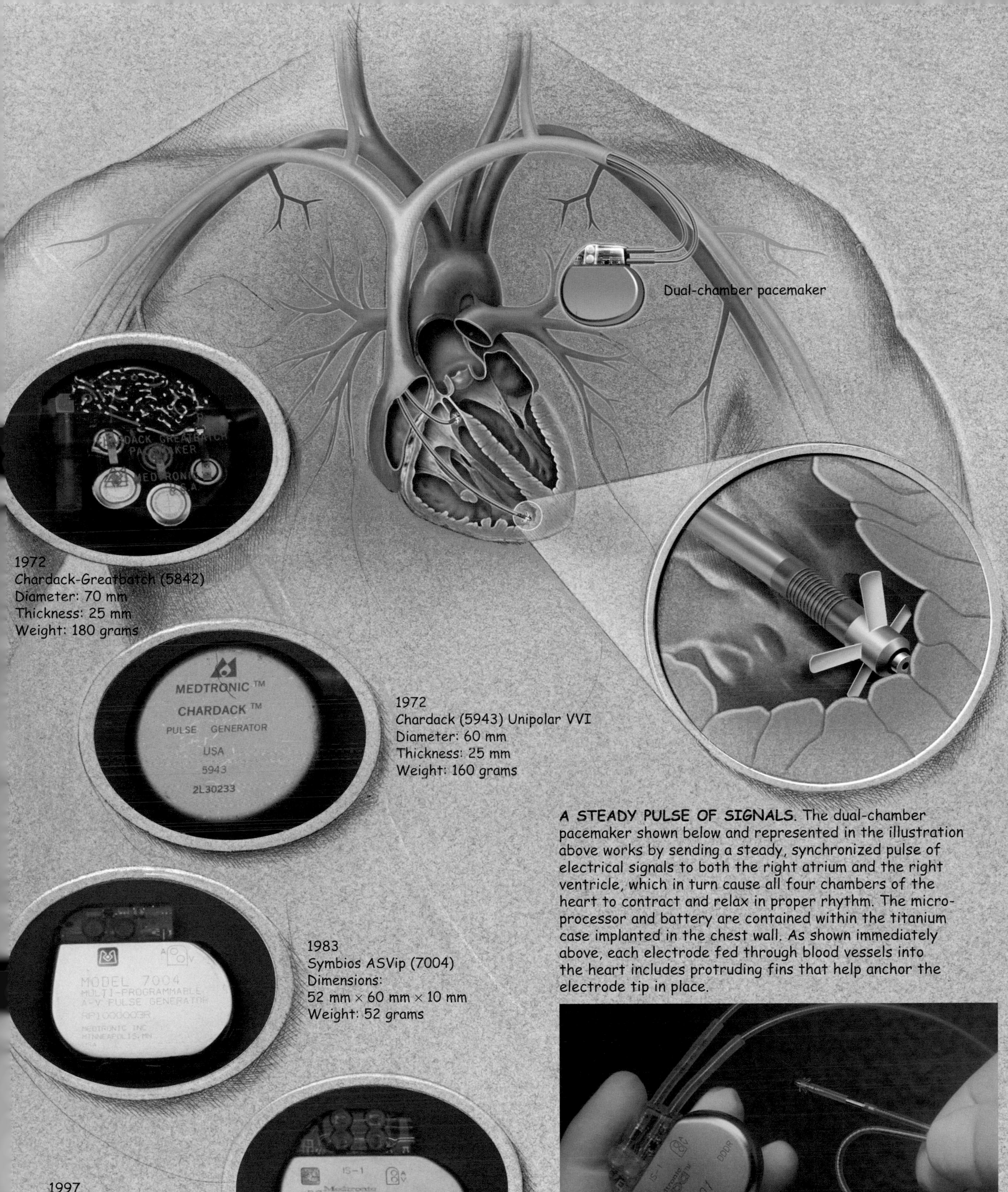

1972
Chardack-Greatbatch (5842)
Diameter: 70 mm
Thickness: 25 mm
Weight: 180 grams

1972
Chardack (5943) Unipolar VVI
Diameter: 60 mm
Thickness: 25 mm
Weight: 160 grams

1983
Symbios ASVip (7004)
Dimensions:
52 mm × 60 mm × 10 mm
Weight: 52 grams

1997
Medtronic Kappa 401 DR
(KDR401)
Dimensions:
46.1 mm × 52.2 mm × 7.6 mm
Weight: 34 grams

A STEADY PULSE OF SIGNALS. The dual-chamber pacemaker shown below and represented in the illustration above works by sending a steady, synchronized pulse of electrical signals to both the right atrium and the right ventricle, which in turn cause all four chambers of the heart to contract and relax in proper rhythm. The microprocessor and battery are contained within the titanium case implanted in the chest wall. As shown immediately above, each electrode fed through blood vessels into the heart includes protruding fins that help anchor the electrode tip in place.

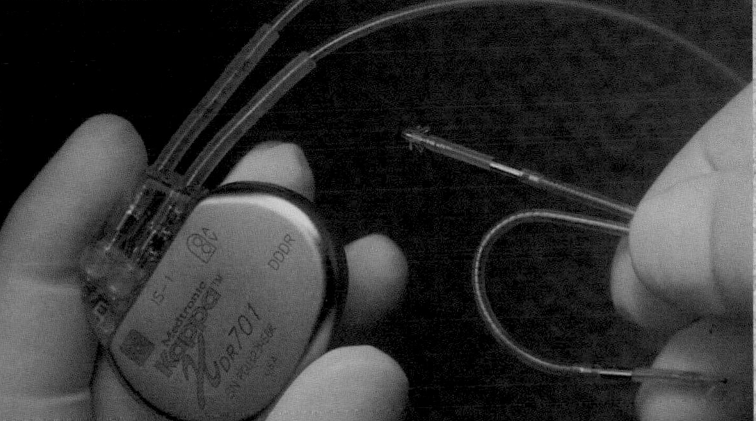

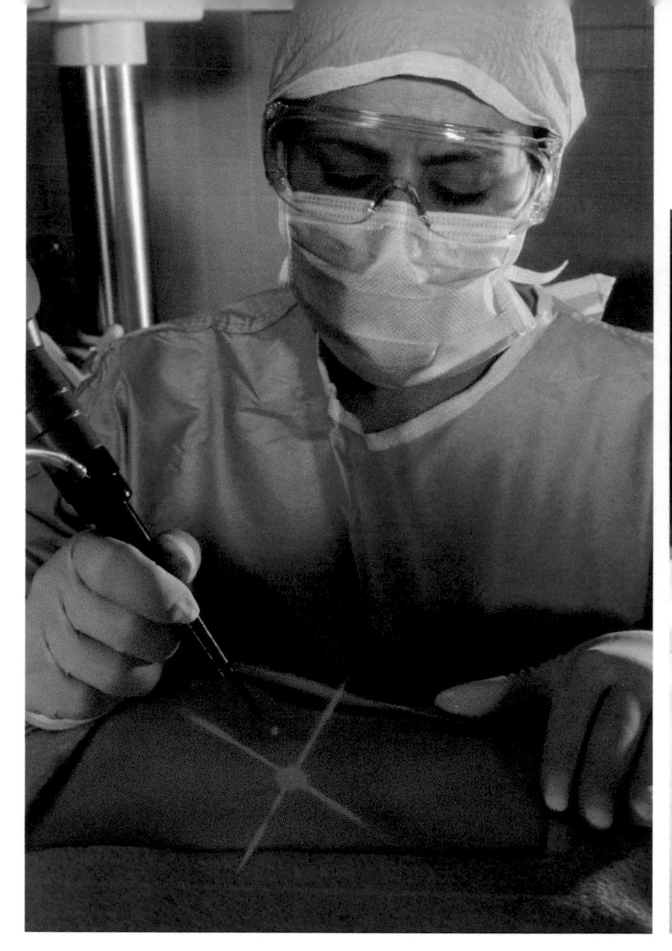

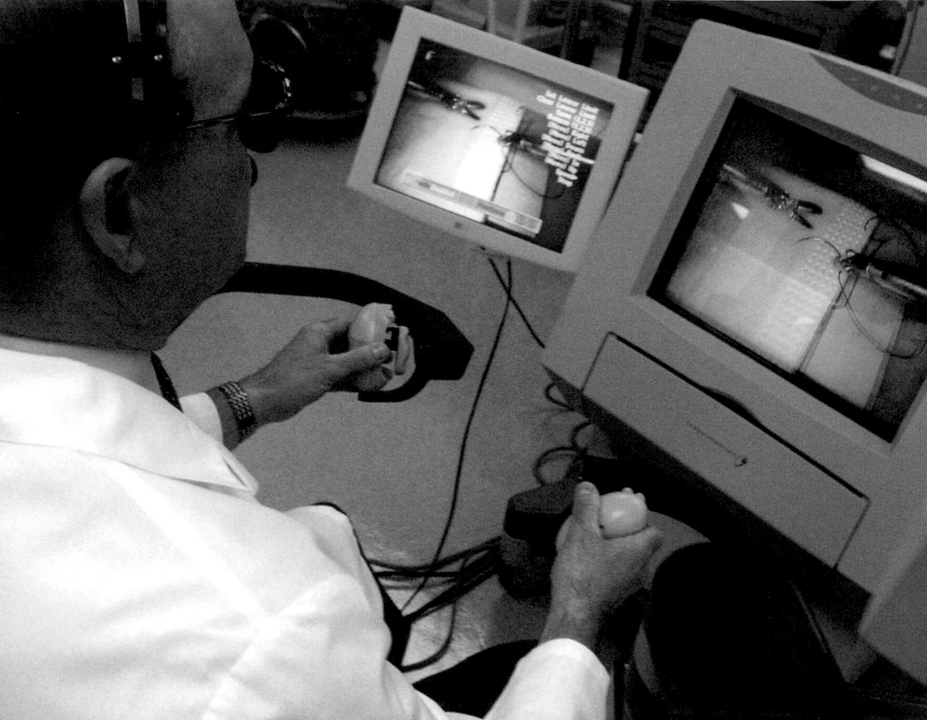

The medical uses of lasers are widespread and varied. They are one of the chief tools of eye surgery and frequently are used to burn away growths (above left). Surgery involving robots is also a burgeoning field, allowing surgeons to manipulate a remote pair of robotic "hands" in minimally invasive procedures (above right).

his own equipment, often adapting such unlikely materials as guitar strings, strips of vinyl insulation, and in one case an automobile speedometer cable to create more effective interventional tools.

Adaptation was nothing new in medicine, and physicians always seemed ready to find new uses for technology's latest offspring. Lasers are perhaps the best case in point. Not long after its invention, the laser was taken up by the medical profession and became one of the most effective surgical tools of the 20th century's last 3 decades. Lasers are now a mainstay of eye surgery and are also routinely employed to create incisions elsewhere in the body, to burn away growths, and to cauterize wounds. Set to a particular wavelength, lasers can destroy brain tumors without damaging surrounding tissue. They have even been used to target and destroy viruses in the blood.

As surgeons recognized the benefits of minimally invasive procedures, which dramatically reduce the risk of infection and widen the range of treatment techniques, they also became aware that they themselves were now a limiting factor. Even with the assistance of operating microscopes attached to laparoscopic tools, surgeons often couldn't move their hands precisely enough. Then in the 1990s researchers began to realize what had long seemed a futuristic dream—using computer-controlled robots to perform operations. Beginning in 1995 Seattle surgeon Frederic Moll, with the help of an electrical engineer named Robert Younge, developed one of the first robotic surgeon prototypes—a combination of sensors, actuators, and microprocessors that translated a surgeon's hand movements into more fine-tuned actions of robotic arms holding microinstruments. Since then other robotics-minded physicians and inventors have created machines that automate practically every step of such procedures as closed-chest heart surgery, with minimal human involvement.

The list of health care technologies that have benefited from engineering insights and accomplishments continues to grow. Indeed, at the end of the century bioengineering seemed poised to be fully integrated into biological and medical research. It seemed possible that advances in understanding the genetic underpinnings of life might ultimately lead to cures for huge numbers of diseases and inherited ills—either by reengineering the human body's own cells or genetically disabling invading organisms. Certainly engineering techniques—particularly computerized analyses—had already helped identify the complexities of the code. The next step, intervening by replacing or correcting or otherwise manipulating genes and their components, seemed in the offing. Although the promise has so far remained unfulfilled, engineering solutions will continue to play a vital role in many of medicine's next great achievements.

Perspective

Wilson Greatbatch

President
Greatbatch Enterprises, Inc.

One can trace the beginning of health care technologies to the first electrocardiograph—a string galvanometer for recording heart voltages—invented in 1903 by Willem Einthoven in the Netherlands. In the late 1920s the advent of vacuum tubes to amplify the electrocardiograph allowed patients to get their hands and feet out of the saline buckets that were necessary to get a good contact with the ECG machine.

Some 20 years later, from 1949 to 1951, I was employed by the Cornell Psychology Department at its animal behavior farm in Varna, New York. My job involved instrumenting 100 sheep and goats for heart function, blood pressure, EEG, and body motion. All this, of course, was still being done with vacuum tube equipment.

Then came the invention of the transistor by Bell Laboratories. Once the transistor was readily available, I knew I could make an implantable pacemaker. Like most breakthrough developments, the implantable pacemaker was the result of a series of efforts, starting with Albert Hyman, who built a hand-crank pacemaker in the 1930s. In the early 1950s Paul M. Zoll built an external pacemaker that plugged into a wall socket, and Earl Bakken built a wearable pacemaker in 1958. That same year Åke Senning in Stockholm implanted the first pacemaker in a human, but the device was not successful. (We define "success" as an implanted device that performs satisfactorily for at least one year.)

By 1958 I had blocked out a transistorized pacemaker circuit and began looking for a surgeon interested in working with me. I met my surgeon partners, William Chardack, chief of surgery, and Andrew Gage at the Buffalo Veterans Administration Hospital. I had $2,000 in cash and enough to feed my family for 2 years, so I quit my job and gave the family money to my wife. Between 1958 and 1960, I built 50 pacemakers in the barn behind my house. We put 40 of them into animals and 10 into patients. In 1960 Chardack, Gage, and I achieved the first successful implanted cardiac pacemaker.

One of my current projects involves eliminating the dangerous interactions between magnetic resonance imaging (MRI) and implantable pacemakers and defibrillators. The powerful electromagnetic fields generated by the MRI machine can travel down the wire leads, possibly damaging the pacemaker circuitry and perhaps scarring the heart. We have worked with a client (Biophan, Inc.) to design, build, and test a pacemaker that eliminates the wire lead. An electronic pulse generator drives a laser, which conducts a light pulse to the heart, where the light pulse is converted back into an electrical impulse to drive the heart. Work is also under way to replace the lithium/silver vanadium battery in the implantable cardiac defibrillator with a hybrid battery that will provide long service life and decrease the charge time on capacitors.

I see the future of bioengineering as largely driven by developments in microtechnology, nanotechnology, and light technology. Researchers are already building motors visible only through a microscope. We are investigating chemical and electrical sensors that operate entirely on the photonic level, with no electronic instrumentation other than that needed to drive a laser. These low-power requirements raise the possibility of building and operating equipment on power generated by the human body itself.

Health care technology and bioengineering are interdisciplinary subjects. They require ideas, people, and intelligence from all the physical sciences, all the natural sciences, all the medical sciences, and all the photonic sciences. This varied group of people, converging under the subject of bioengineering, will shape the future, leading us to the greatest engineering achievements of the 21st century.

HEALTH TECHNOLOGIES
TIMELINE

For most of the 19th century doctors cared for their patients much as they had in prior eras. Most medical tools fit easily into the doctor's little black bag, and diagnoses were based more on instinct than on science. But then in 1895 German physicist Wilhelm C. Roentgen accidentally discovered a form of electromagnetic radiation that could pass through the body and leave an image of its bones or organs on a photographic plate. The birth of the X ray sparked a revolution. Diagnostic tools such as the electrocardiograph, CAT scan, and MRI followed, as did the development of artificial and transplanted organs and joints and myriad other surgical devices and techniques designed to keep the body functioning. These advancements and the discovery of antibiotics and other life-saving drugs contributed to increasing the life span of people throughout the developed world—on average nearly 30 years longer than their ancestors a century ago.

1903 Dutch physician and physiologist Willem Einthoven develops the first electrocardiograph machine, a simple, thin, lightweight quartz "string" galvanometer, suspended in a magnetic field and capable of measuring small changes in electrical potential as the heart contracts and relaxes. After attaching electrodes to both arms and the left leg of his patient, Einthoven is able to record the heart's wave patterns as the string deflects, obstructing a beam of light whose shadow is then recorded on a photographic plate or paper. In 1924 Einthoven is awarded the Nobel Prize in medicine for his discovery.

1927 Harvard medical researcher Philip Drinker, assisted by Louis Agassiz Shaw, devises the first modern practical respirator using an iron box and two vacuum cleaners. Dubbed the iron lung, his finished product—nearly the length of a small car—encloses the entire bodies of its first users, polio sufferers with chest paralysis. Pumps raise and lower the pressure within the respirator's chamber, exerting a pull-push motion on the patients' chests. Only their heads protrude from the huge cylindrical steel drum.

1930s Albert S. Hyman, a practitioner cardiologist in New York City, invents an artificial pacemaker to resuscitate patients whose hearts have stopped. Working with his brother Charles, he constructs a hand-cranked apparatus with a spring motor that turns a magnet to supply an electrical impulse. Hyman tests his device on several small laboratory animals, one large dog, and at least one human patient before receiving a patent, but his invention never receives acceptance from the medical community.

1933 Working on rats and dogs at Johns Hopkins University, William B. Kouwenhoven and neurologist Orthello Langworthy discover that while a low-voltage shock can cause ventricular fibrillation, or arrhythmia, a second surge of electricity, or countershock, can restore the heart's normal rhythm and contraction. Kouwenhoven's research in electric shock and his study of the effects of electricity on the heart lead to the development of the closed-chest electric defibrillator and the technique of external cardiac massage today known as cardiopulmonary resuscitation, or CPR.

1945 Willem J. Kolff successfully treats a dying patient in his native Holland with an "artificial kidney," the first kidney dialysis machine. Kolff's creation is made of wooden drums, cellophane tubing, and laundry tubs and is able to draw the woman's blood, clean it of impurities, and pump it back into her body. Kolff's invention is the product of many years' work, and this patient is his first long-term success after 15 failures. In the course of his work with the artificial kidney, Kolff notices that blue, oxygen-poor blood passing through the artificial kidney becomes red, or oxygen-rich, leading to later work on the membrane oxygenator.

1948 Kevin Touhy receives a patent for a plastic contact lens designed to cover only the eye's cornea, a major change from earlier designs. Two years later George Butterfield introduces a lens that is molded to fit the cornea's contours rather than lie flat atop it. As the industry evolves, the diameter of contact lenses gradually shrinks.

1951 Charles Hufnagel, a professor of experimental surgery at Georgetown University, develops an artificial heart valve and performs the first artificial valve

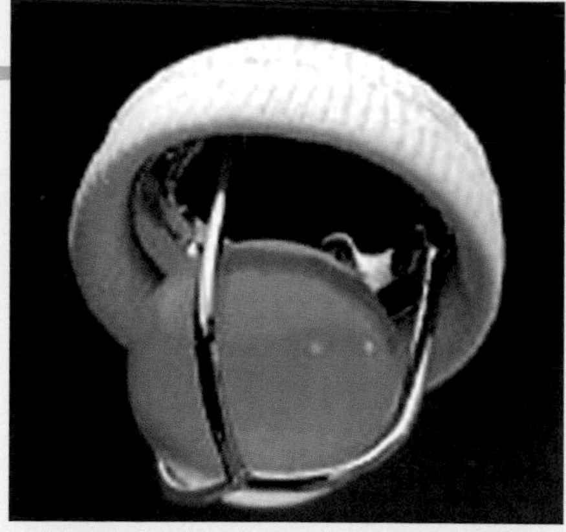

implantation surgery in a human patient the following year. The valve—a methacrylate ball in a methacrylate aortic-size tube—does not replace the leaky valve but acts as an auxiliary. The first replacement valve surgeries are performed in 1960 by two surgeons who develop their ball-in-cage designs independently. In Boston, Dwight Harken develops a double-cage design in which the outer cage separates the valve struts from the aortic wall. At the University of Oregon, Albert Starr, working with electrical engineer Lowell Edwards, designs a silicone ball inside a cage made of stellite-21, an alloy of cobalt, molybdenum, chromium, and nickel. The Starr-Edwards heart valve *(above)* is born and is still in use today.

1952 Paul M. Zoll of Boston's Beth Israel Hospital, in conjunction with the Electrodyne Company, develops the first successful cardiac pacemaker. The bulky device, worn externally on the patient's belt, plugs into an electric wall socket and stimulates the heart through two metal electrodes placed on the patient's bare chest. Five years later doctors begin implanting electrodes into chests. Around the same time a battery-powered external machine is developed by Earl Bakken and C. Walton Lillehei.

1953 Philadelphia physician John H. Gibbon performs the first successful open-heart bypass surgery on 18-year-old Cecelia Bavolek, whose heart and lung functions are supported by a heart-lung machine developed by Gibbon. The device is the culmination of two decades of research and experimentation and heralds a new era in surgery and medicine. Today coronary bypass surgery is one of the most common operations performed.

1954 A team of doctors at Boston's Peter Bent Brigham Hospital successfully performs the first human kidney transplant. Led by Joseph E. Murray, the physicians remove a healthy kidney from the donor, Ronald Herrick, and implant it in his identical twin brother, Richard, who is dying of renal disease. Since the donor and recipient are perfectly matched, the operation proves that in the absence of the body's rejection response, which is stimulated by foreign tissue, human organ transplants can succeed.

Late 1950s English surgeon John Charnley applies engineering principles to orthopedics and develops the first artificial hip replacement procedure, or arthroplasty. In 1962 he devises a low-friction, high-density polythene suitable for artificial hip joints and pioneers the use of methyl methacrylate cement for holding the metal prosthesis, or implant, to the shaft of the femur. Charnley's principles are subsequently adopted for other joint replacements, including the knee and shoulder.

1960 Buffalo, New York, electrical engineer Wilson Greatbatch develops the first totally internal pacemaker using two commercial silicon transistors. Surgeon William Chardack implants the device into 10 fatally ill patients. The first lives for 18 months, another for 30 years.

1963 Francis L'Esperance, of the Columbia-Presbyterian Medical Center, begins working with a ruby laser photocoagulator to treat diabetic retinopathy, a complication of diabetes and a leading cause of blindness in the United States. In 1965 he begins working with Bell researchers Eugene Gordon and Edward Labuda to design an argon laser for eye surgery. (They learn that the blue-green light of the argon laser is more readily absorbed by blood vessels than the red light of the ruby laser.) In early 1968, after further refinements and careful experiments, L'Esperance begins using the argon-ion laser to treat patients with diabetic retinopathy.

1971 Bausch & Lomb licenses Softlens, the first soft contact lens. The new product is the result of years of research by Czech scientists Otto Wichterle and Drahoslav Lim and is based on their earlier invention of a "hydrophilic" gel, a polymer material that is compatible with living tissue and therefore suitable for eye implants. Soft contacts allow more oxygen to reach the eye's cornea than do hard plastic lenses.

1972 Computerized axial tomography, popularly known as CAT or CT scan, is introduced as the most important development in medical filming since the X ray some 75 years earlier. (See *Imaging*)

1978 Graeme Clarke in Australia carries out the first cochlear implant surgery. Advances in integrated circuit technology enable him to design a multiple electrode receiver-stimulator unit about the size of a quarter.

Late 1970s Advances in fiber-optics technology give surgeons a view into joints and other surgical sites through an arthroscope, an instrument the diameter of a pencil, containing a small lens and light system, with a video camera at the outer end. Used initially as a diagnostic tool prior to open surgery, arthroscopic surgery, with its minimal incisions and generally shorter recovery time, is soon widely used to treat a variety of joint problems.

1980s Robert Langer, professor of chemical and biochemical engineering at MIT, develops the foundation of today's controlled drug delivery technology. Using pellets of degradable and nondegradable polymers such as polyglycolic acid, he fashions a porous structure that allows the slow diffusion of large molecules. Such structures are turned into a dime-size chemotherapy wafer to treat brain cancer after surgery. Placed at the site where a tumor has been removed, the wafer slowly releases powerful drugs to kill any remaining cancer cells. By confining the drug to the tumor site, the wafer minimizes toxic effects on other organs.

1981 The first commercial MRI (magnetic resonance imaging) scanner arrives on the medical market. (See *Imaging*, page 160.)

1982 Seattle dentist Barney Clark receives the first permanent artificial heart, a silicone and rubber device designed by many collaborators, including Robert Jarvik, Don Olsen, and Willem Kolff. William DeVries of the University of Utah heads the surgical transplant team. Clark survives for 112 days with his pneumatically driven heart.

1985 The Food and Drug Administration approves Michel Mirowski's implantable cardioverter defibrillator (ICD), an electronic device to monitor and correct abnormal heart rhythms, and specifies that patients must have survived two cardiac arrests to qualify for ICD implantation. Inspired by the death from ventricular fibrillation of his friend and mentor Harry Heller, Mirowski has conceived and developed his invention almost single-handedly. It weighs 9 ounces and is roughly the size of a deck of cards.

1987 France's Alim-Louis Benabid, chief of neurosurgery at the University of Grenoble, implants a deep-brain electrical stimulation system into a patient with advanced Parkinson's disease. The experimental treatment is also used for dystonia, a debilitating disorder that causes involuntary and painful muscle contractions and spasms, and is given when oral medications fail.

New York City ophthalmologist Steven Trokel performs the first laser surgery on a human cornea, after perfecting his technique on a cow's eye. Nine years later the first computerized excimer laser—Lasik—designed to correct the refractive error myopia, is approved for use in the United States. The Lasik procedure has evolved from both the Russian-developed radial keratotomy and its laser-based successor photorefractive keratectomy.

1990 Researchers begin the Human Genome Project, coordinated by the U.S. Department of Energy and the National Institutes of Health, with the goal of identifying all of the approximately 30,000 genes in human DNA and determining the sequences of the three billion chemical base pairs that make up human DNA. The project catalyzes the multibillion-dollar U.S. biotechnology industry and fosters the development of new medical applications, including finding genes associated with genetic conditions such as familial breast cancer and inherited colon cancer. A working draft of the genome is announced in June 2000.

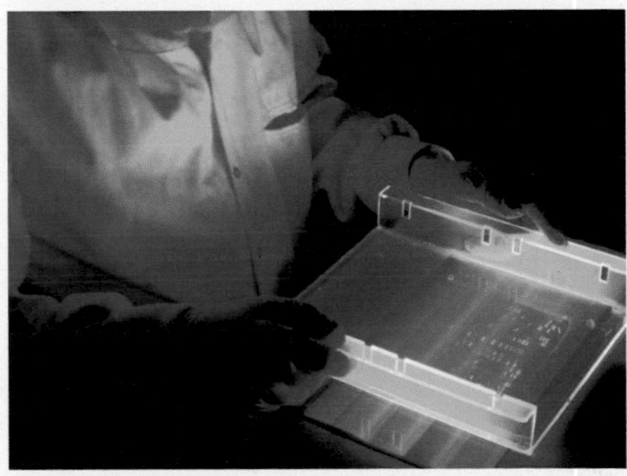

Using ultraviolet light, a researcher examines a gel containing stained DNA fragments.

PETROLEUM AND
PETROCHEMICAL
TECHNOLOGIES

If coal was king in the 19th century, oil was the undisputed emperor of the 20th. Refined forms of petroleum, or "rock oil," became—in quite literal terms—the fuel on which the 20th century ran, the lifeblood of its automobiles, aircraft, farm equipment, and industrial machines.

The captains of the oil industry were among the most successful entrepreneurs of any century, reaping huge profits from oil, natural gas, and their byproducts and building business empires that soared to capitalism's heights. Oil even became a factor in some of the most complex geopolitical struggles in the last quarter of the 20th century, ones still playing out today.

Oil has touched all our lives in other ways as well. Transformed into petrochemicals, it is all around us, in just about every modern manufactured thing, from the clothes we wear and the medicines we take to the materials that make up our computers, countertops, toothbrushes, running shoes, car bumpers, grocery bags, flooring tiles, and on and on and on. Indeed, the products from petrochemicals have played as great a role in shaping the modern world as gasoline and fuel oils have in powering it.

It seems at first a chicken-and-egg sort of question: Which came first—the gas pump or the car pulling up to it? Gasoline was around before the invention of the internal combustion engine but for many years was considered a useless byproduct of the refining of crude oil to make kerosene, a standard fuel for lamps through much of the 19th century. Oil refining of the day—and into the first years of the 20th century—relied on a relatively simple distillation process that separated crude oil into portions, called fractions, of different hydrocarbon compounds (molecules consisting of varying arrangements of carbon and hydrogen atoms) with different boiling points. Heavier kerosene, with more

From crude oil pumped out of the earth (opposite), engineers and scientists during the 20th century learned not only how to extract fuel for automobiles and heating but also how to transform that black viscous stuff into the materials that help keep our children dry and prolong the freshness of fresh foods (left).

carbon atoms per molecule and a higher boiling point, was thus easily separated from lighter gasoline, with fewer atoms and a lower boiling point, as well as from other hydrocarbon compounds and impurities in the crude oil mix. Kerosene was the keeper; gasoline and other compounds, as well as natural gas that was often found alongside oil deposits, were often just burned off.

Then in the first 2 decades of the 20th century horseless carriages in increasing droves came looking for fuel. Researchers had found early on that the internal combustion engine ran best on light fuels like gasoline, but distillation refining just didn't produce enough of it—only about 20 percent gasoline from a given amount of crude petroleum. Even as oil prospectors extended the range of productive wells from Pennsylvania through Indiana and into the vast oil fields of Oklahoma and Texas, the inherent inefficiency of the existing refining process was almost threatening to hold back the automotive industry with gasoline shortages.

The problem was solved by a pair of chemical engineers at Standard Oil of Indiana—company vice president William Burton and Robert Humphreys, head of the lab at the Whiting refinery, the world's largest at the time. Burton and Humphreys had tried and failed to extract more gasoline from crude by adding chemical catalysts, but then Burton had an idea and directed Humphreys to add pressure to the standard heating process used in distillation. Under both heat and pressure, it turned out that heavier molecules of kerosene, with up to 16 carbon atoms per molecule, "cracked" into lighter molecules such as those of gasoline, with 4 to 12 carbons per molecule. Thermal cracking, as the process came to be called, doubled the efficiency of refining, yielding 40 percent gasoline. Burton was issued a patent for the process in 1913, and soon the pumps were keeping pace with the ever-increasing automobile demand.

In the next decades other chemical engineers improved the refining process even further. In the 1920s Charles Kettering and Thomas Midgley, who would later develop Freon (see *Air Conditioning and Refrigeration*), discovered that adding a form of lead to gasoline made it burn smoothly, preventing the unwanted detonations that caused engine knocking. Tetraethyl lead was a standard ingredient of almost all gasolines until the 1970s, when environmental concerns led to the development of efficiently burning gasolines that didn't require lead.

Standard Oil engineers William Burton (below left) and Robert Humphreys (below right) developed the process now known as thermal cracking, which doubled the efficiency of oil refining.

Catalytic cracking plants allowed refineries to convert crude oil into gasoline using heat and chemical catalysts. In the mid-1960s this plant in California produced more than 200,000 gallons of gasoline a day.

Another major breakthrough was catalytic cracking, the challenge that had escaped Burton and Humphreys. In the 1930s a Frenchman named Eugene Houdry perfected a process using certain silica and alumina-based catalysts that produced even more gasoline through cracking and didn't require high pressure. In addition, catalytic cracking produced forms of gasoline that burned more efficiently.

Different forms of all sorts of things were coming out of refineries, driven in part by the demands of war. Houdry had also invented a catalytic process for crude oil that yielded butadiene, a hydrocarbon compound with some interesting characteristics. In the years before and during World War II it became one of two key ingredients in the production of synthetic rubber, an especially vital commodity as the war in the Pacific cut off supplies of natural rubber. The stage was now set for a revolution in petrochemical technology. As the war drove up demands for both gasoline and heavier aviation fuels, supplies of byproduct compounds—known as feedstocks—were increasing. At the same time, chemical engineers working in research labs were finding potential new uses for just those feedstocks, which they were beginning to see as vast untapped sources of raw material.

Throughout the 1920s and 1930s and into the 1940s chemical companies in Europe and the United States, working largely with byproducts of the distillation of coal tar, announced the creation of a wide assortment of new compounds with a variety of characteristics that had the common property of being easily molded—and thus were soon known simply as plastics. Engineering these new compounds for specific attributes was a matter of continual experimentation with chemical processes and combinations of different molecules. Many of the breakthroughs involved the creation of polymers—larger, more complex molecules consisting of smaller molecules chemically bound together, usually through the action of a catalyst. Sometimes the results would be a surprise, yielding a material with unexpected characteristics or fresh insights into what might be possible. Among the most important advances was the discovery of a whole class of plastics that could be remolded after heating, an achievement that would ultimately lead to the widespread recycling of plastics.

Three of the most promising new materials—polystyrene, polyvinyl chloride (PVC), and polyethylene—were synthesized from the same hydrocarbon: ethylene, a relatively rare byproduct of standard petroleum refinery processes. But there, in those ever-increasing feedstocks, were virtually limitless quantities of ethylene just

In the 1940s and 1950s the byproducts of oil refining began to yield synthetic materials such as polyethylene, shown below left being extruded and fed into a cooling bath; nylon, stronger and more durable for parachutes than natural silk; and polyvinyl chloride, or vinyl, used for everything from grocery store plastic wrap to wiring insulation, car door panels, and piping (opposite).

Offshore oil drilling platforms, like these off the coast of California, are made of steel and reinforced concrete and engineered to withstand so-called 100-year storm conditions.

waiting to be cracked. And here also was a moment of serendipity: readily available raw material, a wide range of products to be made from it, and a world of consumers coming out of years of war eager to start the world afresh, preferably with brand-new things.

Plastics and their petrochemical cousins, synthetic fibers, filled the bill. From injection-molded polystyrene products like combs and cutlery, PVC piping, and the ubiquitous polyethylene shopping bags and food storage containers to the polyesters, the acrylics, and nylon, all were within consumers' easy reach. Indeed, synthetic textiles became inexpensive enough to eventually capture half of the entire fiber market. All credit was owed to the ready feedstock supplies.

But those supplies were not as limitless as they had once seemed. With demand for petroleum—both as a fuel and in its many other synthesized forms—skyrocketing, America and other Western countries turned more and more to foreign sources, chiefly in the Middle East. At the same time, oil companies continued to search for and develop new sources, including vast undersea deposits in the Gulf of Mexico and later the North Sea. Offshore drilling presented a whole new set of challenges to petroleum engineers, who responded with some truly amazing constructions, including floating platforms designed to withstand hurricane-force winds and waves. One derrick in the North Sea called "Troll" stands in 1,000 feet of water and rises 1,500 feet above the surface. It is, with the Great Wall of China, one of only two human-made structures visible from the Moon.

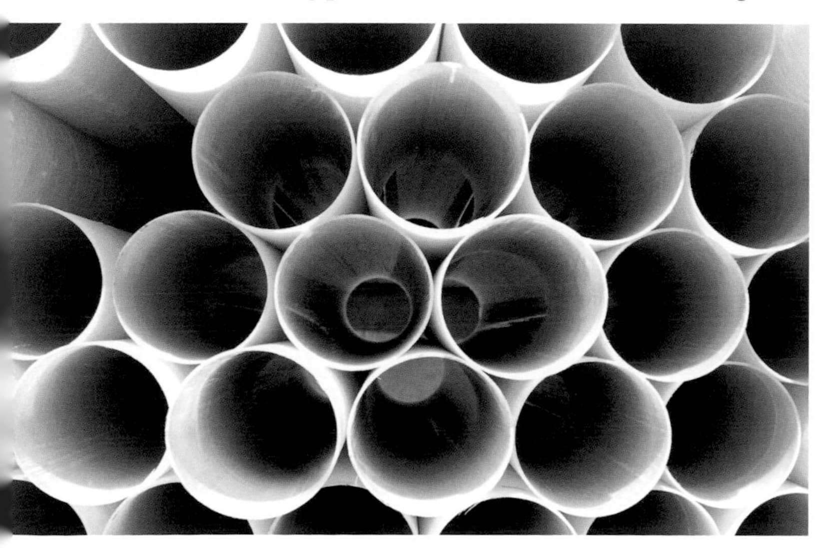

One way or another, the oil continued to flow. In 1900 some 150 million barrels of oil were pumped worldwide. By 2000 world production stood at 22 billion barrels—a day. But a series of crises in the 1970s, including the Arab oil embargo of 1973 and an increasing awareness of the environmental hazards posed by fossil fuels, brought more changes to the industry. Concern over an assured supply of fossil fuel encouraged prospectors, for instance, to develop new techniques for finding oil, including using the seismic waves produced artificially by literally thumping the ground to create three-dimensional images that brought hidden underground deposits into clear view and greatly reduced the fruitless drilling of so-called dry holes. Engineers developed new types of drills that not only reached deeper into the earth—some extending several miles below the surface—but could also tunnel horizontally for thousands of feet, reaching otherwise inaccessible deposits. Known reserves were squeezed as dry as they could be with innovative processes that washed oil out with injected water or chemicals and induced thermal energy. Refineries continued to find better ways to crack crude oil into more and better fuels and even developed other techniques such as reforming, which did the opposite of cracking, fashioning just-right molecules from smaller bits and pieces. And perhaps most significantly of all,

DEEP SEA OIL RIGS

PUMPING BLACK GOLD ON THE HIGH SEAS

The Coast Guard designates them as vessels, and they do indeed float and can be moved around. But today's offshore oil rigs serve primarily as stable platforms in the hunt for, and the harvesting of, new under-the-seabed sources of natural gas and oil. Earlier generations of these gargantuan structures were quite literally offshore, within sight of land; the latest versions ply their trade 150 miles out or even farther, drilling into a seabed that can be a mile or more below the surface.

Offshore exploration and production begins with high-tech seismic surveys in which sound waves are bounced off deep layers beneath the seabed and the reflected patterns are analyzed by supercomputers to create three-dimensional maps of potentially productive fields. Then come exploratory "vessels" such as the one illustrated below, drilling rigs that can be moved efficiently from place to place above the field, test-drilling for ideal development locations. Once the best sites are found, more complex structures such as the one depicted opposite get to work, not only drilling but also performing initial processing before pumping the crude oil and gas through undersea pipelines to onshore refineries.

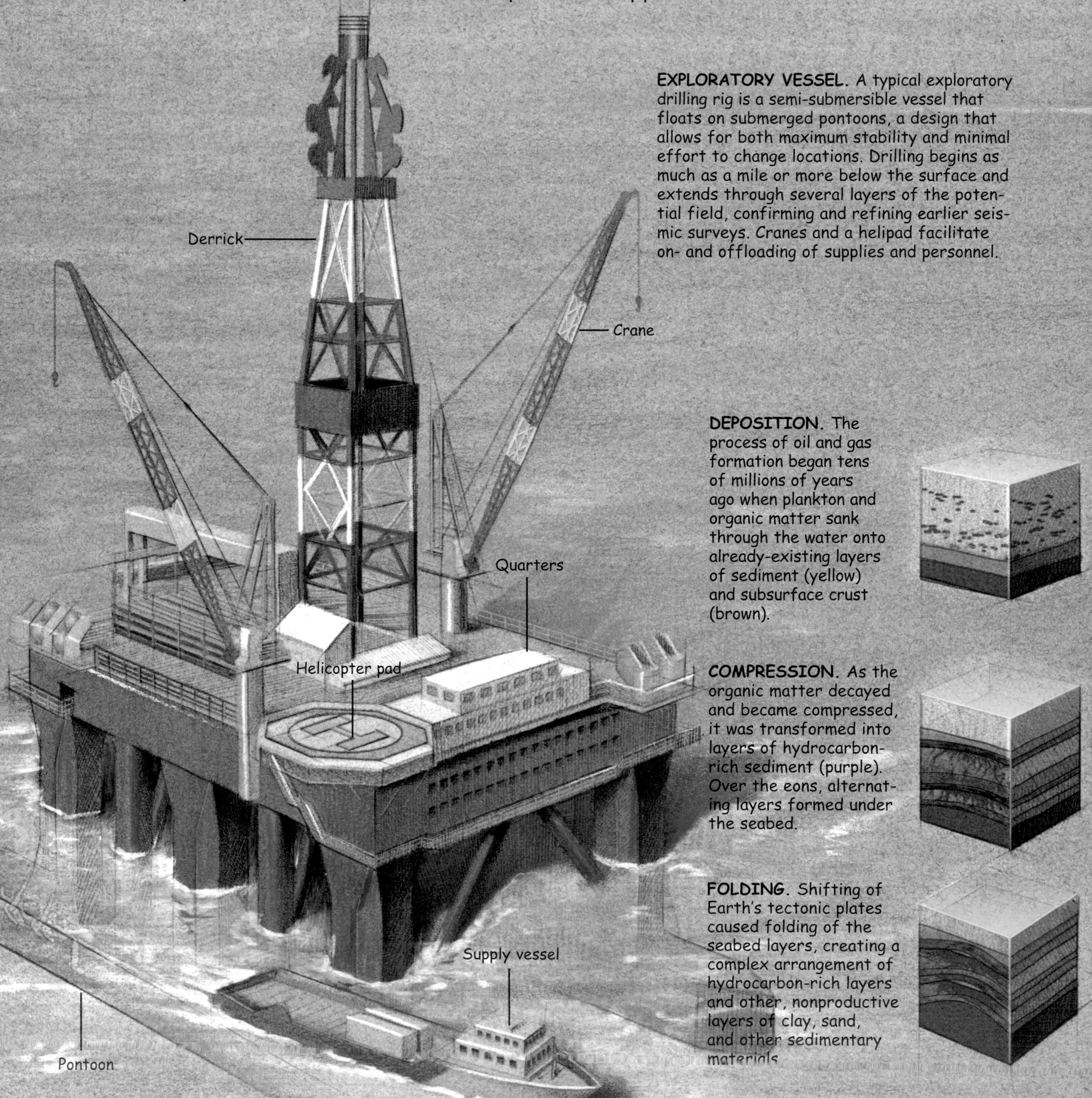

EXPLORATORY VESSEL. A typical exploratory drilling rig is a semi-submersible vessel that floats on submerged pontoons, a design that allows for both maximum stability and minimal effort to change locations. Drilling begins as much as a mile or more below the surface and extends through several layers of the potential field, confirming and refining earlier seismic surveys. Cranes and a helipad facilitate on- and offloading of supplies and personnel.

DEPOSITION. The process of oil and gas formation began tens of millions of years ago when plankton and organic matter sank through the water onto already-existing layers of sediment (yellow) and subsurface crust (brown).

COMPRESSION. As the organic matter decayed and became compressed, it was transformed into layers of hydrocarbon-rich sediment (purple). Over the eons, alternating layers formed under the seabed.

FOLDING. Shifting of Earth's tectonic plates caused folding of the seabed layers, creating a complex arrangement of hydrocarbon-rich layers and other, nonproductive layers of clay, sand, and other sedimentary materials.

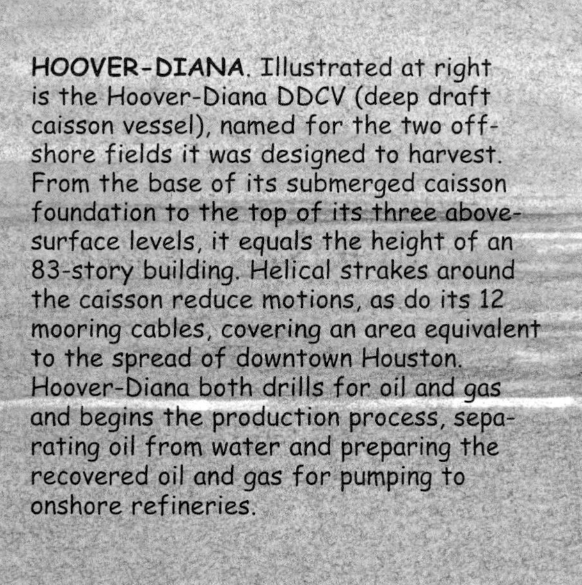

HOOVER-DIANA. Illustrated at right is the Hoover-Diana DDCV (deep draft caisson vessel), named for the two offshore fields it was designed to harvest. From the base of its submerged caisson foundation to the top of its three above-surface levels, it equals the height of an 83-story building. Helical strakes around the caisson reduce motions, as do its 12 mooring cables, covering an area equivalent to the spread of downtown Houston. Hoover-Diana both drills for oil and gas and begins the production process, separating oil from water and preparing the recovered oil and gas for pumping to onshore refineries.

UNDERSEA WELLS. The Hoover-Diana DDCV receives oil and gas from multiple subsea wells in the Diana reservoir, about 15 miles away, as well as from wells drilled directly beneath it in the Hoover field. Because they lie about a mile beneath the surface, the Diana wells are maintained only by remotely operated vehicles (ROVs), which are themselves controlled by technicians aboard Hoover-Diana. Subsea manifolds feed recovered oil and gas back to the production platform through miles-long pipelines and up the caisson in pipes called risers

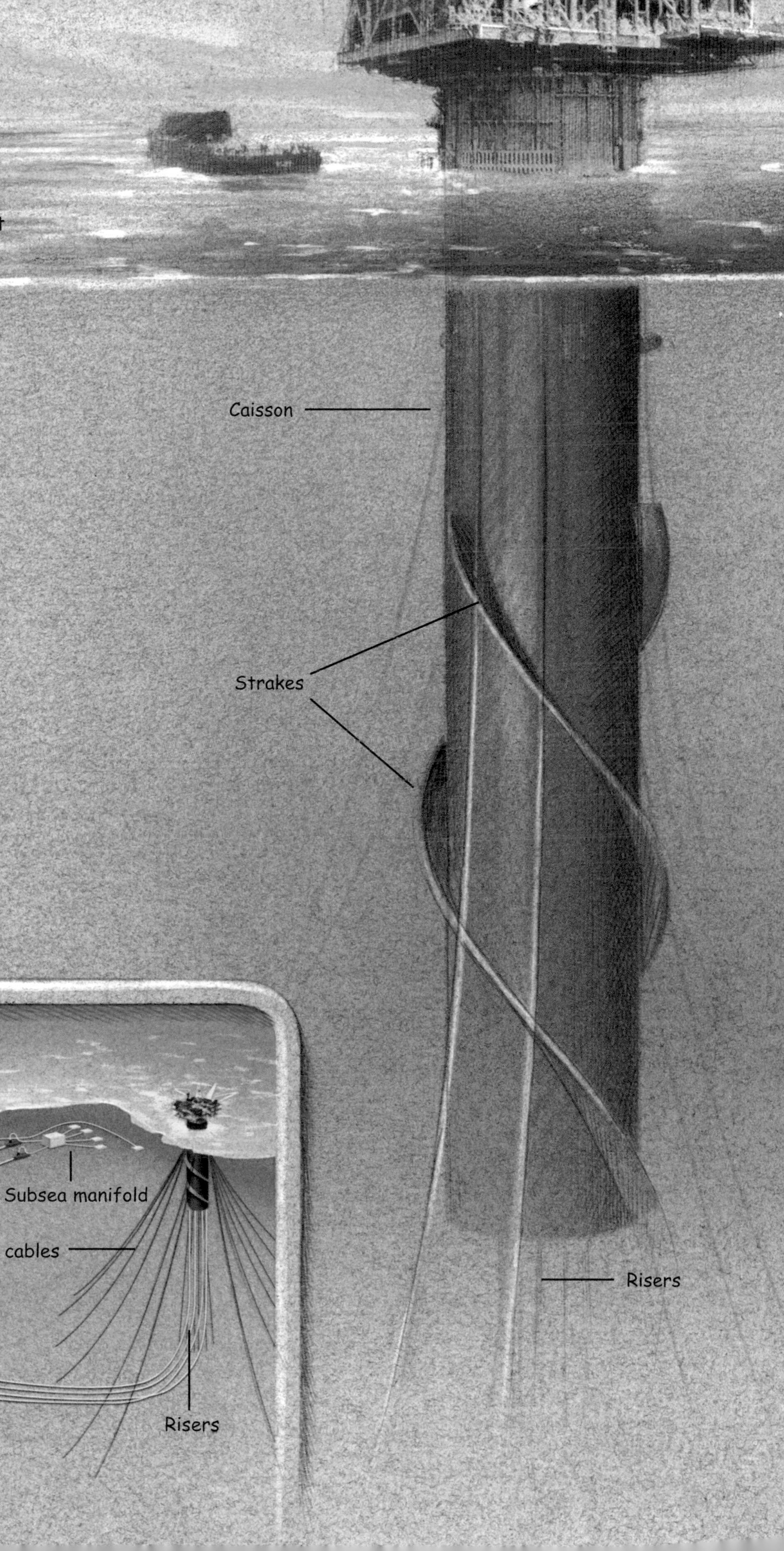

Even as Eugene Houdry's catalytic converter (above) is improving air quality by removing harmful pollutants from auto exhaust, the petrochemical industry has found a way to turn recycled plastic bottles (below) into lightweight but warm synthetic fleece clothing for people of all ages.

natural gas—so often found with oil deposits—was finally recognized as a valuable fuel in its own right, becoming an economically significant energy source beginning in the 1960s and 1970s.

Initially attractive because it was cheap and relatively abundant, natural gas also held the advantage of being cleaner burning and far less damaging to the environment, factors that became increasingly important with the passage of the Clean Air Act in the 1970s. Indeed, natural gas has replaced crude oil as the most important source of petrochemical feedstocks.

Petrochemical and automotive engineers had already responded to environmental concerns in a variety of ways. As early as the 1940s German émigré Vladimir Haensel invented a type of reforming refining process called platforming that used very small amounts of platinum as a catalyst and produced high-octane, efficient-burning fuel without the use of lead. Haensel's process, which was eventually recognized as one of the most significant chemical engineering technologies of the past 50 years, made the addition of lead to gasoline no longer necessary. Today, more than 85 percent of the gasoline produced worldwide is derived from platforming.

Also well ahead of the environmental curve was Eugene Houdry, who had developed catalytic cracking; in 1956 he invented the catalytic converter, a device that removed some of the most harmful pollutants from automobile exhaust and that ultimately became standard equipment on every car in the United States. Other engineers also developed methods for removing more impurities, such as sulfur, during refining, making the process itself a cleaner affair. For its part, natural gas was readily adopted as an alternative to home heating oil and has also been used in some cities as the fuel for fleets of buses and taxicabs, reducing urban pollution. Environmental concerns have also affected the other side of the petrochemical business, leading to sophisticated processes for recycling existing plastic products.

Somewhere around the middle of the 20th century, petroleum replaced coal as the dominant fuel in the United States, and petroleum processing technologies allowed petrochemicals to replace environmentally harmful coal tar chemistry. The next half-century saw this dominance continue and even take on new forms, as plastics and synthetic fibers entered the consumer marketplace. Despite increasingly complex challenges, new generations of researchers and engineers have continued to keep the black gold bonanza in full swing.

Perspective

Lee R. Raymond

Chairman and CEO
Exxon Mobil Corporation

My first major academic interest was chemical engineering, which I studied as an undergraduate at the University of Wisconsin. I liked the subject so well that I decided to pursue a doctorate in it, but this time at the University of Minnesota. It was the top-ranked graduate school and also a bit closer to where I grew up in Watertown, South Dakota.

I've always been amused when people act surprised about where I grew up, since they seem to think that South Dakota, Minnesota, and Wisconsin are like Siberia. But I'm proud of my roots. It's my feeling that people who come out of America's heartland have strong beliefs in fundamental values, in education, and in a commitment to do a good job.

Perhaps the thing that appealed most to me about being an engineer was a curiosity about how things are designed and built. I'm fascinated by technology and by the research that underpins the incredible technical advances we see all around us.

After getting my doctorate, I decided to take a research job with what was then the Standard Oil Company (New Jersey) because it seemed like a good way to quickly broaden my experience. At the time I joined Jersey, I thought I would eventually be returning to academic life. That was 40 years ago, and I just never made the trip back to academia.

One reason is that I found a company that both satisfied my curiosity and fulfilled my abiding interest in technology. I was able to work in a company committed to R&D, a place where I was exposed to stimulating colleagues and many areas of research and, above all, a place where what we worked on had practical applications of benefit to people. The experience has been exhilarating, and even though it's been a long time since I have been asked to do practical engineering, I am still drawn to being able to work in a place that puts technology at the forefront of its activities. I have also found management in the petroleum industry to be both challenging and rewarding. I derive my greatest sense of accomplishment from watching people develop and grow in competence, in seeing them take on difficult challenges and master them.

Of course, even though I am no longer a practicing engineer, I see many aspects of the energy business where the skills and perspectives of an engineer are vital. For example, I think I can sense the sorts of projects that are likely to be achievable and those that lie beyond what is doable, at least in the short to medium term. An engineering background also has helped in assessing areas of public policy where the science and technology that some people are enthusiastic about may not yet be mature enough to rely on and yet in other areas it is within reach.

A science and engineering background gives me enormous faith and confidence in the power of technology. No one who has lived their entire professional life in an industry like petroleum can escape a sense of awe at what has been achieved technically and at the benefits that have come from that technological power. My experiences have also given me a huge sense of optimism about the innovations we are likely to see in the future.

TIMELINE

When retired railroad conductor Edwin Drake struck oil in 1859 in Titusville, Pennsylvania, he touched off the modern oil industry. For the next 40 years the primary interest in oil was as a source of kerosene, used for lighting lamps. Then came the automobile and the realization that the internal combustion engine ran best on gasoline, a byproduct of the process of extracting kerosene from crude oil. As the demand grew for gasoline to power not only cars but also internal combustion engines of all kinds, chemical engineers honing their refining techniques discovered a host of useful byproducts of crude—and the petrochemical industry was born. Oil had truly become black gold.

1901 North America's first oil gusher *(right)* blows at the Spindletop field near Beaumont in southeastern Texas, spraying more than 800,000 barrels of crude into the air before it can be brought under control. The strike boosts the yearly oil output in the United States from 2,000 barrels in 1859 to more than 65 million barrels by 1901.

1913 Chemical engineers William Burton and Robert Humphreys of Standard Oil patent a method of oil refining that significantly increases gasoline yields. Known as thermal cracking, the chemists discover that by applying both heat and pressure during distillation, heavier petroleum molecules can be broken down, or cracked, into gasoline's lighter molecules. The discovery is a boon to the new auto industry, whose fuel of choice is gasoline.

German organic chemist Friedrich Bergius develops a high-pressure hydrogenation process that transforms heavy oil and oil residues into lighter oils, boosting gasoline production. In 1926 IG Farben Industries, where Carl Bosch had been developing similar high-pressure processes, acquires the patent rights to the Bergius process. Bergius and Bosch share a Nobel Prize in 1931.

1920s By using fractional distillation, two German coal researchers create synthetic gasoline. Known as the Fischer-Tropsch method, the gasoline is produced by combining either coke and steam or crushed coal and heavy oil, then exposing the mixture to a catalyst to form synthetic gasoline. The process plays a critical role in helping to meet the increasing demand for gasoline as automobiles come into widespread use and later for easing gasoline shortages during World War II.

1921 Charles Kettering of General Motors and his assistants, organic chemists Thomas Midgley, Jr., *(right)* and T. A. Boyd, discover that adding lead to gasoline eliminates engine knock. Until the 1970s, when environmental concerns forced its removal, tetraethyl lead was a standard ingredient in gasoline.

1928 By mounting a derrick and drilling outfit onto a submersible barge, Texas oilman Louis Giliasso creates an efficient portable method of offshore drilling. The transportable barge allows a rig to be erected in as little as a day, which makes for easier exploration of the Texas and Louisiana coastal wetlands. More permanent offshore piers and platforms had been successfully operating since the late 1800s off the coast of California near Santa Barbara, where oil seepage in the Pacific had been reported by Spanish explorers as early as 1542.

1930s U.S. refineries take advantage of a new process of alkalinization and fine-powder fluid-bed production that increases the octane rating of aviation gasoline to 100. This becomes important in the success of the Royal Air Force and the U.S. Army Air Force in World War II.

1936 French scientist Eugene Houdry introduces catalytic cracking. By using silica and alumina-based catalysts, he demonstrates not only that more gasoline can be produced from oil without the use of high pressure but also that it has a higher octane rating and burns more efficiently.

1920s-1940s An assortment of new compounds derived from byproducts of the oil-refining process enter the market. Three of the most promising new materials

—synthesized from the hydrocarbon ethylene—are polystyrene, a brittle plastic known also as styrofoam; polyvinyl chloride, used in plumbing fixtures and weather-resistant home siding; and polyethylene, which is flexible, inexpensive, and widely used in packaging. New synthetic fibers and resins are also introduced, including nylon *(below)*, acrylics, and polyester, and are used to make everything from clothing and sports gear to industrial equipment, parachutes, and plexiglass.

1942 The first catalytic cracking unit is put on-stream in Baton Rouge, Louisiana, by Standard Oil, New Jersey.

1947 German-born American chemical engineer Vladimir Haensel invents platforming, a process for producing cleaner-burning high-octane fuels using a platinum catalyst to speed up certain chemical reactions. Platforming eliminates the need to add lead to gasoline.

A consortium of oil companies led by Kerr-McGee drills the world's first commercial oil well out of sight of land in the Gulf of Mexico, 10.5 miles offshore and 45 miles south of Morgan City, Louisiana. Eleven oil fields are mapped in the gulf by 1949, with 44 exploratory wells in operation.

1955 The first jack-up oil-drilling rig is designed for offshore exploration. The rig features long legs that can be lowered into the seabed to a depth of 500 feet, allowing the platform to be raised to various heights above the level of the water.

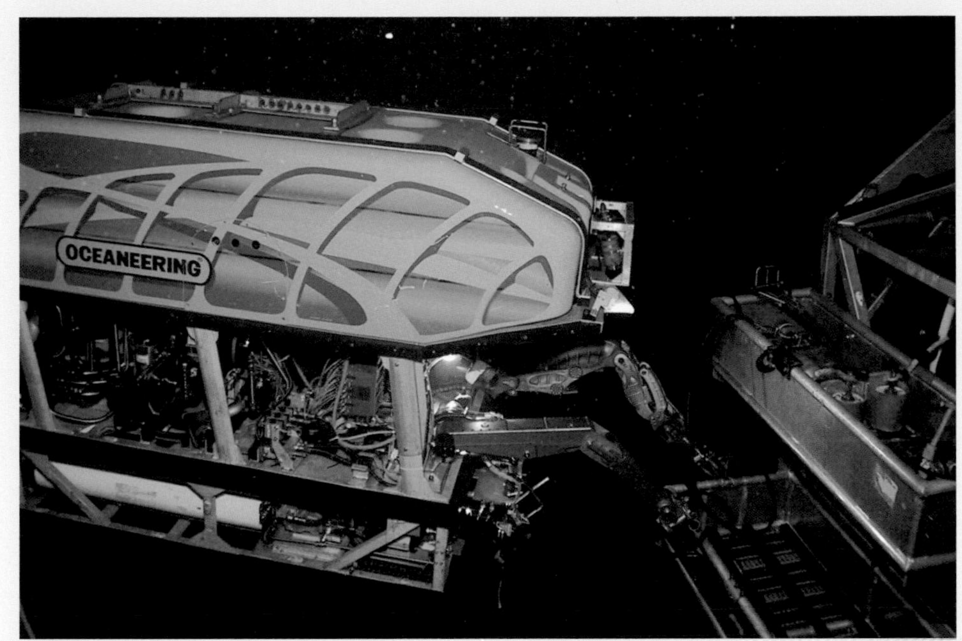

This heavy work-class subsea robot performs intricate tasks in water depths down to 10,000 feet or more.

1960s Synthetic oils are in development to meet the special lubricating requirements of military jets. Mobil Oil and AMSOIL are leaders in this field; their synthetics contain such additives as polyalphaolefins, derived from olefin, one of the three primary petrochemical groups. Saturated with hydrogen, olefin-carbon molecules provide excellent thermal stability. Following on the success of synthetic oils in military applications, they are introduced into the commercial market in the 1970s for use in automobiles.

1970s The introduction of digital seismology in oil exploration increases accuracy in locating underground pools of oil. The technique of using seismic waves to look for oil is based on determining the time interval between the sending of a sound wave (generated by an explosion, an electric vibrator, or a falling weight) and the arrival of reflected or refracted waves at one or more seismic detectors. Analysis of differences in arrival times and amplitudes of the waves tells seismologists what kinds of rock the waves have traveled through.

Teleco, Inc., of Greenville, South Carolina, and the U.S. Department of Energy introduce mud pulse telemetry, a system of relaying pressure pulses through drilling mud to convey the location of the drill bit. Mud pulse telemetry is now an oil industry standard, saving millions of dollars in time and labor.

1980s Remotely operated vehicles (ROVs) are developed for subsea oil work. Controlled from the surface, ROVs vary from beachball-size cameras to truck-size maintenance robots (*above*).

1990s The combined efforts of private industry, the Department of Energy, and national laboratories such as Argonne and Lawrence Livermore result in the introduction of several new tools and techniques designed to reduce the costs and risks of drilling, including reducing potential damage to the geological formation and improving environmental protection. Among such tools are the near-bit sensor, which gathers data from just behind the drill bit and transmits it to the surface, and carbon dioxide/sand fracturing stimulation, a technique that allows for nondamaging stimulation of a natural gas formation.

2000 The *Hoover-Diana*, a 63,000-ton deep-draft caisson vessel, goes into operation in the Gulf of Mexico. A joint venture by Exxon Mobil and BP, it is a production platform mounted atop a floating cylindrical concrete tube anchored in 4,800 feet of water. The entire structure is 83 stories high, with 90 percent of it below the surface. Within half a year it is producing 20,000 barrels of oil and 220 million cubic feet of gas a day. Two pipelines carry the oil and gas to shore.

LASERS AND FIBER OPTICS

If necessity is the mother of invention, the odds of a breakthrough in telecommunications were rising fast as the 20th century passed its midpoint. Most long-distance message traffic was then carried by electrons traveling along copper or coaxial cables, but the flow was pinched and expensive, with demand greatly outstripping supply. Over the next few decades, however, the bottlenecks in long-haul communications would be cleared away by a radically new technology.

Its secret was light—a very special kind of radiance produced by devices called lasers and channeled along threads of ultrapure glass called optical fibers. Today, millions of miles of the hair-thin strands stretch across continents and beneath oceans, knitting the world together with digital streams of voice, video, and computer data, all encoded in laser light.

When the basic ideas behind lasers occurred to Columbia University physicist Charles Townes in 1951, he wasn't thinking about communications, much less the many other roles the devices would someday play in such fields as manufacturing, health care, consumer electronics, merchandising, and construction. He wasn't even thinking about light. Townes was an expert in spectroscopy—the study of matter's interactions with electromagnetic energy—and what he wanted was a way to generate extremely short-wavelength radio waves or long-wavelength infrared waves that could be used to probe the structure and behavior of molecules. No existing instrument was suitable for the job, but early one spring morning as he sat on a park bench wrestling with the problem, he suddenly recognized that molecules themselves might be enlisted as a source.

All atoms and molecules exist only at certain characteristic energy levels. When an atom or molecule shifts from one level to another, its electrons emit or absorb photons—packets of electromagnetic energy with a telltale wavelength (or frequency) that may range from very long radio waves to ultrashort gamma rays, depending on the size of the energy shift. Normally the leaps up and down the energy ladder don't yield a surplus of photons,

but Townes saw possibilities in a distinctive type of emission described by Albert Einstein back in 1917.

If an atom or molecule in a high-energy state is "stimulated" by an impinging photon of exactly the right wavelength, Einstein noted, it will create an identical twin—a second photon that perfectly matches the triggering photon in wavelength, in the alignment of wave crests and troughs, and in the direction of travel. Normally, there are more molecules in lower-energy states than in higher ones, and the lower-energy molecules absorb photons, thus limiting the radiation intensity. Townes surmised that under the right conditions the situation might be reversed, allowing the twinning to create amplification on a grand scale. The trick would be to pump energy into a substance from the outside to create a general state of excitement, then keep the self-duplicating photons bouncing back and forth in a confined space to maximize their numbers.

Not until 1954 did he and fellow researchers at Columbia prove it could be done. Using an electric field to direct excited molecules of ammonia gas into a thumb-sized copper chamber, they managed to get a sustained output of the desired radio waves. The device was given the name maser, for microwave amplification by stimulated emission of radiation, and it proved valuable for spectroscopy, the strengthening of extremely faint radio signals, and a few other purposes. But Townes would soon create a far bigger stir, teaming up with his physicist brother-in-law Arthur Schawlow to show how stimulated emission might be achieved with photons at the much shorter wavelengths of light—hence the name laser, with the "m" giving way to "l." In a landmark paper published in 1958 they explained that light could be reflected back and forth in the energized medium by means of two parallel mirrors, one of them only partly reflective so that the built-up light energy could ulti-

In a few short decades, lasers have progressed from small experimental devices (opposite) to the copper vapor laser on the mammoth Starfire Optical Range telescope (left and previous pages), which projects a beam into the upper atmosphere to create an artificial "star." By monitoring changes in the shape of the star due to air turbulence, the telescope's adaptive optics can compensate.

Charles Townes (below left) carried out initial work on masers and later, with brother-in-law Arthur Schawlow (below right) on lasers, publishing their findings in 1958.

mately escape. Six years later Townes received a Nobel Prize for his work, sharing it with a pair of Soviet scientists, Aleksandr Prochorov and Nikolai Gennadievich Basov, who had independently covered some of the same ground.

The first functioning laser—a synthetic ruby crystal that emitted red light—was built in 1960 by Theodore Maiman, an electrical engineer and physicist at the Hughes Research Laboratories. That epochal event set off a kind of evolutionary explosion. Over the next few decades lasers would take forms as big as a house and as small as a grain of sand. Along with ruby, numerous other solids were put to work as a medium for laser excitation. Various gases proved viable too, as did certain dye-infused liquids and some of the electrically ambivalent materials known as semiconductors. Researchers also developed many ways to excite a laser medium into action, pumping in the necessary energy with flash lamps, other lasers, electricity, and even chemical reactions.

As for the laser light itself, it soon came in a broad range of wavelengths, from infrared to ultraviolet, with the output delivered as either pulses or continuous beams. All laser light has the same highly organized nature, however. In the language of science, it is practically monochromatic (of essentially the same wavelength), coherent (the crests and troughs of the waves perfectly in step, thus combining their energy), and highly directional. The result is an extremely narrow and powerful beam, far less inclined to spread and weaken than a beam of ordinary light, which is composed of a jumble of wavelengths out of step with one another.

Lasers have found applications almost beyond number. In manufacturing, infrared carbon dioxide lasers cut and heat-treat metal, trim computer chips, drill tiny holes in tough ceramics, silently slice through textiles, and pierce the openings in baby bottle nipples. In construction the narrow, straight beams of lasers guide the laying of pipelines, drilling of tunnels, grading of land, and alignment of buildings. In medicine, detached retinas are spot-welded back in place with an argon laser's green light, which passes harmlessly through the central part of the eye but is absorbed by the blood-rich tissue at the back. Medical lasers are also used to make surgical incisions while simultaneously cauterizing blood vessels to minimize bleeding, and they allow doctors to perform exquisitely precise surgery on the brain and inner ear.

Many everyday devices have lasers at their hearts. A CD or DVD player, for example, reads the digital contents of a rapidly spinning disc by bouncing laser light off minuscule irregularities stamped onto the disc's surface. Barcode scanners in supermarkets play a laser beam over a printed pattern of lines and spaces to extract price information and keep track of inventory.

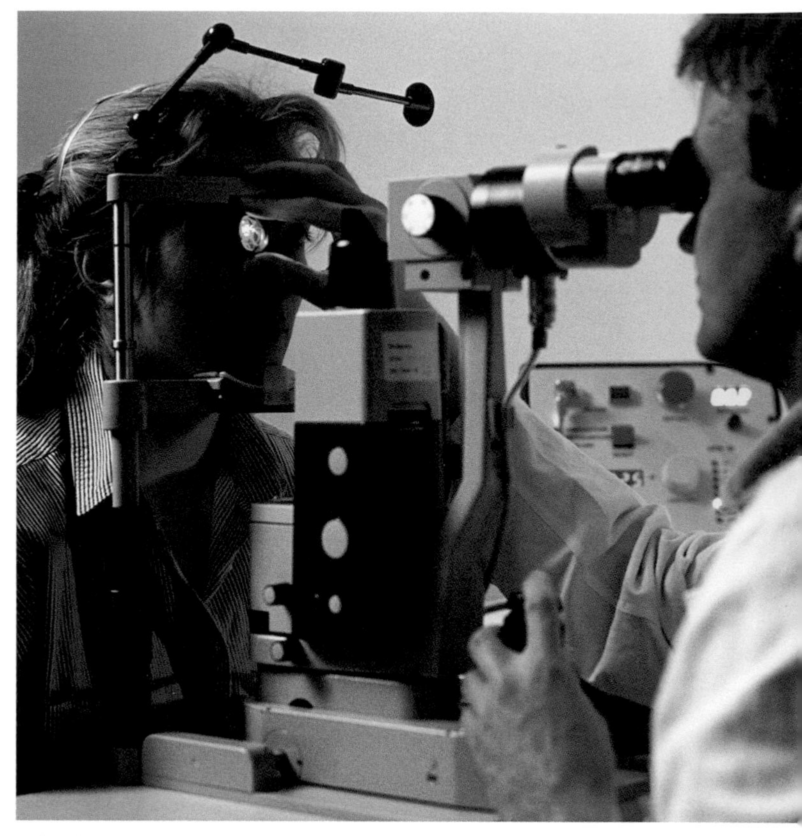

The diverse applications of laser technology include repairing detached retinas (above) and attempting to create nuclear fusion, as seen opposite at the Lawrence Livermore National Laboratory.

Monitoring the slightest movement in the San Andreas Fault is a major occupation in Parkfield, California, where a laser system bounces light off a network of distant reflectors and can detect movements of less than 1 millimeter across a distance of 3 or 4 miles.

Pulsed lasers are no less versatile than their continuous-beam brethren. They can function like optical radar, picking up reflections from objects as small as air molecules, enabling meteorologists to detect wind direction or measure air density. The reflections can also be timed to measure distances—in some cases, very great indeed. A high-powered pulsed laser, aimed at mirrors that astronauts placed on the lunar surface, was used to determine the distance from Earth to the Moon to within 2 inches. The pulses of some lasers are so brief—a few quadrillionths of a second—that they can visually freeze the lightning-fast movements of molecules in a chemical reaction. And superpowerful laser pulses may someday serve as the trigger for controlled fusion, the long-sought thermonuclear process that could provide humankind with almost boundless energy.

Whatever the future holds, the laser's status as a world-changing innovation has already been secured by its role in long-distance communications. But that didn't happen without some pioneering on another frontier—fiber optics. At the time lasers emerged, the ability of flexible strands of glass to act as a conduit for light was a familiar phenomenon, useful for remote viewing and a few other purposes. Such fibers were considered unsuitable for communications, however, because any data encoded in the light were quickly blurred by chaotic internal reflections as the waves traveled along the channel. Then in 1961 two American researchers, Will Hicks and Elias Snitzer, directed laser beams through a glass fiber made so thin—just a few microns—that the light waves would follow a single path rather than ricocheting from side to side and garbling a signal in the process.

This was a major advance, but practical communication with light was blocked by a more basic difficulty. As far as anyone knew, conventional glass simply couldn't be made transparent enough to carry light far. Typically, light traveling along a fiber lost about 99 percent of its energy by the time it had gone just 30 feet. Fortunately for the future of fiber optics, a young Shanghai-born electrical engineer named Charles Kao was convinced that glass could do much better.

Working at Standard Telecommunications Laboratories in England, Kao collected and analyzed samples from glassmakers and concluded that the energy loss was mainly due to impurities such as water and minerals, not the basic glass ingredient of silica itself. A paper he published with colleague George Hockham in 1966 predicted that optical fibers could be made pure enough to carry signals for miles. The challenges of manufacturing such stuff were formidable, but in 1970 a team at Corning Glass Works succeeded in creating a fiber hundreds of yards long that performed just as Kao and Hockham had foreseen. Continuing work at Corning and AT&T Bell Labs developed the manufacturing processes necessary to produce miles of high-quality fiber.

At about the same time, researchers were working hard on developing a light source to partner with optical fibers. Their efforts were focused on semiconductor lasers, sand-grain-sized mites that could be coupled to the end of a thread of glass. Semiconducting materials are solid compounds that conduct electricity imperfectly. When a tiny sandwich of differing materials is electrically energized, laser action takes place in the junction region, and the polished ends of the materials act as mirrors to confine the light photons while they multiply prolifically.

Three traits were essential in a semiconductor laser tailored to telecommunications. It would have to generate a continuous beam rather than pulses. It would need to function at room temperature and operate for hundreds of thousands of hours without failure. Finally, the laser's output would have to be in the infrared range, optimal for transmission down a fiber of silica glass. In 1967 Morton Panish and Izuo Hayashi of Bell Labs spelled out the basic requirements in materials and design. Two other Bell Labs researchers, J. R. Arthur and A. Y. Cho, subsequently found a way to create an ultrathin layer of material at the center of the semicon-

LASERS AND FIBER OPTICS

TWIN TECHNOLOGIES FOR HIGH VOLUME COMMUNICATIONS

Considering its future as one of the most versatile tools ever devised, the birth of the laser in the late 1950s was a quiet event. Nor did any great fanfare greet a pioneering study, at about the same time, of how light behaves when piped through a very narrow channel of glass. But the seeds of a spectacular synergy had been planted. Within 2 decades, lasers and optical conduits for their light would open the way to almost unimaginable communications plenitude: In theory, if not yet in practice, a single laser beam traveling along an optical fiber can carry the world's entire present-day traffic of telephone, video and data interchanges, with plenty of room to spare.

Before the two technologies could be effectively partnered, however, some major hurdles had to be cleared. One was the fabrication of sufficiently transparent optical fibers, drawn to filament fineness from rods of ultrapure silica glass. Another was the development of semiconductor lasers—long-lasting, able to switch on and off very rapidly, and tiny enough to be coupled to the end of a fiber. An important recent innovation is an all-optical way of regenerating a signal on long-haul routes, achieved by inserting stretches of special fibers that function as lasers themselves. The amount of information that potentially can be handled by fiber-optic technology continues to grow at an astonishing rate, posing a challenge of a most unusual kind: figuring out how to use it all.

STIMULATED EMISSION IN A LASER CAVITY. Ordinary light of the kind from lightbulbs or the Sun is produced when atoms or molecules release excess energy by emitting it in particles called photons. Laser light, by contrast, is produced when a photon encounters an atom or molecule in a state of high energy, stimulating it to emit a twin photon, identical in wavelength, phase and direction of travel. In the sequence shown at right, photons travel back and forth through an energized laser medium—a so-called resonant cavity, mirrored at both ends. Each pass through the medium generates more perfectly matching photons, amplifying the light's intensity in a kind of chain reaction.

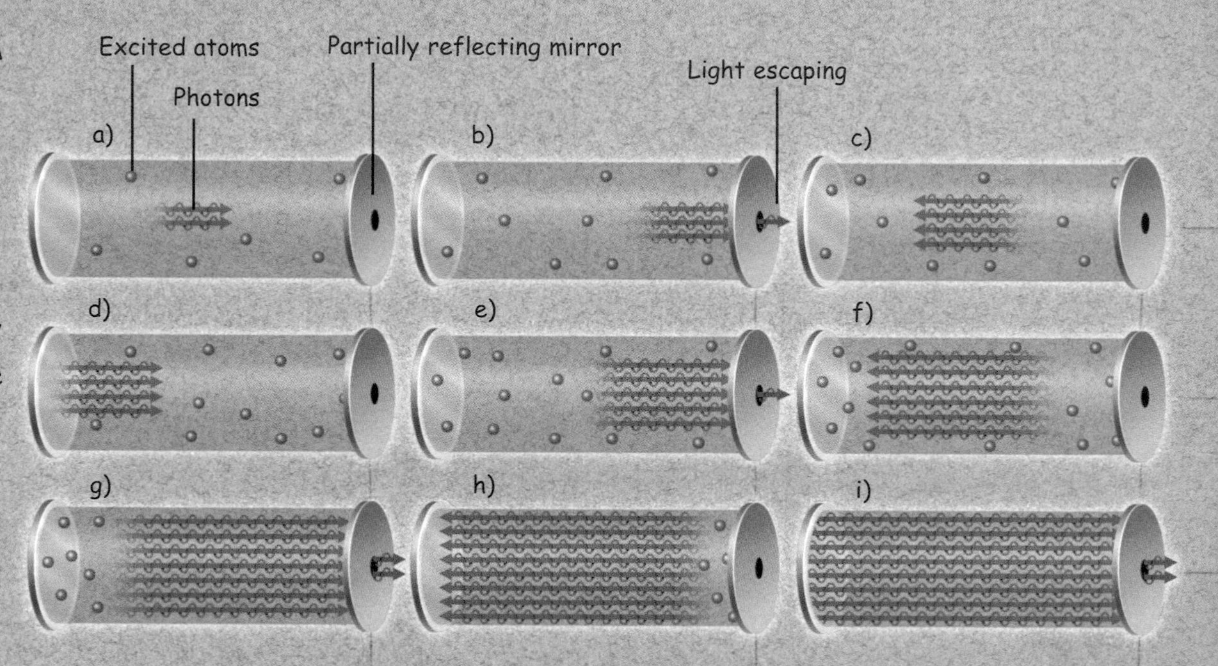

One end of a typical laser's cavity is fully reflective; the other is partially reflective and allows some photons to escape. As seen above, a swarm of photons grows with each back-and-forth pass. Finally, the laser medium reaches equilibrium and the cavity is saturated with stimulated emission—the limit of its amplification.

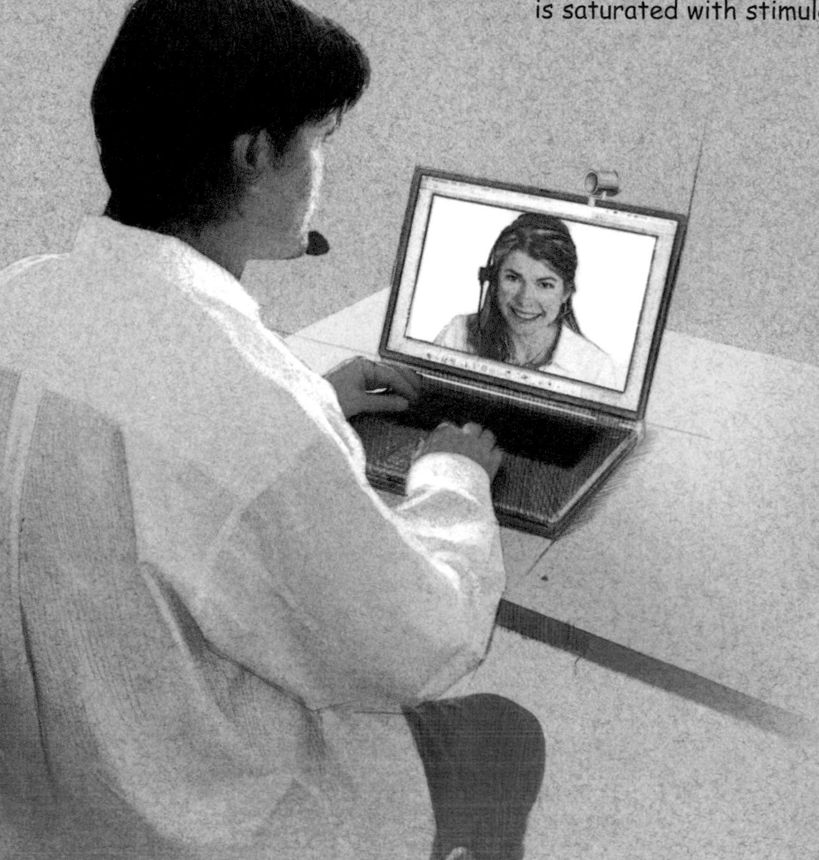

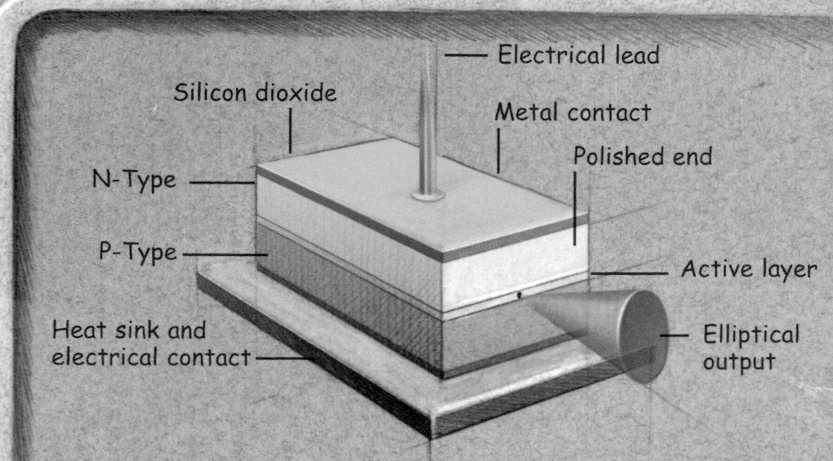

SEMICONDUCTOR DIODE LASER. The lasers used for fiber-optic communications are sand-grain-size semiconductor sandwiches energized by an electrical current. Infrared light is created in a very thin central layer where electrons and holes—areas of electron deficiency—combine and annihilate each other, emitting photons. The ends of the semiconductor crystal are polished to bounce the photons back and forth, and one end is fully reflective. Light emerges from the other end with an elliptical cross-section, a function of the rectangular emission face.

WINDOWS OF LIGHT TRANSMISSION. Even the purest silica glass contains some impurities that absorb or scatter pulses of light as they travel along a fiber, but the degree of optical loss varies with different wavelengths. The first optical-fiber systems were designed to operate at wavelengths around 850 nanometers, a region chosen because the losses are relatively low and also because technology for generating and detecting the light already existed. As new technology came along, systems shifted to longer wavelengths lending themselves to better performance—first 1310 nanometers, then 1550 nanometers, a wavelength offering a theoretical minimum of loss in silica glass and also suitable for the all-optical signal-regenerating methods developed for long-haul routes. Short-range systems may use light around 660 nanometers—in the visible rather than infrared region of the spectrum, and generated by light-emitting diodes (LEDs) rather than laser diodes.

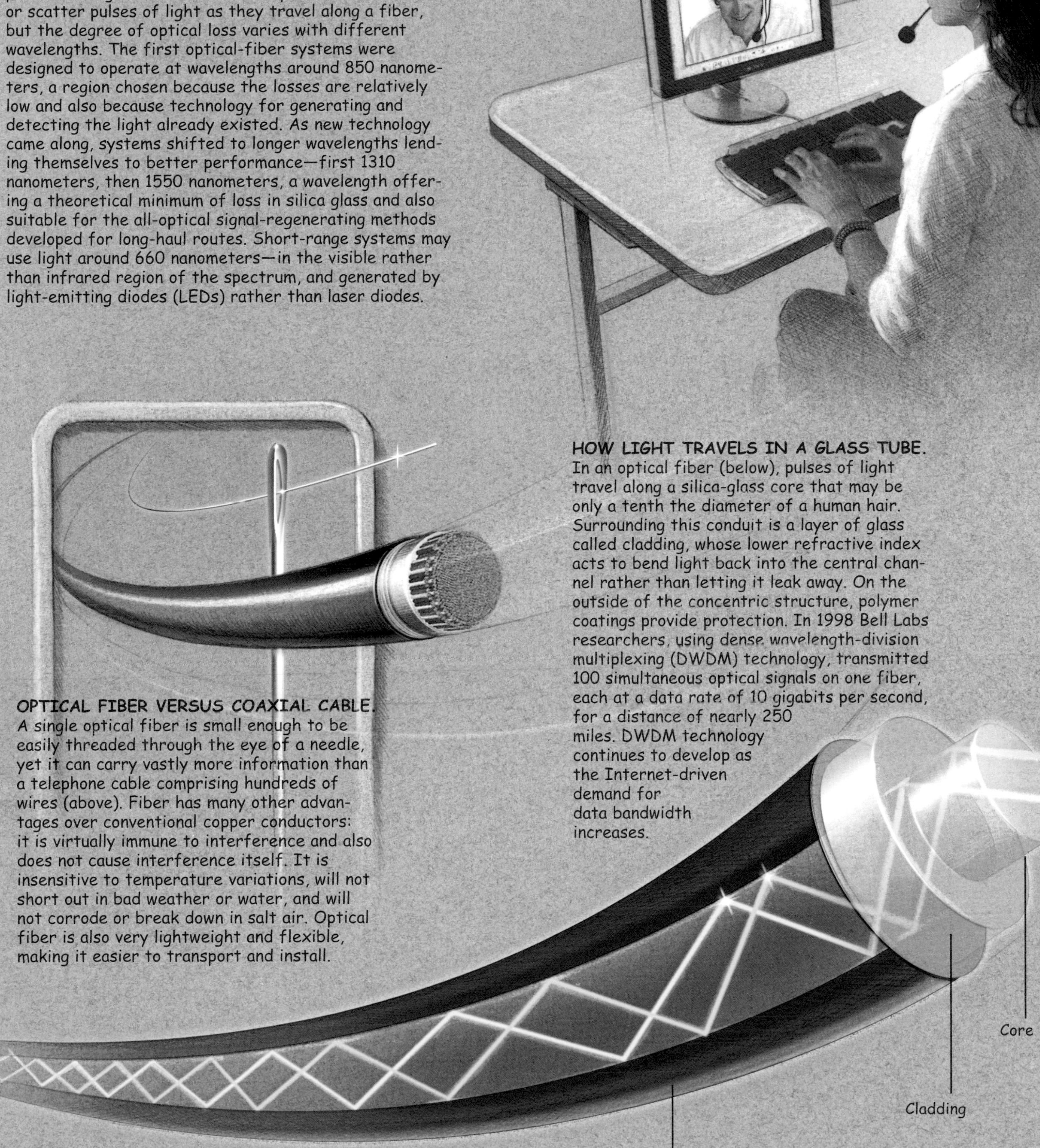

HOW LIGHT TRAVELS IN A GLASS TUBE. In an optical fiber (below), pulses of light travel along a silica-glass core that may be only a tenth the diameter of a human hair. Surrounding this conduit is a layer of glass called cladding, whose lower refractive index acts to bend light back into the central channel rather than letting it leak away. On the outside of the concentric structure, polymer coatings provide protection. In 1998 Bell Labs researchers, using dense wavelength-division multiplexing (DWDM) technology, transmitted 100 simultaneous optical signals on one fiber, each at a data rate of 10 gigabits per second, for a distance of nearly 250 miles. DWDM technology continues to develop as the Internet-driven demand for data bandwidth increases.

OPTICAL FIBER VERSUS COAXIAL CABLE. A single optical fiber is small enough to be easily threaded through the eye of a needle, yet it can carry vastly more information than a telephone cable comprising hundreds of wires (above). Fiber has many other advantages over conventional copper conductors: it is virtually immune to interference and also does not cause interference itself. It is insensitive to temperature variations, will not short out in bad weather or water, and will not corrode or break down in salt air. Optical fiber is also very lightweight and flexible, making it easier to transport and install.

In the late 1980s technicians release the first fiber-optic cable (above left) that crossed the Atlantic Ocean, from Long Beach, New Jersey, to Marseilles, France. A decade later fiber optic lines spanned both the Atlantic and Pacific Oceans, produced by manufacturing plants (above right).

ductor sandwich that produced laser light with unprecedented efficiency.

By the mid-1970s all the necessary ingredients for fiber-optic communications were ready, and operational trials got under way. The first commercial service was launched in Chicago in 1977, with 1.5 miles of underground fiber connecting two switching stations of the Illinois Bell Telephone Company. Improvements in both lasers and fibers would keep coming after that, further widening light's already huge advantage over other methods of communication.

Any transmission medium's capacity to carry information is directly related to frequency—the number of wave cycles per second, or hertz. The higher the frequency, the more wave cycles per second, and the more information can be packed into the transmission stream. Light used for fiber-optic communications has a frequency millions of times higher than radio transmissions and 100 billion times higher than electric waves traveling along copper telephone wires. But that's just the beginning. Researchers have learned how to send multiple light streams along a fiber simultaneously, each carrying a huge cargo of information on a separate wavelength. In theory, more than a thousand distinct streams can ride along a single glass thread at the same time.

Toward the 20th century's end, one of the few lingering constraints was removed by a device that is both laser and fiber. For all the marvelous transparency of silica glass, light inevitably weakens as it travels along, requiring amplification from time to time. In the early years of fiber optics, the necessary regeneration was done by devices that converted the light signals into electricity, boosted them, and then changed them back into light again. This limited the speed of transmission because the electronic amplifier was slower than the fiber. But the 1990s saw the appearance of vastly superior amplifiers that are lasers themselves. These optical amplifiers consist of short stretches of fiber, doped with the element erbium and optically energized by an auxiliary "pump" laser. The erbium-doped amplifiers revive the fading photons every 50 miles or so without the need for electrical conversion. The amplification can occur for a relatively broad range of wavelengths, allowing roughly 40 different wavelengths to be amplified simultaneously.

For the most part the devices that switch messages from one fiber to another (as from one router to another on the Internet) still must convert a message from light to electricity and back again. Yet even as researchers and engineers actively pursue the development of all-optical switches, this last bottleneck scarcely hampers the flow of information carried on today's fiber-optic systems. Flashing incessantly between cities, countries, and continents, the prodigious torrent strains the gossamer web not at all.

Perspective

Charles H. Townes

Professor
Department of Physics
University of California, Berkeley

The laser invention happened because I wanted very much to be able to make an oscillator at frequencies as high as the infrared in order to extend the field of microwave spectroscopy in which I was working. I had tried several ideas, but none worked very well. At the time I was also chairman of a committee for the navy that was examining ways to obtain very short-wave oscillators. In 1951, on the morning before the last meeting of this committee in Washington, I woke up early worrying over our lack of success. I got dressed and stepped outside to Franklin Park, where I sat on a bench admiring the azaleas and mulling over our problem.

Why couldn't we think of something that would work at high frequencies? I went through the possibilities, including, of course, molecules, which oscillate at high frequencies. Although I had considered molecules before, I had dismissed them because of certain laws of thermodynamics. But suddenly I recognized, "Hey, molecules don't have to obey such a law if they are not in equilibrium." And I immediately took a piece of paper out of my pocket and wrote equations to see if selection of excited molecules by molecular beam methods could produce enough molecules to provide a feedback oscillator. Wow! It looked possible.

I went back to my hotel and told Art Schawlow about the idea, since he was staying at the same place. Back at Columbia University, I wrote the idea carefully in my notebook and had Schawlow witness it in preparation for the possibility of a patent. (In my previous career at Bell Labs, I had produced several patents in the field of radar and hence was familiar with patent requirements.) Soon my students and I began building the first maser—not yet producing light but demonstrating the principles. Its extension to waves as short as light came a few years later, after much excitement over the maser and as a result of my continued collaboration with Schawlow, then at Bell Labs. An essential element in this discovery, I believe, was my experience in both engineering and physics: I knew both quantum mechanics and the workings and importance of feedback oscillators.

When Schawlow and I first distributed our paper on how to make a laser, a number of friends teased me with the comment, "That's an invention looking for an application. What can it do?" To me, communications with a potentially large bandwidth and beam directionality seemed an obvious application. But the use of fibers did not occur to me, and that's what really changed communications, especially with the development of low-loss materials.

Lasers combine optics and electronics. And so in addition to their revolutionary role in communications, lasers by now have found a wealth of applications—in medicine, manufacturing, measurements and control, computing, possibly nuclear power, and much new science. Thirteen Nobel prizes have been awarded for work utilizing lasers or masers as scientific tools.

Both lasers and fiber optics are fields that can be expected to grow and develop further, including in ways still not foreseen. Consider that all of the separate principles and ideas involved in the invention of masers and lasers were known and understood by someone in the scientific or technical community at least as early as the mid-1930s. Yet it took 25 more years for these ideas to be put together to make a laser. What might we be missing or overlooking now?

LASERS AND FIBER OPTICS
TIMELINE

The slender glass filaments that carry telephone conversations and high-speed computer data as lightwaves are enclosed in layers of reflective cladding (white) and protective buffering.

From surgical instruments and precision guides in construction to barcode scanners and compact disc readers, lasers are integral to many aspects of modern life and work. But perhaps the farthest-flung contribution of the 20th century's combination of optics and electronics has been in telecommunications. With the advent of highly transparent fiber-optic cable in the 1970s, very high-frequency laser signals now carry phenomenal loads of telephone conversations and data across the country and around the world.

1917 Albert Einstein proposes the theory of stimulated emission—that is, if an atom in a high-energy state is stimulated by a photon of the right wavelength, another photon of the same wavelength and direction of travel will be created. Stimulated emission will form the basis for research into harnessing photons to amplify the energy of light.

1954 Charles Townes, James Gordon, and Herbert Zeiger at Columbia University develop a "maser" (for microwave amplification by stimulated emission of radiation), in which excited molecules of ammonia gas amplify and generate radio waves. The work caps 3 years of effort since Townes's idea in 1951 to take advantage of high-frequency molecular oscillation to generate short-wavelength radio waves.

1958 Townes and physicist Arthur Schawlow publish a paper showing that masers could be made to operate in optical and infrared regions. The paper explains the concept of a laser (light amplification by stimulated emission of radiation)—that light reflected back and forth in an energized medium generates amplified light.

1960 Theodore Maiman, a physicist and electrical engineer at Hughes Research Laboratories, invents an operable laser using a synthetic pink ruby crystal as the medium. Encased in a "flash tube" *(below)* and bookended by mirrors, the laser successfully produces a pulse of light. Prior to Maiman's working model, Columbia University doctoral student Gordon Gould also designs a laser, but his patent application is initially denied. Gould finally wins patent recognition nearly 30 years later.

Bell Laboratories researcher and former Townes student Ali Javan and his colleagues William Bennett, Jr., and Donald Herriott (*left to right below*) invent a continuously operating helium-neon gas laser.

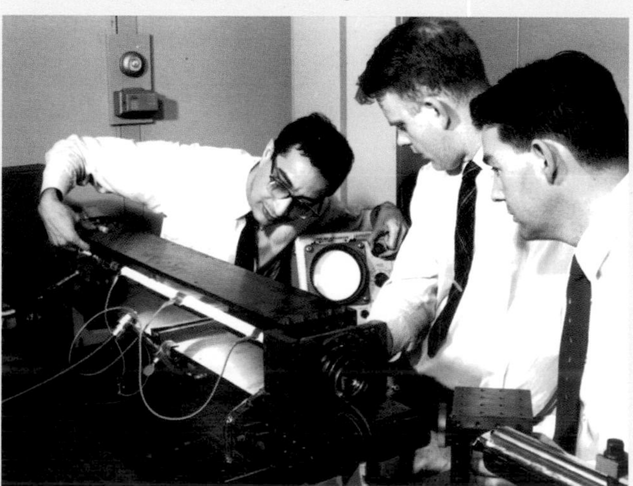

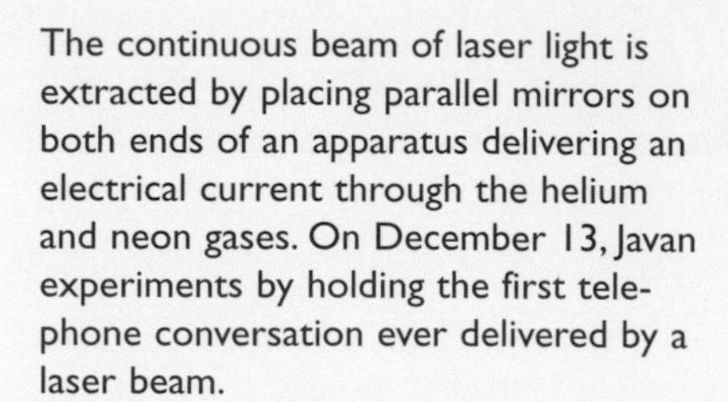

The continuous beam of laser light is extracted by placing parallel mirrors on both ends of an apparatus delivering an electrical current through the helium and neon gases. On December 13, Javan experiments by holding the first telephone conversation ever delivered by a laser beam.

1961 In the first medical use of the ruby laser, Charles Campbell of the Institute of Ophthalmology at Columbia-Presbyterian Medical Center and Charles Koester of the American Optical Corporation use a prototype ruby laser photocoagulator to destroy a human patient's retinal tumor.

Industry researchers Elias Snitzer and Will Hicks demonstrate a laser beam directed through a thin glass fiber. The fiber's core is small enough that the light follows a single path, but most scientists still consider fibers unsuitable for communications because of the high loss of light across long distances.

1962 Three groups—at General Electric, IBM, and MIT's Lincoln Laboratory—simultaneously develop a gallium arsenide laser that converts electrical energy directly into infrared light and that much later is used in CD and DVD players as well as computer laser printers.

1963 Physicist Herbert Kroemer proposes the idea of heterostructures, combinations of more than one semiconductor built in layers that reduce energy requirements for lasers and help them work more efficiently. These heterostructures will later be used in cell phones and other electronic devices.

1966 Charles Kao and George Hockham of Standard Telecommunications Laboratories in England publish a landmark paper demonstrating that optical fiber can transmit laser signals with much reduced loss if the glass strands are pure enough. Researchers immediately focus on ways to purify glass.

1970 Corning Glass Works scientists Donald Keck, Peter Schultz, and Robert Maurer report the creation of optical fibers that meet the standards set by Kao and Hockham. The purest glass ever made, it is composed of fused silica from the vapor phase and exhibits light loss of less than 20 decibels per kilometer (1 percent of the light remains after traveling 1 kilometer). By 1972 the team creates glass with a loss of 4 decibels per kilometer. Also in 1970, Morton Panish (*below right*) and Izuo Hayashi (*below left*) of Bell

Laboratories, along with a group at the Ioffe Physical Institute in Leningrad, demonstrate a semiconductor laser that operates continuously at room temperature. Both breakthroughs will pave the way toward commercialization of fiber optics.

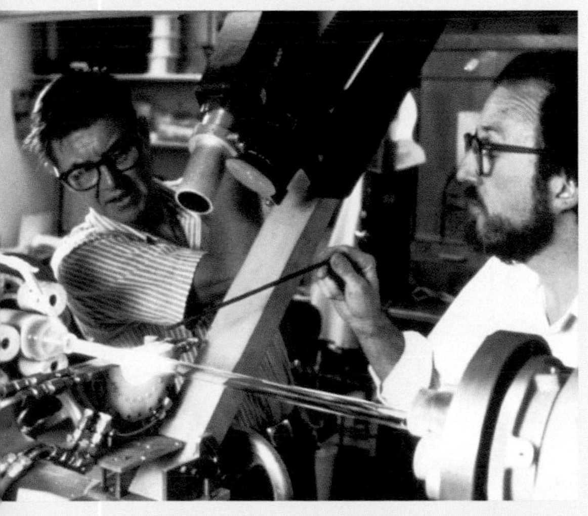

1973 John MacChesney (*above left*) and Paul O'Connor (*above right*) at Bell Laboratories develop a modified chemical vapor deposition process that heats chemical vapors and oxygen to form ultratransparent glass that can be mass-produced into low-loss optical fiber. The process still remains the standard for fiber-optic cable manufacturing.

1975 Engineers at Laser Diode Labs develop the first commercial semiconductor laser to operate continuously at room temperatures. The continuous-wave operation allows the transmission of telephone conversations.

Standard Telephones and Cables in the United Kingdom installs the first fiber-optic link for interoffice communications after a lightning strike damages equipment and knocks out radio transmission used by the police department in Dorset.

1977 Telephone companies begin trials with fiber-optic links carrying live telephone traffic. GTE opens a line between Long Beach and Artesia, California, whose transmitter uses a light-emitting diode. Bell Labs establishes a similar link for the phone system of downtown Chicago, 1.5 miles of underground fiber that connects two switching stations.

1980 AT&T announces that it will install fiber-optic cable linking major cities between Boston and Washington, D.C. The cable is designed to carry three different wavelengths through graded-index fiber—technology that carries video signals later that year from the Olympic Games in Lake Placid, New York. Two years later MCI announces a similar project using single-mode fiber carrying 400 bits per second.

1987 David Payne at England's University of Southampton introduces fiber amplifiers that are "doped" with the element erbium. These new optical amplifiers are able to boost light signals without first having to convert them into electrical signals and then back into light.

1988 The first transatlantic fiber-optic cable is installed, using glass fibers so transparent that repeaters (to regenerate and recondition the signal) are needed only about 40 miles apart. The shark-proof TAT-8 is dedicated by science fiction writer Isaac Asimov, who praises "this maiden voyage across the sea on a beam of light." Linking North America and France, the 3,148-mile cable is capable of handling 40,000 telephone calls simultaneously using 1.3-micrometer wavelength lasers and single-mode fiber. The total cost of $361 million is less than $10,000 per circuit; the first transatlantic copper cable in 1956 costs $1 million per circuit to plan and install.

1991 Emmanuel Desurvire of Bell Laboratories, along with David Payne and P. J. Mears of the University of Southampton, demonstrate optical amplifiers that are built into the fiber-optic cable itself. The all-optic system can carry 100 times more information than cable with electronic amplifiers.

1996 TPC-5, an all-optic fiber cable that is the first to use optical amplifiers, is laid in a loop across the Pacific Ocean. It is installed from San Luis Obispo, California, to Guam, Hawaii, and Miyazaki, Japan, and back to the Oregon coast and is capable of handling 320,000 simultaneous telephone calls. In 1997 the Fiber Optic Link Around the Globe (FLAG) becomes the longest single-cable network in the world and provides infrastructure for the next generation of Internet applications. The 17,500-mile cable begins in England and runs through the Strait of Gibraltar to Palermo, Sicily, before crossing the Mediterranean to Egypt. It then goes overland to the FLAG operations center in Dubai, United Arab Emirates, before crossing the Indian Ocean, Bay of Bengal, and Andaman Sea; through Thailand; and across the South China Sea to Hong Kong and Japan.

NUCLEAR
TECHNOLOGIES

Beating swords into plowshares—that's how advocates of nuclear technology have long characterized efforts to develop peaceful applications of the atom's energy. In an ongoing controversy, opponents point to the destructive potential and say that, despite the benefits, this is almost always a tool too dangerous to use. Beyond the controversy, however, lies the story of scientific and engineering breakthroughs that unfolded over a remarkably short period of time—with unprecedented effects on the world, for both good and ill.

Although a cloud of potential doom has shadowed the future since the first atomic bomb was tested in the New Mexico desert in July 1945, the process that led to that moment also paved the way for myriad technologies that have improved the lives of millions around the world.

It all began with perhaps the most famous formula in the history of science—Albert Einstein's deceptively simple mathematical expression of the relationship between matter and energy. $E=mc^2$, or energy equals mass multiplied by the speed of light squared, demonstrated that under certain conditions mass could be converted into energy and, more significantly, that a very small amount of matter was equivalent to a very great deal of energy. Einstein's formula, part of his work on relativity published in 1905, gained new significance in the 1930s as scientists in several countries were making a series of discoveries about the workings of the atom. The culmination came in late 1938, when Lise Meitner, an Austrian physicist who had recently escaped Nazi Germany and was living in Stockholm, got a message from longtime colleagues Otto Hahn and Fritz Strassmann in Berlin. Meitner had been working with them on an experiment involving bombarding uranium atoms with neutrons, and Hahn and Strassman were reporting a puzzling result. The product of the experiment seemed to be barium, a much lighter element. Meitner and her nephew, physicist Otto Frisch, recognized that what had occurred was the splitting of the uranium atoms, a process Meitner and Frisch were the first to call "fission." Italian physicist Enrico Fermi had achieved the same result several years earlier, also without realizing exactly what he had done. Among other things, fission converted some of the original atom's mass into energy, an amount Meitner and Frisch were able to calculate accurately using Einstein's formula. The news spread quickly through the scientific community and soon reached a much wider audience. On January 29, 1939, the *New York Times*, misspeaking slightly, headlined the story about the discovery: "Atomic Explosion Frees 200,000,000 Volts."

Fermi knew that when an atom splits it releases other neutrons, and he was quick to realize that under the right conditions those neutrons could go on to split other atoms in a chain reaction. This would lead to one of two things: a steady generation of energy in the form of heat or a huge explosion. If each splitting atom caused one released neutron to split another atom, the chain reaction was said to be "critical" and would create a steady release of heat energy. But if each fission event released two, three, or more neutrons that went on to split other atoms, the chain reaction was deemed "supercritical" and would rapidly cascade into an almost instantaneous, massive, explosive release of energy—a bomb. In the climate of the times, with the world on the brink of war, there was little doubt in which direction the main research effort would turn. Fermi, who had emigrated to the United States, became part of the top-secret American effort known as the Manhattan Project, which, in an astonishingly short period of time from its beginnings in 1942, turned fission's potential into the reality of the world's first atomic bombs.

Otto Hahn and Fritz Strassmann used the equipment shown opposite to discover the first chemical evidence of nuclear fission products in 1938. Today the energy unleashed by fission in nuclear power plants like the one at California's Diablo Canyon (previous pages), supplies electricity to millions of people around the world.

The Manhattan Project, headed by General Leslie Groves of the Army Corps of Engineers, included experimental facilities and manufacturing plants in several states, from Tennessee to Washington. Dozens of top-ranking physicists and engineers took part. One of the most significant breakthroughs was achieved by Fermi himself, who in 1942 created the first controlled, self-sustaining nuclear chain reaction in a squash court beneath the stands of the University of Chicago stadium. To do it, he had built the world's first nuclear reactor, an achievement that would ultimately lead to the technology that now supplies a significant proportion of the world's energy. But it was also the first practical step toward creating a bomb.

Key insights by (left to right above) Lise Meitner, Otto Hahn, and Otto Frisch and by Enrico Fermi (below) enabled researchers in midcentury to harness the energy released by splitting the nucleus of the atom.

Fermi recognized that the key to both critical and supercritical chain reactions was the fissionable fuel source. Only two potential fuels were known: uranium-235 and what was at the time still a hypothetical isotope, plutonium-239. (An isotope is a form of a given element with a different number of neutrons. The number refers to the combined total of protons and neutrons in the nucleus.) Uranium-235 exists in only 0.7 percent of natural uranium ore; the other 99.3 percent is uranium-238, a more stable isotope that tends to absorb neutrons rather than split and that can keep chain reactions from even reaching the critical stage. Plutonium-239 is created when an atom of uranium-238 absorbs a single neutron.

Manhattan Project engineers first set about enriching uranium, using chemical processes to increase the proportion of fissionable uranium-235 up to levels that could produce supercritical chain reactions. Most of this work was done in Oak Ridge, Tennessee, where Fermi was also involved in building another nuclear reactor, to test whether it was really possible to sustain a critical chain reaction that would produce plutonium-239 from the original uranium fuel. Plutonium, it

Tucked into a small valley in East Tennessee, Oak Ridge National Laboratory housed part of the Manhattan Project during World War II. Originally known as the Clinton Engineer Works, the site grew to be the fifth largest city in the state, home to 75,000 people (above). Scarcely a year after the war ended, Oak Ridge was producing and shipping radioisotopes for cancer treatment (below).

turned out, was an even more efficient fuel for supercritical chain reactions. Both efforts were successes and went on to provide the raw material for the first and only atomic bombs ever used in war—the Hiroshima bomb of uranium-235 enriched to 70 percent, and the Nagasaki bomb, which had a plutonium core, both ignited by implosion.

Bomb development ultimately led to thermonuclear weapons, in which the fusion of hydrogen atoms releases far greater amounts of energy. The first atomic bomb tested in New Mexico yielded the equivalent of 18 kilotons of TNT; thermonuclear hydrogen bombs yield up to 10 megatons. The Cold War drove both the United States and the Soviet Union to develop ever more lethal nuclear weapons, all based on the principles worked out and put into action by the scientists and engineers of the Manhattan Project. Although the consequences of their actions remain highly controversial, the brilliance of their technological achievements is undimmed.

Fermi's reactor in Tennessee opened the door to the first peacetime use of nuclear technology. When a fissionable material splits, it can produce any of a variety of radioisotopes, unstable isotopes whose decay emits radiation that can be dangerous—as in the fallout of a nuclear bomb. In a reactor, the radiation is contained, and scientists had already discovered that, if properly handled, radioisotopes could have beneficial uses, particularly in medicine. Cancer cells, for example, are especially sensitive to radiation damage because they divide so rapidly, and doctors were learning to use small targeted doses of radiation to destroy tumors. So reaction was swift in the summer of 1946 when Oak Ridge published a list of the radioisotopes its reactor was producing in the June issue of *Science*. By early August the lab was sending its first radioisotope shipment to Brainard Cancer Hospital in St. Louis, Missouri.

The field of nuclear medicine is now an integral part of health care throughout the world. Doctors use dozens of different radioisotopes in both diagnostic and therapeutic procedures, creating images of blood vessels, the brain, and other internal organs (see *Imaging*), and helping to destroy harmful growths. Radiation continues to be a mainstay of cancer treatment and has evolved to include not just targeted beams of radiation but also the implantation of small radioactive pellets and the use of so-called radiopharmaceuticals, drugs that deliver appropriate doses of radiation to specific tissues. Because even

One of the first peacetime applications of the work done during the war was the treatment of cancer, using focused beams of radioisotopes from a linear accelerator (below right) to destroy cancerous tumors in patients lying in the next room.

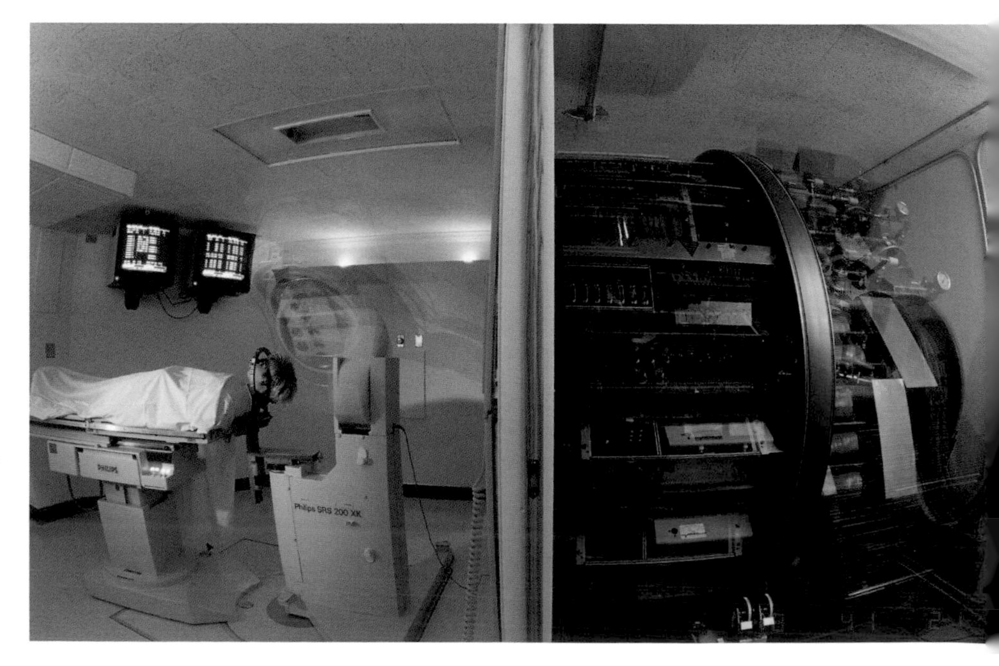

a small amount of radiation is easily detectable, researchers have also developed techniques using radioisotopes as a kind of label to tag and trace individual molecules. This labeling has proved particularly effective in the study of genetics by making it possible to identify individual DNA "letters" of the genetic code.

From the start, of course, researchers had known that another use for atomic energy was as a power source. After World War II the U.S. government was quick to realize that potential as well. In 1946 President Truman signed a law that created the Atomic Energy Commission, whose mandate included not only the development of atomic weapons but also the exploration of other applications. One of these was to power navy ships, and in 1948 Captain (later Admiral) Hyman Rickover was assigned the task of developing a reactor that could serve as the power plant for a submarine. Rickover, who had been part of the Manhattan Project, would become known as "the father of the nuclear navy." Under his leadership, engineers at the Westinghouse Bettis Atomic Power Laboratory in Pennsylvania designed the first pressurized-water reactor (PWR), which ultimately became the dominant type of power plant reactor in the United States. Rickover's team pioneered new materials and reactor designs, established safety and control standards and operating procedures, and built and tested full-scale propulsion prototypes. The final result was the USS *Nautilus*, commissioned in 1954 as the world's first nuclear-powered vessel. Six years later the USS *Triton* became the first submarine to circumnavigate the globe while submerged. Soon a fleet of nuclear submarines was patrolling the world's oceans, able to stay submerged for months at a time and go for years without refueling because of their nuclear power source. Masterpieces of engineering, nuclear submarines and aircraft carriers have operated without accident for nearly 6 decades.

Even before the *Nautilus* was finished, nuclear power plants were about to come into their own. On December 20, 1951, near the town of Arco, Idaho, engineers from Argonne National Laboratory started up a reactor that was connected to a steam turbine generator. When the chain reaction reached criticality, the heat of the nuclear fuel turned water into steam, which drove the generator and cranked out 440 volts, enough electricity to power four lightbulbs. It was the first time a nuclear reaction had created usable power. A few years later Arco became the world's first community to get its entire power supply from a nuclear reactor when the town's power grid was temporarily connected to the reactor's turbines.

Arco had been an experiment, but by 1957 a commercially viable nuclear power plant was operating in the western Pennsylvania town of Shippingport. It was one of the first practical manifestations of President Eisenhower's Atoms for Peace Program, established in 1953 specifically to promote commercial applications of atomic energy. Nuclear power plants of various designs were soon supplying significant percentages of energy needs throughout the developed world. There was certainly no question about the advantages. One ton of nuclear fuel produces the energy equivalent of 2 million to 3 million tons of fossil fuel. Looked at another way,

Nearly half a century after the event, Arco, Idaho, still proclaims its status as the first city to be lit by atomic power, a milestone that occurred in July 1955.

1 kilogram of coal generates 3 kilowatt-hours of electricity; 1 kilogram of oil generates 4 kilowatt-hours; and 1 kilogram of uranium generates up to 7 million kilowatt-hours. Also, unlike coal- and oil-burning plants, nuclear plants release no air pollutants or the greenhouse gases that contribute to global warming. Currently, some 400 nuclear plants provide electricity around the world, including 20 percent of energy in the United States, 80 percent in France, and more than 50 percent in Japan.

But the nuclear chain reaction still carries inherent dangers, which were made frighteningly apparent during the reactor accidents at Pennsylvania's Three Mile Island plant in 1979 and Ukraine's Chernobyl plant in

NUCLEAR SUBMARINES

NO LIMITS BENEATH THE WAVES

Less than 13 years after the first sustained nuclear chain reaction, the U.S. Navy launched the submarine *Nautilus*, the first of what was soon a fleet of nuclear-powered vessels—including aircraft carriers—that could cruise for as long as 10 years without refueling. The *Nautilus*, which could travel at top speed submerged for days at a time, vastly outstripped older, diesel-powered subs, whose fresh-air-guzzling engines could sustain top speed under water for no more than 30 minutes. One of the first of *Nautilus*'s many records was covering some 13,700 miles, from August to October of 1958, without breaking the surface once.

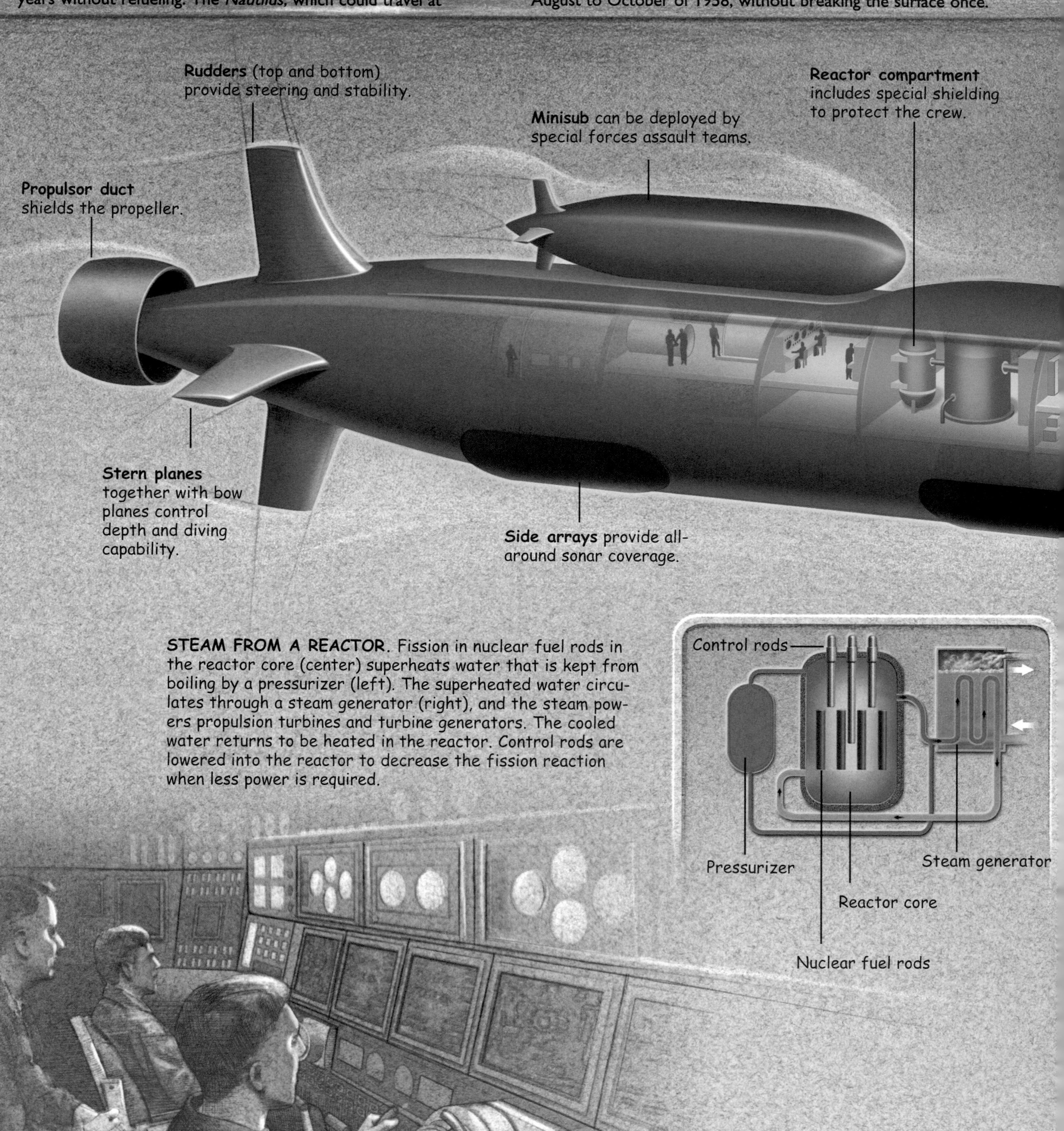

STEAM FROM A REACTOR. Fission in nuclear fuel rods in the reactor core (center) superheats water that is kept from boiling by a pressurizer (left). The superheated water circulates through a steam generator (right), and the steam powers propulsion turbines and turbine generators. The cooled water returns to be heated in the reactor. Control rods are lowered into the reactor to decrease the fission reaction when less power is required.

As shown below, the reactor in a nuclear vessel superheats water in one closed-loop system that in turn creates steam in a secondary system; the steam drives turbines that power the craft's propeller and also generates electricity to run the rest of the submarine's equipment.

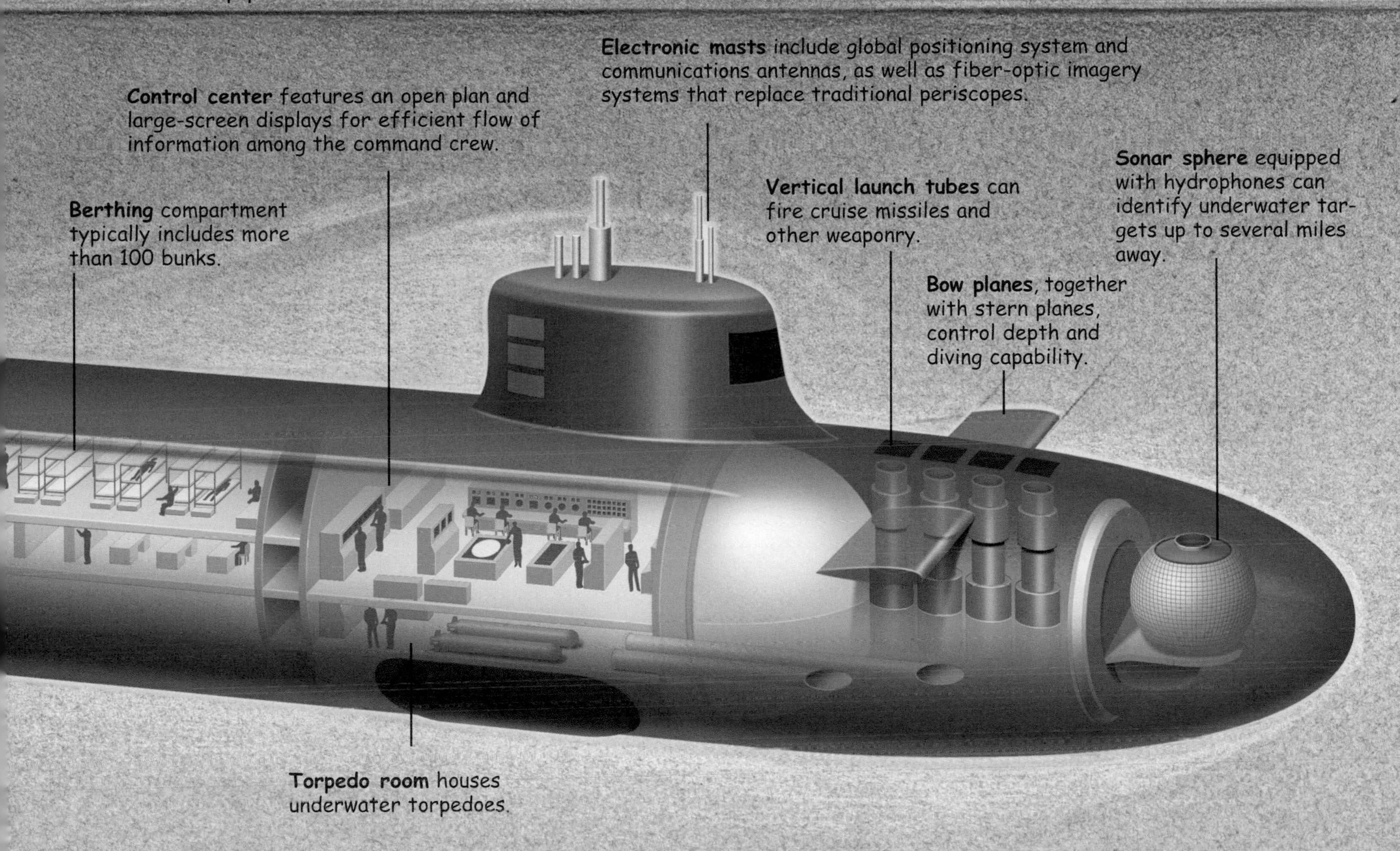

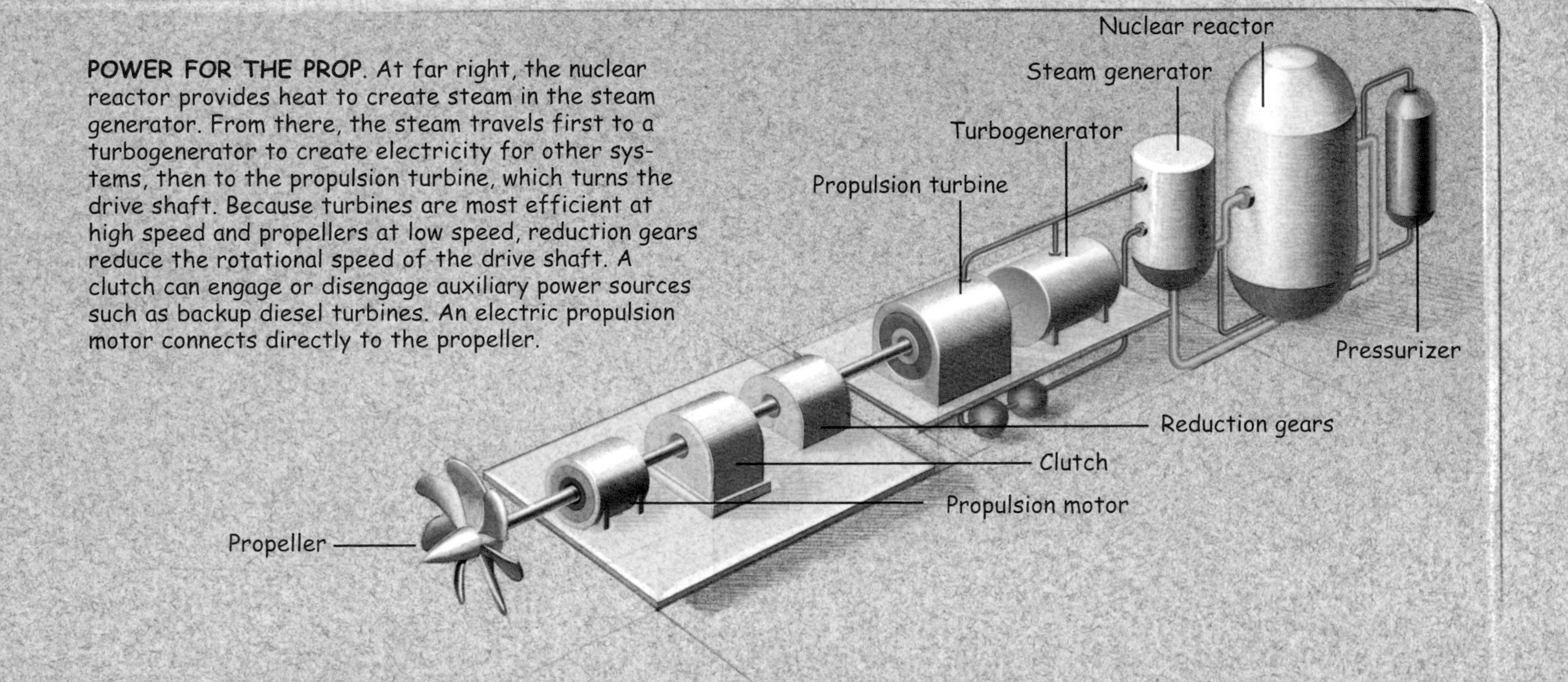

POWER FOR THE PROP. At far right, the nuclear reactor provides heat to create steam in the steam generator. From there, the steam travels first to a turbogenerator to create electricity for other systems, then to the propulsion turbine, which turns the drive shaft. Because turbines are most efficient at high speed and propellers at low speed, reduction gears reduce the rotational speed of the drive shaft. A clutch can engage or disengage auxiliary power sources such as backup diesel turbines. An electric propulsion motor connects directly to the propeller.

1986. In each case, radiation was released into the atmosphere, a small amount at Three Mile Island but a tragically large amount at Chernobyl. Human error played a significant role in both events, but Chernobyl also revealed the need to improve safeguards in future reactor designs.

Although public sentiment in the United States turned against nuclear power for a number of years after the Three Mile Island accident, the international growth of nuclear power continued virtually unabated, with an additional 350 nuclear plants built worldwide in the past 2 decades—almost doubling the previous total. A strong incentive for continuing to improve nuclear technology is the fact that it may offer a solution to global warming and reduce the free release of emissions such as sulfur oxides and nitrous oxides as well as trace metals. In addition, engineers in other countries and the United States have continued to refine reactor designs to improve safety. Most recently, designs have been proposed for reactors that are physically incapable of going supercritical and causing a catastrophic meltdown of the reactor's radioactive core, and such designs are ready to be moved beyond the drawing board. Nations with nuclear power plants continue to wrestle with the problem of disposing of nuclear waste—spent nuclear fuel and fission products—which can remain radioactively lethal for thousands, and even tens of thousands, of years. In the United States, most power plants store their own nuclear waste onsite in huge pools of water, while the longer-term option of a national repository, deep within the bedrock of Yucca Mountain in Nevada, continues to be debated. Other countries reprocess waste, extracting every last particle of fissionable fuel. And plans are also afoot to convert nuclear material from obsolete weapons—particularly those of the former Soviet Union—into usable nuclear fuel. Even though no new nuclear plants have been ordered in the United States since 1977, most existing facilities have requested extensions of their operating licenses—in part because of the many advantages of nuclear power over other forms of energy.

Still, developments in nuclear technology remain controversial. A case in point is the irradiation of food, approved by the Food and Drug Administration in 1986 but only slowly gaining public acceptance. Irradiation involves subjecting foods to high doses of radiation, which kills harmful bacteria on spices, fruits, and vegetables and in raw meats, preventing foodborne illnesses and dramatically reducing spoilage. No residual radiation remains in the food, but—despite laboratory evidence to the contrary—critics have expressed concerns that the process may cause other chemical changes that could give rise to toxic or carcinogenic substances. Nevertheless, as its benefits become more and more obvious, irradiaton has come into wider use.

In a typical industrial irradiator used for food products, the radiation source (cobalt-60) is housed in a room with concrete walls 2 meters thick. Food to be irradiated moves in and out of the room by a conveyor system. When workers must enter the room, the source is lowered to the bottom of a storage pool, where water absorbs the radiation energy. The gamma rays emitted by the cobalt-60 to kill bacteria penetrate the food so rapidly that no heat is produced. Because the cobalt-60 rods slowly decay to non-radioactive nickel, no radioactive waste remains.

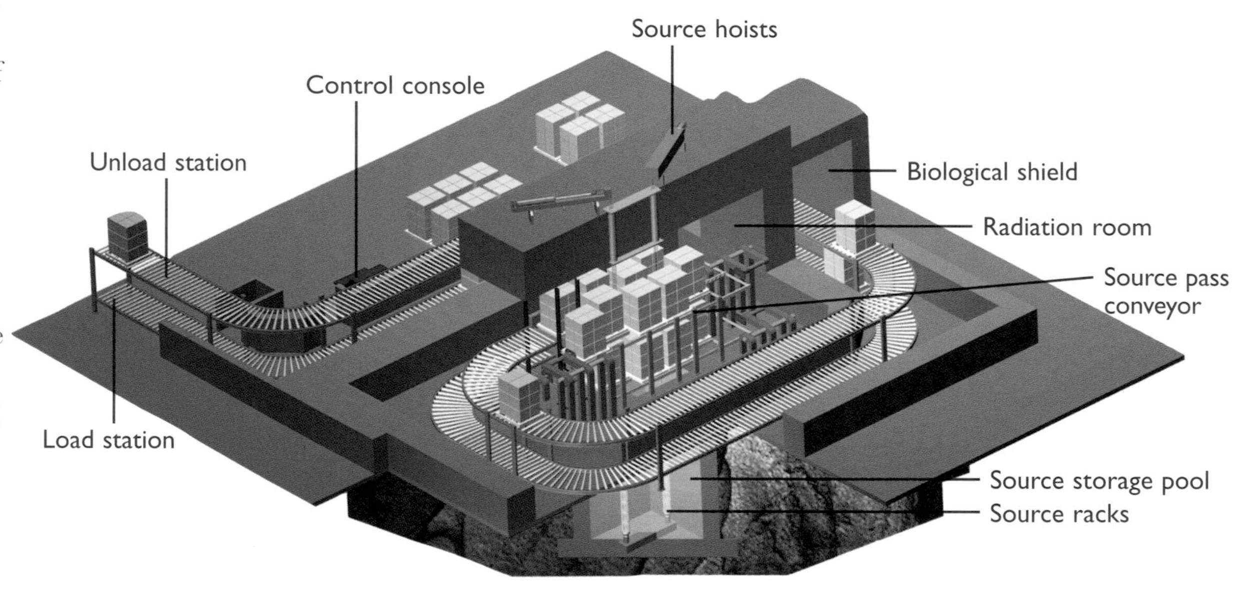

In 1970 many nations signed a nuclear nonproliferation treaty in an effort to limit the spread of nuclear weapons. That issue remains front and center in the news, even as engineers keep working to make peaceful uses of nuclear power safer. It may well be that harnessing the tremendous power of the atom will continue to be a story of both swords and plowshares.

Perspective

Shirley Ann Jackson

President
Rensselaer Polytechnic Institute

Throughout its relatively brief history the use of nuclear energy has been marked by contrast and sometimes controversy. We have had to weigh the answers to some hard questions: Does the global threat of nuclear weapons overshadow the benefits of peaceful nuclear technologies? Against this backdrop, how do we create a regulatory regime that allows the promise of civilian use of nuclear technology to be realized? When I was appointed chairman of the U.S. Nuclear Regulatory Commission (NRC) in 1995, I quickly found that these contrasts and questions were best understood in their historical context.

In the United States, beginning with the creation of the Atomic Energy Commission (AEC) in 1946, the development and regulation of nuclear energy have evolved along two streams: military uses (i.e., weapons) and civilian or "peaceful" uses (e.g., nuclear power and nuclear medicine). The secrecy necessary initially to protect weapons development gradually extended to the civilian nuclear power industry, sometimes leading to public distrust and misunderstanding that would linger for half a century. Meanwhile, other nations were racing to acquire the secrets of nuclear science as well. After President Eisenhower's 1953 "Atoms for Peace" speech to the United Nations, the International Atomic Energy Agency was created as a vehicle for offering peaceful nuclear technology to the entire world while preventing, through multilateral treaties, the proliferation of weapons technology.

Two decades later in the United States, the Energy Reorganization Act of 1974 abolished the AEC, replacing it with the Nuclear Regulatory Commission, which would focus on protecting public health and safety. The act also created the Energy Research and Development Administration (ERDA) to focus on energy research and development and on the federal government's nuclear energy defense activities. In 1977 ERDA became the U.S. Department of Energy. Then came two pivotal events—the nuclear accidents at Three Mile Island in 1979 and Chernobyl in 1986. The accident at Three Mile Island, resulting in a minor radiation leak, led Congress to reorganize the NRC again, changing aspects of the regulatory and management structure to improve the focus on safety.

These history lessons helped shape my priorities at the NRC: public transparency, a vigorous international presence, and smarter regulation. The commission began taking dramatic steps to give the public an active role in NRC deliberations. We stepped up support for international nuclear safety and security programs, and, together with senior regulators from eight other countries, I founded the International Nuclear Regulators Association.

To promote smarter investments in safety, I pushed for regulation that would take advantage of the insights gained through "probabilistic risk assessments"—exhaustive engineering analyses that ranked the relative risks associated with systems, structures, and components throughout a given nuclear plant. Operations involving high risk—for example, maintenance on a system that would ensure emergency cooling water during an accident—required more stringent quality assurance measures. Other operations of less risk could be handled in a less stringent way. This is the essence of risk-informed operation (and regulation). This risk ranking, when combined with standard emergency analysis and operational history, made regulation more cost effective, by directing the greatest investment toward areas of greatest vulnerability.

The current state of nuclear technology has benefited dramatically from persistent efforts on each of these fronts. Not all of the challenges of nuclear technology have been solved. The terrorist attacks of September 2001 raised the ante for nuclear security, and the year 2002 brought new challenges to international efforts to prevent the spread of nuclear weapons. But great strides have been made: Nuclear power has become economically competitive while simultaneously operating more safely than ever before, and public confidence in nuclear technology—and appreciation of its many benefits—is on the rise.

NUCLEAR TECHNOLOGIES TIMELINE

Even though the ancient Greeks correctly theorized that everything was made up of simple particles, which they called atoms, it wasn't until the beginning of the 20th century that scientists realized the atom could be split. Nuclear physicists such as Britain's Joseph John Thomson and Denmark's Niels Bohr mapped out the atom's elementary building blocks (the electron, proton, and neutron) and paved the way for the discovery of nuclear fission—the process that transformed the atom into a new and powerful source of energy. Today atomic energy generates clean, low-cost electricity, powers some of the world's largest ships, and assists in the development of the latest health care techniques.

1905 German-born physicist Albert Einstein introduces his special theory of relativity, which states that the laws of nature are the same for all observers and that the speed of light is not dependent on the motion of its source. The most celebrated result of his work is the mathematical formula $E=mc^2$, or energy equals mass multiplied by the speed of light squared, which demonstrates that mass can be converted into energy. Einstein wins the Nobel Prize in physics in 1921 for his work on the photoelectric effect.

1932 English physicist and Nobel laureate James Chadwick exposes the metal beryllium to alpha particles and discovers the neutron, an uncharged particle. It is one of the three chief subatomic particles, along with the positively charged proton and the negatively charged electron. Alpha particles, consisting of two neutrons and two protons, are positively charged, and are given off by certain radioactive materials. His work follows the contributions of New Zealander Ernest Rutherford, who demonstrated in 1919 the existence of protons. Chadwick also studies deuterium, known as heavy hydrogen, an isotope of hydrogen used in nuclear reactors.

British physicist John Cockcroft teams with Ernest Walton of Ireland to split the atom with protons accelerated to high speed. Their work wins them the Nobel Prize in physics in 1951.

1937 The Westinghouse Corporation builds the 5-million-volt Van de Graaff generator. Named for its inventor, physicist Robert Van de Graaff, the generator gathers and stores electrostatic charges. Released in a single spark and accelerated by way of a magnetic field, the accumulated charge, equivalent to a bolt of lightning, can be used as a particle accelerator in atom smashing and other experiments.

1939 Physicists Otto Hahn and Fritz Strassmann of Germany, along with Lise Meitner of Austria and her nephew Otto Frisch, split uranium atoms in a process known as fission. The mass of some of the atoms converts into energy, thus proving Einstein's original theory.

1939-1945 The U.S. Army's top-secret atomic energy program, known as the Manhattan Project, employs scientists in Los Alamos, New Mexico, under the direction of physicist J. Robert Oppenheimer *(above)*, to develop the first transportable atomic bomb. Other Manhattan Project teams at Hanford, Washington, and Oak Ridge, Tennessee, produce the plutonium and uranium-235 necessary for nuclear fission.

1942 Italian-born physicist and Nobel winner Enrico Fermi and his colleagues at the University of Chicago achieve the first controlled, self-sustaining nuclear chain reaction in which neutrons released during the splitting of the atom continue splitting atoms and releasing more neutrons. Fermi's team builds a low-powered reactor, insulated with blocks of graphite, beneath the stands at the university's stadium. In case of fire, teams of students stand by, equipped with buckets of water.

1945 To force the Japanese to surrender and end World War II, the United States drops atomic bombs on Hiroshima, an important army depot and port of embarkation, and Nagasaki, a coastal city where the Mitsubishi torpedoes used in the attack on Pearl Harbor were made.

1946 The U.S. Congress passes the Atomic Energy Act to establish the Atomic Energy Commission, which replaces the Manhattan Project. The commission is charged with overseeing the use of nuclear technology in the postwar era.

The U. S. Army's Oak Ridge facility in Tennessee ships the first nuclear-reactor-produced radioisotopes for peacetime civilian use to Brainard Cancer Hospital in St. Louis.

1948 The U.S. government's Argonne National Laboratory, operated in Illinois by the University of Chicago, and the Westinghouse Corporation's Bettis Atomic

Some of the scientists instrumental in developing the atom bomb gather in 1945 for the opening of the Institute of Nuclear Studies and Institute of Metals at the University of Chicago. Seated left to right: W. H. Zachariasen, Harold C. Urey, Cyril Smith, Enrico Fermi, and Samuel K. Allison. Standing left to right: Edward Teller, Thorfin Hogness, Walter Zinn, Clarence Zener, Joseph E. Mayer, Philip W. Schutz, R. H. Christ, and Carl Eckhart.

Power Laboratory in Pittsburgh, announce plans to commercialize nuclear power to produce electricity for consumer use.

1951 Experimental Breeder Reactor I at the Idaho National Engineering and Environmental Laboratory (INEEL) produces the world's first usable amount of electricity from nuclear energy. When neutrons released in the fission process convert uranium into plutonium, they generate, or breed, more fissile material, thus producing new fuel as well as energy. No longer in operation, the reactor is now a registered national historic landmark and is open to the public for touring.

1953 BORAX-I, the first of a series of Boiling Reactor Experiment reactors, is built at INEEL. The series is designed to test the theory that the formation of steam bubbles in the reactor core does not cause an instability problem. BORAX-I proves that steam formation is, in fact, a rapid, reliable, and effective mechanism for limiting power, capable of protecting a properly designed reactor against "runaway" events. In July 1955, BORAX-III becomes the first nuclear power plant in the world to provide an entire town with all of its electricity. When power from the reactor is cut in, utility lines supplying conventional power to the town of Arco, Idaho (population 1,200), are disconnected. The community depends solely on nuclear power for more than an hour.

1954 The U.S. Congress passes the Atomic Energy Act of 1954, amending the 1946 act to allow the Atomic Energy Commission to license private companies to use nuclear materials and also to build and operate nuclear power plants. The act is designed to promote peaceful uses of nuclear energy through private enterprise, implementing President Dwight D. Eisenhower's Atoms for Peace Program.

1955 The USS *Nautilus SSN 571*, the world's first nuclear-powered submarine, gets under way on sea trials. The result of the efforts of 300 engineers and technicians working under the direction of Admiral Hyman Rickover, "father of the nuclear navy," it is designed and built by the Electric Boat Company of Groton, Connecticut, and outfitted with a pressurized-water reactor built by the Westinghouse Corporation's Bettis Atomic Power Laboratory. In 1958 the *Nautilus* is the first ship to voyage under the North Pole.

USS Nautilus, *the first nuclear-powered submarine, enters New York Harbor during routine training operations.*

1957 The International Atomic Energy Agency is formed with 18 member countries to promote peaceful uses of nuclear energy. Today it has 130 members.

The first U.S. large-scale nuclear power plant begins operation in Shippingport, Pennsylvania. Built by the federal government but operated by the Duquesne Light Company in conjunction with the Westinghouse Bettis Atomic Power Laboratory, the pressurized-water reactor supplies power to the city of Pittsburgh and much of western Pennsylvania. In 1977 the original reactor is replaced by a more efficient light-water breeder reactor.

1962 The first advanced gas-cooled reactor is built at Calder Hall in England. Intended originally to power a naval vessel, the reactor is too big to be installed aboard ship and is instead successfully used to supply electricity to British consumers. A smaller pressurized-water reactor, supplied by the United States, is then installed on Britain's first nuclear-powered submarine, the HMS *Dreadnaught.*

1966 The Advanced Testing Reactor at the Idaho National Engineering and Environmental Laboratory begins operation for materials testing and isotope generation.

1969 The Zero Power Physics Reactor (ZPPR), a specially designed facility for building and testing a variety of types of reactors, goes operational at Argonne National Laboratory-West in Idaho. Equipped with a large inventory of materials from which any reactor could be assembled in a few weeks, ZPPR operates at very low power, so the materials do not become highly radioactive and can be reused many times. Nuclear reactors can be built and tested in ZPPR for about 0.1% of the capital cost of construction of the whole power plant.

1974 The Energy Reorganization Act of 1974 splits the Atomic Energy Commission into the Energy Research and Development Administration (ERDA) and the Nuclear Regulatory Commission (NRC). ERDA's responsibilities include overseeing the development and refinement of nuclear power, while the NRC takes up the issue of safe handling of nuclear materials.

1979 The nuclear facility at Three Mile Island near Harrisburg, Pennsylvania, experiences a major failure when a water pump in the secondary cooling system of the Unit 2 pressurized-water reactor malfunctions. A jammed relief valve then causes a buildup of heat, resulting in a partial meltdown of the core but only a minor release of radioactive material into the atmosphere.

1986 The Chernobyl nuclear disaster occurs in Ukraine during unauthorized experiments when four pressurized-water reactors overheat, releasing their water coolant as steam. The hydrogen formed by the steam causes two major explosions and a fire, releasing radioactive particles into the atmosphere that drift over much of the European continent.

1990s The U.S. Naval Nuclear Propulsion Program pioneers new materials and develops improved material fabrication techniques, radiological control, and quality control standards.

2000 The fleet of more than 100 nuclear power plants in the United States achieve world record reliability benchmarks, operating annually at more than 90 percent capacity for the last decade—the equivalent of building 10 gigawatt nuclear power plants in that period. In the 21 years since the Three Mile Island accident, the fleet can claim the equivalent of 2,024.6 gigawatt-years of safe reactor operation, compared to a total operational history of fewer than 253.9 gigawatt-years before the accident. Elsewhere in the world, nuclear power energy production grows, most notably in China, Korea, Japan, and Taiwan, where more than 28 gigawatts of nuclear power plant capacity is added in the last decade of the century.

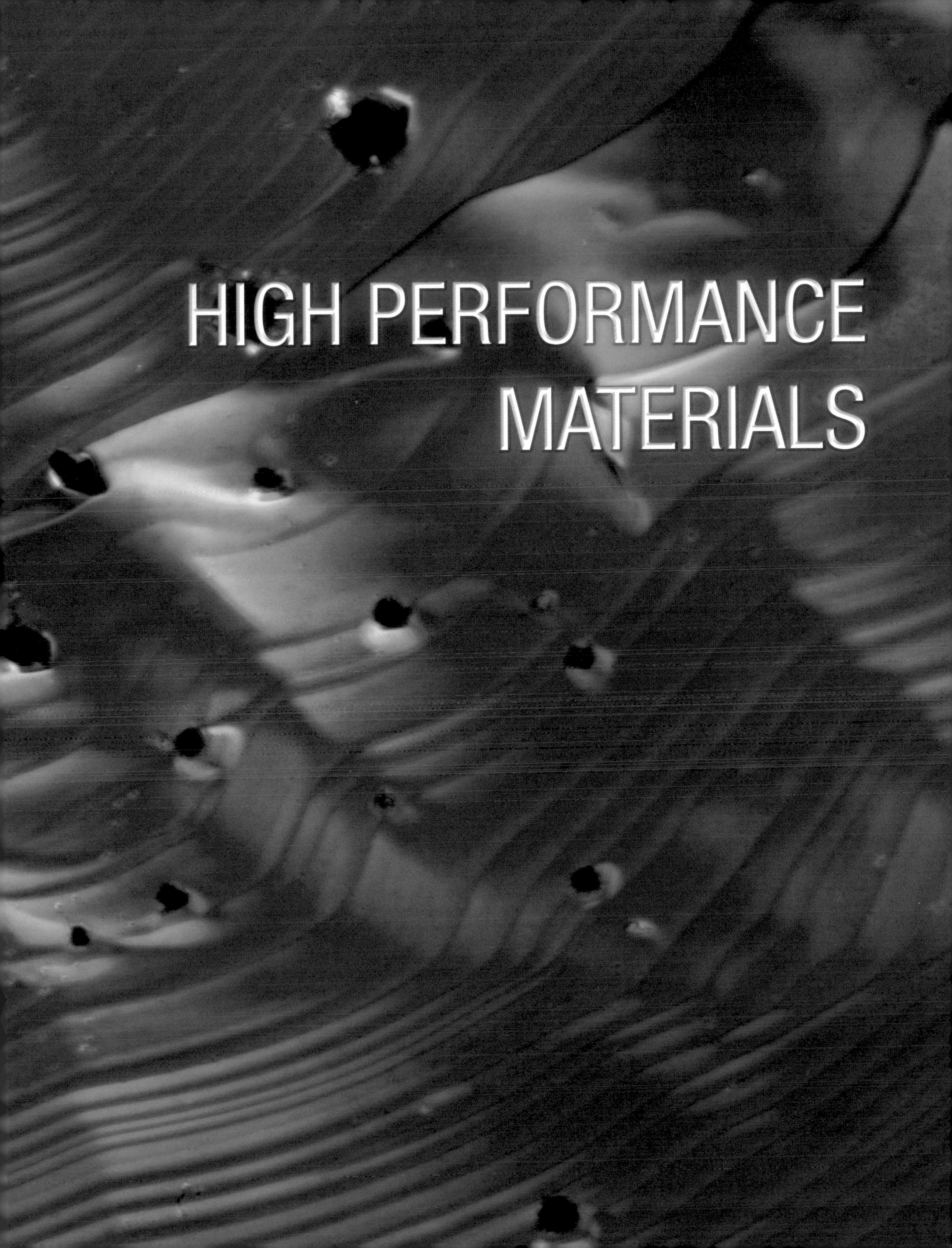
HIGH PERFORMANCE
MATERIALS

"All hail, King Steel," wrote Andrew Carnegie in a 1901 paean to the monarch of metals, praising it for working "wonders upon the earth." A few decades earlier a British inventor named Henry Bessemer had figured out how to make steel in large quantities, and Carnegie and other industry titans were now producing millions of tons of it each year, to be used for the structural framing of bridges and skyscrapers, the tracks of sprawling railway networks, the ribs and plates of steamship hulls, and a multitude of other applications extending from food cans to road signs.

In the decades to come, however, there would be many more claimants to wonder-working glory—among them other metals, polymers, ceramics, blends called composites, and the electrically talented group known as semiconductors. Over the course of the 20th century, virtually every aspect of the familiar world, from clothing to construction, would be profoundly changed by new materials. High performance materials would also make possible some of the century's most dazzling technological achievements: airplanes and spacecraft, microchips and magnetic disks, lasers and the fiber-optic highways of the Internet. And behind all that lies another, less obvious, wonder—the ability of scientists and engineers to customize matter for particular applications by manipulating its composition and microstructure: they start with a design requirement and create a material that answers it.

Of the various families of metals represented among high performance materials, steel still stands supreme in both versatility and volume of production. Hundreds of alloys are made by adding chromium, nickel, manganese, molybdenum, vanadium, or other metals to the basic steel recipe of iron plus a small but critical amount of carbon. Some of these alloys are superstrong or ultrahard; some are almost impervious to corrosion; some can withstand constant flexing; some possess certain desired electrical or magnetic properties. Highly varied microstructures can be produced by processing the metal in various ways.

Until well into the 20th century, new steel alloys were concocted mainly by trial-and-error cookery, but steelmakers at least had the advantage of long experience—3 millennia of it, in fact. That wasn't the case with aluminum, the third most common element in Earth's crust, yet never seen in pure form until 1825. It was heralded as a marvel—light, silvery, resistant to corrosion—but the metal was so difficult to separate from its ore that it remained a rarity until the late 19th century, when a young American, Charles Martin Hall, found that electricity could pull aluminum atoms apart from tight-clinging oxygen partners. Extensive use was still blocked by the metal's softness, limiting it to such applications as jewelry and tableware. But in 1906 a German metallurgist named Alfred Wilm, by happy chance, discovered a strengthening method. He made an alloy of aluminum with a small amount of copper and heated it to a high temperature, then quickly cooled it. At first the aluminum was even softer than before, but within a few days it became remarkably strong, a change caused by the formation of minute copper-rich particles in the alloy, called precipitation hardening. This lightweight material became invaluable in aviation and other transportation applications.

By the 1920s Leo Baekeland (right) had developed a plastic that was being used to make everything from electrical insulation to billiard balls and toothbrushes. And by the 1940s lightweight aluminum sheet was being used for aircraft fuselages (below).

In recent decades other high performance metals have found important roles in aircraft. Titanium, first isolated in 1910 but not produced in significant quantities until the 1950s, is one of them. It is not only light and resistant to corrosion but also can endure intense heat, a requirement for the skin of planes traveling at several times the speed of sound. But even titanium can't withstand conditions inside the turbine of a jet engine, where temperatures may be well above 2,000° F. Turbine blades are instead made of nickel- and cobalt-based materials known as superalloys, which remain strong in fierce heat while spinning at tremendous speed. To ensure they have the maximum possible resistance to high-temperature deformation, the most advanced of these blades are grown from molten metal as single crystals in ceramic molds.

Another major category of high performance materials is that of synthetic polymers, commonly known as plastics. Unknown before the 20th century, they are now ubiquitous and immensely varied. The first of the breed was created in 1907 by a Belgium-born chemist named Leo Baekeland (*right*). Working in a suburb of New York City, he spent years experimenting with mixtures of phenol (a distillate of coal tar) and formaldehyde (a wood-alcohol distillate). Eventually he discovered that, under controlled heat and pressure, the two liquids would react to yield a thick brownish resin. Further heating of the resin produced a powder, which became a useful varnish if dissolved in alcohol. And if the powder was remelted in a mold, it rapidly hardened and held its shape. Bakelite, as the hard plastic was called, was an excellent electrical insulator. It was tough; it wouldn't burn; it didn't crack or fade; and it was unaffected by most solvents. By the 1920s the translucent, amber-colored plastic was everywhere—in pipe stems and toothbrushes, billiard balls and fountain pens, combs and ashtrays. It was "the material of a thousand purposes," *Time* magazine said.

The basic steel being used early in the 20th century (opposite) today comes as hundreds of alloys with carbon, chromium, nickel, and other metals added to make the alloys superstrong or virtually corrosion-proof, like the block of stainless steel being cut by an industrial laser at left. Synthetic silicon carbide crystals (previous pages) are valued for use in high-tech ceramics, semiconductors, abrasives, hard saws, and armor.

THE MIRACLE OF ADVANCED MATERIALS

FROM THE EVERYDAY TO THE EXTRAORDINARY

If any factor can be deemed indispensable to 20th-century technological progress and the prospects for more of the same, the synthesis of high performance materials fits the role. In realms as rarefied as rocketry or as prosaic as packaging, tailor-made materials answer a wide range of special needs—great strength, an ability to endure fierce heat, resistance to corrosion, flexibility, lightness, transparency, a knack for swift shuffling of electrical charges, and many other critical engineering requirements. The most wild-eyed alchemist of yesteryear could never have envisioned today's reality of textiles made of fibers a hundred times finer than a human hair, sturdy replacements for bone, or sand-grain-sized crystals that emit light.

Designing matter for a particular set of desired properties is exacting work, often involving such tools as electron microscopes and powerful computers, and the challenges of manufacturing the materials reliably and affordably can be no less formidable. Even so, the menu of new alloys, polymers, ceramics, and blends called composites has grown so vast that, as noted in the scene on these pages, they are now embedded everywhere.

IN THE COCKPIT. Flight information is supplied by liquid-crystal displays, which control light transmission by orienting molecules with an electrical field.

Cockpit windshields are made of specially hardened polycarbonate, acrylic, or laminated glass designed to endure rapid changes in mechanical stress and can be heated to counteract icing without sacrificing low weight and high visual quality.

CLOTHING. Strong, lightweight, breathable fabrics such as polyester are made from polymers—extremely long molecules that may comprise a million or more atoms. One polymer fiber, Lycra spandex, can stretch to five times its resting length without permanent deformation.

CELL PHONE. As cellular networks move to higher frequencies for increased capacity, the silicon-based electronics of cell phones are being replaced by so-called compound semiconductors such as gallium arsenide, which can operate at greater speeds. Compound semiconductors are used in the low-noise receivers of cell phone handsets.

BIONIC JOINTS. Prosthetic devices such as the hip replacement shown left need to be strong enough to handle the body's weight and the forces of movement, resistant to constant wear and corrosion, and not susceptible to bodily rejection. They consist of metal—stainless steel, titanium, or cobalt and chrome alloys—coated with durable polyethylene, a polymer.

SUNGLASSES. Lenses darken when embedded molecules of silver chloride or silver halide change shape in reaction to ultraviolet (UV) rays in sunlight. In the absence of UV rays, the reverse chemical reaction takes place, the molecules snap back to their original shape, and the lenses lighten.

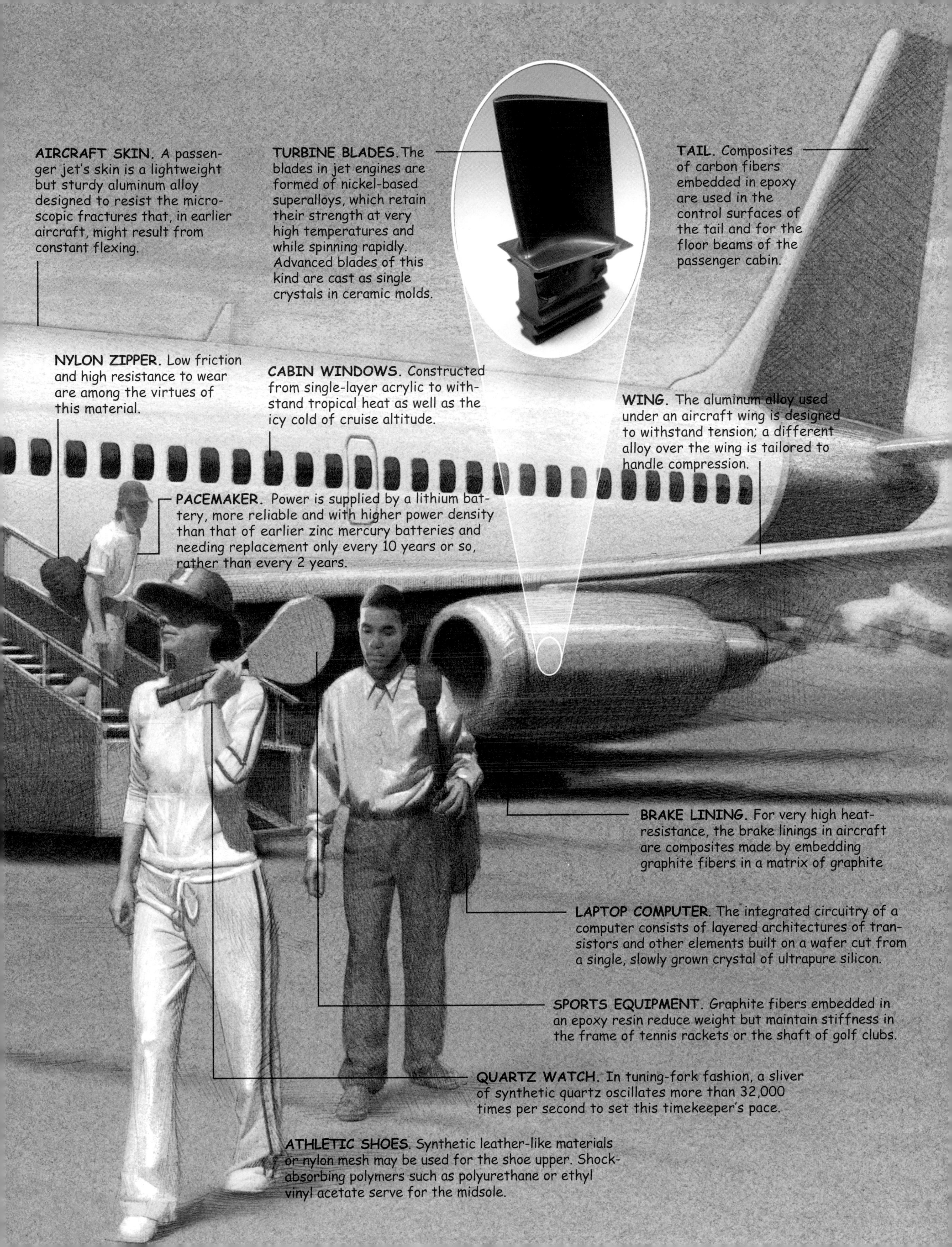

AIRCRAFT SKIN. A passenger jet's skin is a lightweight but sturdy aluminum alloy designed to resist the microscopic fractures that, in earlier aircraft, might result from constant flexing.

TURBINE BLADES. The blades in jet engines are formed of nickel-based superalloys, which retain their strength at very high temperatures and while spinning rapidly. Advanced blades of this kind are cast as single crystals in ceramic molds.

TAIL. Composites of carbon fibers embedded in epoxy are used in the control surfaces of the tail and for the floor beams of the passenger cabin.

NYLON ZIPPER. Low friction and high resistance to wear are among the virtues of this material.

CABIN WINDOWS. Constructed from single-layer acrylic to withstand tropical heat as well as the icy cold of cruise altitude.

WING. The aluminum alloy used under an aircraft wing is designed to withstand tension; a different alloy over the wing is tailored to handle compression.

PACEMAKER. Power is supplied by a lithium battery, more reliable and with higher power density than that of earlier zinc mercury batteries and needing replacement only every 10 years or so, rather than every 2 years.

BRAKE LINING. For very high heat-resistance, the brake linings in aircraft are composites made by embedding graphite fibers in a matrix of graphite

LAPTOP COMPUTER. The integrated circuitry of a computer consists of layered architectures of transistors and other elements built on a wafer cut from a single, slowly grown crystal of ultrapure silicon.

SPORTS EQUIPMENT. Graphite fibers embedded in an epoxy resin reduce weight but maintain stiffness in the frame of tennis rackets or the shaft of golf clubs.

QUARTZ WATCH. In tuning-fork fashion, a sliver of synthetic quartz oscillates more than 32,000 times per second to set this timekeeper's pace.

ATHLETIC SHOES. Synthetic leather-like materials or nylon mesh may be used for the shoe upper. Shock-absorbing polymers such as polyurethane or ethyl vinyl acetate serve for the midsole.

Advanced materials make up objects some of us may handle frequently, such as graphite composite frames for tennis rackets (above) and soft contact lenses made of acrylic polymers that absorb water (above right). Meanwhile, research continues on so-called nanotechnologies, such as a wire just 10 atoms wide (red bar below) that could be used in computers operating at the limits of miniaturization.

use in building panels, bathtubs, boat hulls, and other marine products. Since then, many metals, polymers, and ceramics have been exploited as both matrix and reinforcement. In the 1960s, for instance, the U.S. Air Force began seeking a material that would be superior to aluminum for some aircraft parts. Boron had the desired qualities of lightness and strength, but it wasn't easily formed. The solution was to turn it into a fiber that was run through strips of epoxy tape; when laid in a mold and subjected to heat and pressure, the strips yielded strong, lightweight solids—a tail section for the F-14 fighter jet, for one. While an elegant solution, boron fibers were too expensive to find wide use, highlighting the critical interplay between cost and performance that drives materials applications.

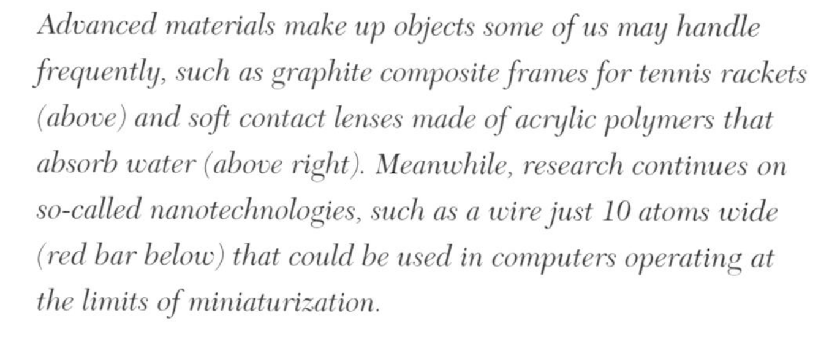

Many composites are strengthened by graphite fibers. They may be embedded in a matrix of graphite to produce a highly heat-resistant material—the lining for aircraft brakes, for example—or the matrix can be an epoxy, as with composite shafts for golf clubs or frames for tennis rackets. Other sorts of composites abound in the sports world. Skis can be reinforced with Kevlar fibers; the handlebars of some lightweight racing bikes are made of aluminum reinforced with aluminum oxide particles. Ceramic-matrix composites find use in a variety of hostile environments, ranging from outer space to the innards of an automobile engine.

Tens of thousands of materials are now available for various engineering purposes, and new ones are constantly being created. Sometimes the effort is grandly scaled — measured in vast tonnages of a metal or polymer, for instance—but many a recent triumph is rooted in exquisite precision and control. This is especially the case in the amazing realm of electronics, built on combinations of metals, semiconductors, and oxides in miniaturized geometries—the fingernail-sized microchips of computers or CD players, the tiny lasers and threadlike optical fibers of communications networks, the magnetic particles dispersed on discs and other surfaces to record digital data. Making transistors, for example, begins with the growing of flawless crystals of silicon, since the electrical properties of the semiconductor are sensitive to minuscule amounts of impurities (in some cases, just one atom in a million or less) and to tiny imperfections in their crystalline structure. Similarly, optical fibers are composed of silica glass so pure that if the Pacific Ocean were made of the same material, an observer on the surface would have no difficulty seeing details on the bottom miles below. Such stuff is transforming our lives as dramatically as steel once did, and engineering at the molecular level of matter promises much more of the same.

Perspective

Mary L. Good

Professor and Dean
Donaghey College of Information Science and Systems Engineering
University of Arkansas at Little Rock

We have long identified epochs of human history in terms of the materials exploited—referring, for example, to the Stone Age or the Iron Age. The hallmark of progress in every age has been the way "materials engineers" worked to improve the usefulness of materials, whether extracting coal or iron ore from the earth or creating new materials from combinations, such as iron and carbon to produce steel. For most of history such improvements have been incremental and have depended on experimentation, accidents, and passing on from generation to generation the "art" of materials processing and finishing. However, by the early 1980s, instrumentation, simulation techniques, and the accumulation and analysis of materials databases had moved materials engineering and structural design much closer to the fundamental physics and chemistry of the materials' building blocks. Thus, the concept of "materials by design" began to have some champions, and the potential for creating new materials with designed properties for specific applications no longer was considered "science fiction."

Those of us involved in the invention and improvement of catalytic processes and new catalysts found the idea of molecular design of catalysts with predetermined properties compelling. Catalytic science was driven by "trial and error" experimentation and the ability to determine correlations between performance and composition. In June 1984 the Research Laboratories of UOP (Universal Oil Products) and the Signal Companies, where I was president, were awarded a research contract from the Department of Energy to evaluate the concept of materials by design by assessing the current status of relevant theoretical, computational, and experimental tools. After several workshops with leaders in the field, we prepared an extensive report in 1986 to describe areas of understanding and islands of ignorance. This activity was one of the most stimulating and challenging of my career. At the time, computational theorists had good models for the quantum mechanics and properties of electronic systems, and engineers understood macroengineering design and had a good grasp of finite element analysis and process simulation. However, the understanding of molecular dynamics, atomic structure, and multimolecular pieces, or subunits, was quite primitive.

Fifteen years later I revisited this topic in a paper and a lecture to an international conference. Progress in materials by design in the interim had been profound. Theoretical calculations had progressed to quantum calculations of a few atoms to form the basis of quantum computing; atomic force microscopy could now image individual atoms, molecules, and molecular machines; atomic and molecular reactions on catalyst substrates could be imaged and analyzed directly. In addition, researchers could blend new alloys with totally new properties from existing materials and process them to provide desired physical properties. Others were building analytical chemistry instrumentation on a 1-square-inch chip and designing and producing a variety of micromachines.

Clearly the next 15 years will continue these insights into materials at the atomic and molecular level. The science of nanotechnology—the understanding of materials at the nanometer and molecular size—is now building on these prior excursions into the submicroscopic world. No longer "a science looking for applications," nanotechnology is turning some of these discoveries into real products, ranging from high performance fabrics in which nanostructures are intertwined with conventional synthetic fibers to nanocarbon fibers used to transform properties of polymeric materials. Prototypes of quantum dots, utilizing a few atoms, to be used for the next generation of supercomputers are just one of many examples of products to come. In materials engineering, atomic and molecular materials by design and the nanoproducts they can produce, may very well make the 21st century the "Nano Age"!

HIGH PERFORMANCE MATERIALS
TIMELINE

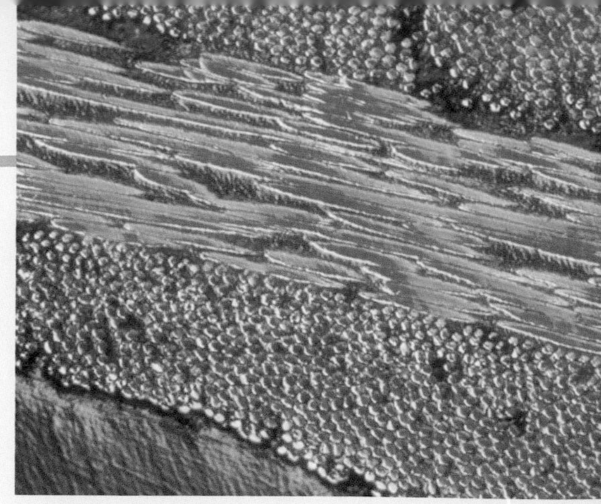

A light micrograph reveals strands of glass fiber in a matrix of polypropylene, a class of flexible but tough synthetic plastic.

Over the millennia human beings have tinkered with substances to devise new and useful materials not ordinarily found in nature. But little prepared the world for the explosion in materials research that marked the 20th century. From automobiles to aircraft, sporting goods to skyscrapers, clothing (both everyday and superprotective) to computers and a host of electronic devices—all bear witness to the ingenuity of materials engineers.

1907 Leo Baekeland, a Belgian immigrant to the United States, creates Bakelite, the first thermosetting plastic. An electrical insulator that is resistant to heat, water, and solvents, Bakelite is clear but can be dyed and machined.

1909 Alfred Wilm, then leading the Metallurgical Department at the German Center for Scientific Research near Berlin, discovers "precipitation hardening," a phenomenon that is the basis for the creation of strong, lightweight aluminum alloys essential to aeronautics and other technologies in need of such materials. Many other materials are also strengthened by precipitation hardening.

1913 Although created earlier in the century by a Frenchman and a German, stainless steel is rediscovered by Harry Brearley in Sheffield, England, and he is credited with popularizing it. Made of iron with about 13 percent chromium and a small portion of carbon, stainless steel does not rust.

1915 Corning research physicist Jesse Littleton cuts the bottom from a glass battery jar produced by Corning, takes it home, and asks his wife to bake a cake in it. The glass withstands the heat during the baking process, leading to the development of borosilicate glasses for kitchenware and later to a wide range of glass products marketed as Pyrex.

1925 A stainless steel containing 18 percent chromium, 8 percent nickel, and 0.2 percent carbon comes into use. Known as 18/8 austenitic grade, it is adopted by the chemical industry starting in 1929. By the late 1930s the material's usefulness at high temperatures is recognized and it is used in the production of jet engines during World War II.

1930 Wallace Carothers and a team at DuPont, building on work begun in Germany early in the century, make synthetic rubber. Called neoprene, the substance is more resistant than natural rubber to oil, gasoline, and ozone, and it becomes important as an adhesive and a sealant in industrial uses.

1930s Engineers at the Owens Illinois Glass Company and Corning Glass Works develop several means to make glass fibers commercially viable. Composed of ingredients that constitute regular glass, the glass fibers produced in the 1930s are made into strands, twirled on a bobbin, and then spun into yarn. Combined with plastics, the material is called fiberglass and is used in automobiles, boat bodies, and fishing rods, and is also made into material suitable for home insulation.

1933 Polyethylene, a useful insulator, is discovered by accident by J. C. Swallow, M. W. Perrin, and Reginald Gibson in Britain. First used for coating telegraph cables, polyethylene is then developed into packaging and liners. Processes developed later render it into linear low-density polyethylene and low-density polyethylene.

1934 Experimenting over 4 years to craft an engineered substitute for silk, Wallace Carothers and his assistant Julian Hill at DuPont ultimately discover a successful process with polyamides. They also learn that their polymer increases in strength and silkiness as it is stretched, thus also discovering the benefits of cold drawing. The new material, called nylon, is put to use in fabrics, ropes, and sutures and eventually also in toothbrushes, sails, carpeting, and more.

1936 The Rohm and Haas Company of Philadelphia presses polymethyl acrylate between two pieces of glass, thereby making a clear plastic sheet of the material. It is the forerunner of what in the United States is called Plexiglass (polyvinyl methacrylate). Far tougher than glass, it is used as a substitute for glass in automobiles, airplanes, signs, and homes.

1938 Annoyed one day that a tank presumably full of tetrafluoroethylene gas is empty, DuPont scientist Roy Plunkett investigates and discovers that the gas had polymerized on the sides of the tank vessel. Waxy and slippery, the coating is also highly resistant to acids, bases, heat, and solvents. At first Teflon is used only in the war effort, but it later becomes a key ingredient in the manufacture of cookware, rocket nose cones, heart pacemakers, space suits, and artificial limbs and joints.

1940s Metallurgists develop nickel-based superalloys that are extremely resistant to high temperatures, pressure, centrifugal force, fatigue, and oxidation. The class of nickel-based superalloys with chromium, titanium, and aluminum makes the jet engine possible, and are eventually used in spacecraft as well as in ground-based power generators.

Scientists in the Netherlands develop ceramic magnets, known as ferrites, that are complex multiple oxides of iron, nickel, and other metals. Such magnets quickly become vital in all high-frequency communications, including the sound recording industry. Nickel-zinc-based ceramic magnets eventually become important as computer memory cores and in televisions and telecommunications equipment.

1945 Scientists in Ohio, Russia, and Japan all develop barium titanate, a ceramic that develops an electrical charge when mechanically stressed (and vice versa). Such ceramics advance the technologies of sound recordings, sonar, and ultrasonics.

1946 As a chemist at DuPont in the 1930s, Earl Tupper develops a sturdy but pliable synthetic polymer he calls Poly T. By 1947 Tupper forms his own corporation and makes nesting Tupperware bowls along with companion airtight lids. Virtually breakproof, Tupperware begins replacing ceramics in kitchens nationwide.

1950s Silicones, a family of chemically related substances whose molecules are made up of silicon-oxygen cores with carbon groups attached, become important as waterproofing sealants, lubricants, and surgical implants.

1952 Corning research chemist S. Donald Stookey discovers a heat treatment process for transforming glass objects into fine-grained ceramics. Further development of this new Pyroceram composition leads to the introduction of CorningWare in 1957.

1953 Karl Zeigler develops a method for creating a high-density polyethylene molecule that can be manufactured at low temperatures and pressures but has a very high melting point. It is made into dishes, squeezable bottles, and soft plastic materials.

DuPont opens a U.S. manufacturing plant to produce Dacron, a synthetic material first developed in Britain in 1941 as polyethylene terephthalate. Because it has a higher melting temperature than other synthetic fibers, Dacron revolutionizes the textiles industry.

1954 Working at General Electric's research laboratories, scientists use a high-pressure vessel to synthesize diamonds, converting a mixture of graphite and metal powder to minuscule diamonds. The process requires a temperature of 4,800°F and a pressure of 1.5 million pounds per square inch, but the tiny diamonds are invaluable as abrasives and cutting points.

Following work done in the late 1940s by Robert Milton and Donald Breck of the Linde Division of Union Carbide Corporation, the company markets two new families of synthetic zeolites (from the Greek for "boiling stone," referring to the visible loss of water that occurs when zeolites are heated) as a new class of industrial materials for separation and purification of organic liquids and gases. As the key materials for "cracking"—that is, separating and reducing the large molecules in crude oil—they revolutionize the petroleum and petrochemical industries. Synthetic zeolites are also put to use in soil improvement, water purification, and radioactive waste treatment, and as a more environmentally friendly replacement in detergents for phosphates.

1955 Building on the work of Karl Ziegler, Giullo Natta in Italy develops a high molecular weight polypropylene that has high tensile strength and is resistant to heat, ushering in an age of "designer" polymers. Polypropylene is put to use in films, automobile parts, carpeting, and medical tools.

1959 British glassmakers Pilkington Brothers announce a revolutionary new process of glass manufacturing developed by engineer Alastair Pilkington. Called "float" glass, it combines the distortion-free qualities of ground and polished plate glass with the less expensive production method of sheet glass. Tough and shatter-resistant, float glass is used in windows for shops and skyscrapers, windshields for automobiles and jet aircraft, submarine periscopes, and eyeglass lenses.

1960s Engineers begin to grow large single crystals of silicon with nearly perfect purity and perfection. The crystals are then sliced into thin wafers, etched, and doped to become semiconductors, the basis for the electronics industry.

Borosilicate glass is developed for encapsulating radioactive waste. Better but more expensive trapping materials are made from crystalline ceramic materials zirconolite and perovskite and from the most widespread material of all for containing radioactivity—carefully designed cements.

1962 Researchers at the Naval Ordnance Laboratory in White Oak, Maryland, discover that a nickel-titanium (Ni-Ti) alloy has so-called shape memory properties, meaning that the metal can undergo deformation yet "remember" its original shape, often exerting considerable force in the process. Although the shape memory effect was first observed in other materials in the 1930s, research now begins in earnest into the metallurgy and practical uses of these materials. Today a number of products using Ni-Ti alloys are on the market, including eyeglass frames that can be bent without sustaining permanent damage, guide wires for steering catheters into blood vessels in the body, and arch wires for orthodontic correction.

1964 Chemists develop acrylic paints, which dry more quickly than previous paints and drip and blister less. They are used for fabric finishes in industry and on automobiles.

British engineer Leslie Phillips makes carbon fiber by stretching synthetic fibers and then heating them to blackness. The result is fibers that are twice as strong as the same weight of steel. Carbon fibers find their way into bulletproof vests, high performance aircraft, automobile tires, and sports equipment.

1970s Amorphous metal alloys are made by cooling molten metal alloys extremely rapidly (more than a million degrees a second), producing a glassy solid with distinctive magnetic and mechanical properties. Such alloys are put to use in signal and power transformers and as sensors.

1977 Researchers Hideki Shirakawa, Alan MacDiarmid, and Alan Heeger announce the discovery of electrically conducting organic polymers. These are developed into light-emitting diodes (LEDs), solar cells, and displays on mobile telephones. The three are awarded the Nobel Prize in chemistry in 2000.

1980s Materials engineers develop "rare earth metals" such as iron neodymium boride, which can be made into magnets of high quality and permanency for use in sensors, computer disk drives, and automobile electrical motors. Other rare earth metals are used in color television phosphors, fluorescent bulbs, lasers, and magneto-optical storage systems with a capacity 15 times greater than that of conventional magnetic disks.

1986-1990s Engineers develop "synthetic skin." One type seeds fibroblasts from human skin cells into a three-dimensional polymer structure, all of which is eventually absorbed into the body of the patient. Another type combines human lower skin tissue with a synthetic epidermal or upper layer.

1990s-present Scientists investi-gate nanotechnology, the manipulation of matter on atomic and molecular scales. Electronic channels only a few atoms thick could lead to molecule-sized machines, extraordinarily sensitive sensors, and revolutionary manufacturing methods.

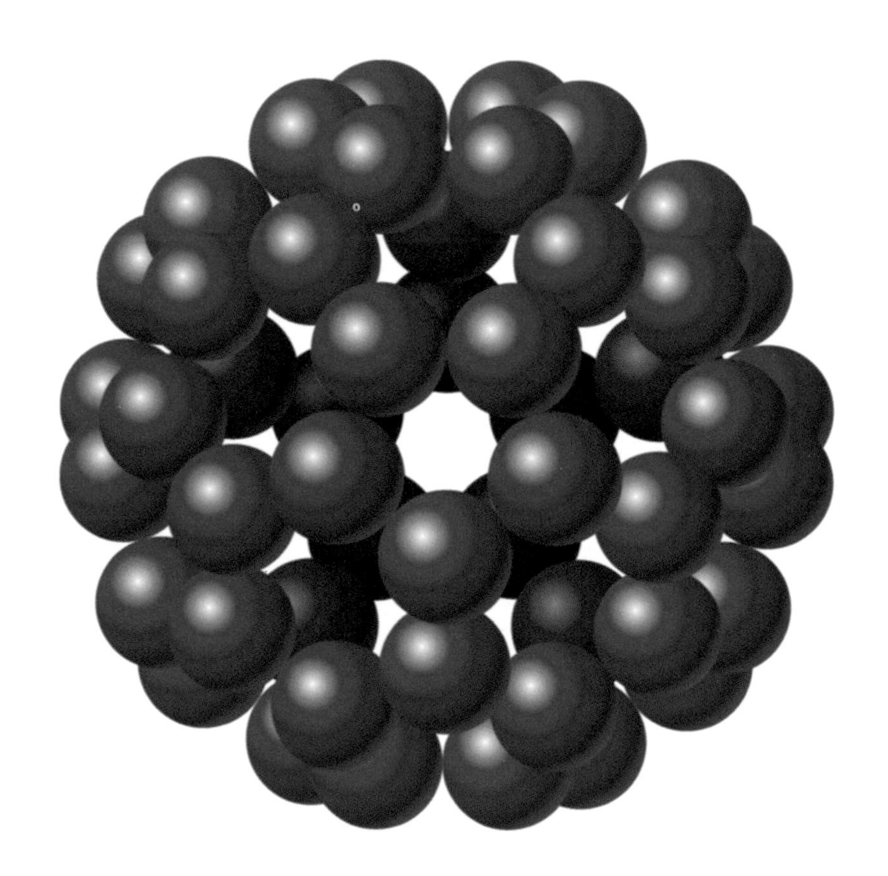

Buckminsterfullerene

AFTERWORD

My first serious attempt at technological prediction began in 1961 in the journal that has published most of my scientific writings—*Playboy* magazine. They were later assembled in *Profiles of the Future* (Bantam Books, 1964), which I am happy to say has just been reissued.

Whatever philosophers or theologians may say, our civilization is largely a product of technology. It is a worthwhile—and sometimes humbling—enterprise to consider what those inventions were. Many of them we take so completely for granted that we forget somebody had to think of them first.

Let us begin with the earliest ones—the wheel, the plough, bridle and harness, metal tools, glass. (I almost forgot buttons—where would we be without those?)

Moving some centuries closer to the present, we have writing, masonry (particularly the arch), moveable type, explosives, and perhaps the most revolutionary of all inventions because it multiplied the working life of countless movers and shakers—spectacles.

The harnessing and taming of electricity, first for communications and then for power, is the event that divides our age from all those that have gone before. I am fond of quoting the remark made by the chief engineer of the British Post Office, when rumors of a certain Yankee invention reached him: "The Americans have need of the Telephone—but we do not. We have plenty of messenger boys." I wonder what he would have thought of radio, television, computers, fax machines—and perhaps above all—e-mail and the World Wide Web. The father of the WWW, Tim Berners-Lee, generously suggested I may have anticipated it in my 1964 short story "Dial F for Frankenstein" (*Playboy* again!).

As I reluctantly approach my 85th birthday I have two main hopes—I won't call them expectations—for the future. The first is carbon 60—better known as Buckminsterfullerene, which may provide us with materials lighter and stronger than any metals. It would revolutionize every aspect of life and make possible the Space Elevator, which will give access to near-Earth space as quickly and cheaply as the airplane has opened up this planet.

The other technological daydream has suddenly come back into the news after a period of dormancy, probably caused by the "cold fusion" fiasco. It seems that what might be called low-energy nuclear reactions may be feasible, and a claim was recently made in England for a process that produces 10 times more energy than its input. If this can be confirmed—and be scaled up—our world will be changed beyond recognition. It would be the end of the Oil Age—which is just as well because we should be *eating* oil, not burning it.

Arthur C. Clarke

ACKNOWLEDGMENTS

This book is based on the *Greatest Engineering Achievements of the 20th Century*, a project designed by the National Academy of Engineering (NAE), the American Association of Engineering Societies, and National Engineers Week to raise public awareness of the role engineering has played in improving our quality of life. Initiated in 1999, it has been a collaborative project with many sectors of the engineering field—industry, academia, non-profits, professional societies, and individual engineers—joining together with the common message of engineering a better quality of life. So we have many people and organizations to thank, but I'd like to explain the origins and intent of the project before I begin those acknowledgments.

In 1999 the NAE invited 60 professional engineering societies to solicit nominations from their members for the *Greatest Engineering Achievements of the 20th Century* project. From the assembled nominations each society submitted its top five to the NAE. A total of 27 societies submitted nominations.

The NAE formed a selection committee comprised of leading engineering experts from academia, industry, government, and a wide range of engineering disciplines. The chief criterion for nominations was the significance that each engineering achievement had on our quality of life during the 20th century. From the initial 105 nominations the committee selected 48 nominations for final consideration. These 48 were grouped into larger categories. For example, specific innovations in building bridges and roads were combined under "highways," and the tractor, combine, and chisel plow were combined under "agricultural mechanization." This reduced the number of nominations to 28, and the committee met in December 1999 to select and rank the top 20, which are now the content for this book.

This book would not have been possible without the underwriting support of Robert A. Pritzker, president and chief executive officer of Colson Associates, Inc., who had the vision and desire to share this information with a wider audience. The United Engineering Foundation provided critical seed money at the beginning of the project.

Project editor Roberta Conlan and authors George Constable and Bob Somerville took on the tremendously challenging job of bringing this book together. They were expert and talented at sifting the facts and human-interest stories and portraying them in easy-to-understand language for nontechnical audiences.

One of the unique features of this book is that it gathers together in one place the thoughts of many of the nation's most prominent engineers. The writers of these "Perspectives" are all NAE members. I cannot thank them enough for helping to make the book infinitely more interesting than a mere history book would have been. I thank them for sharing their stories and insights. Other NAE members are mentioned in the book, as well, especially those who have been awarded the NAE's Draper and Russ Prizes for their technology achievements. A list of the awardees can be found on the NAE Web site <www.nae.edu> at the awards link. I congratulate them again on their achievements. Finally and foremost, I want to thank NAE member Neil Armstrong and Foreign Associate Arthur C. Clarke for reading all the chapters and providing their own unique perspectives at the fore and aft of this book.

In addition to original nominations that provided the content source for the book and Web site, the professional societies contributed immensely to this project in staff and resources. In particular, we thank Tom Price, Pat Natale, Leslie Collins, June Scangarello, Pender McCarter, Jane Howell, Michael Geselowitz, and Chuck Blue.

The Selection Committee, chaired by Guy Stever, also served as reviewers for the material herein. We thank the committee for its many, many hours of service. The Book Development Committee, chaired by Paul Torgersen, helped to frame the book. We thank the committee members for their vision.

We are grateful to a number of other individuals without whose timely assistance the many pieces of this book would not have come together: Carolina Asirifi, Ann Behar, Edward Eckert, Mary Ann Giglio, Julie Jordan, Diane Kaylor, Stephen Hiles, Peter Haynes, Beth Margulies, Lucy Norman, Christine O'Connell, Scott Papillon, Alice Portale, Charles Snearly, Ronald Taylor, Richard Teplitsky, Brooker Thro, and Thomas Way.

To make the information as accurate as possible, we entreated the volunteer efforts of a wide range of experts necessary to review a book with such a diverse number of topics. We gratefully thank our many reviewers for their generous time and effort. They are not responsible for any errors that might remain in the text.

The genesis of the idea of recognizing the impact of engineering on the 20th century and initiating the Great Achievements Project came from NAE senior staff officer Greg Pearson. Robin Drummond Gibbin, senior staff officer, pursued the concept at a meeting of the American Association of Engineering Societies to elicit the support of the professional engineering community and then assumed direction of the project. Executive officer Lance Davis joined the NAE staff shortly after the project began and since then has provided oversight and acted as the "in-house" technical consultant for the project. Maribeth Keitz, senior staff associate, provided staff support to coordinate committee meetings, Web sites, public events, and other essential activities. Finally, thanks go to Carrie Harless and Kim Garcia for their hours of proofreading.

Our intent has been to share with a wide public audience the impact of engineering on society, for truly our quality of life was remarkably changed as a result of the engineering innovations of the 20th century. We hope that you will find this book entertaining, informative, and provocative.

Wm. A. Wulf
President, National Academy of Engineering

PROFESSIONAL ENGINEERING SOCIETY PROJECT PARTNERS

American Association of State Highway and Transportation Officials
American Ceramic Society
American Chemical Society
American Council of Engineering Companies
American Institute of Aeronautics and Astronautics
American Institute of Chemical Engineers
American Institute for Medical and Biological Engineering
American Nuclear Society
American Public Works Association
American Society for Engineering Education
American Society of Agricultural Engineers
American Society of Civil Engineers
American Society of Heating, Refrigerating, and Air-Conditioning Engineers
American Society of Mechanical Engineers
American Society of Naval Engineers
Institute for Operations Research and the Management Sciences
Institute of Electrical and Electronics Engineers
National Association of Corrosion Engineers
National Society of Professional Engineers
Society for Mining, Metallurgy, and Exploration
Society of Automotive Engineers
Society of Fire Protection Engineers
Society of Naval Architects and Marine Engineers
Society of Plastics Engineers
Society of Women Engineers
Structural Engineers Association of California
Tau Beta Pi Association

GREAT ACHIEVEMENTS SELECTION COMMITTEE

H. Guyford Stever (*Committee Chair*)
Former science advisor to President Ford; Former president, Carnegie Mellon University; Former director, National Science Foundation

Harold M. Agnew
Retired president, General Atomics; past director, Los Alamos National Laboratory

Frances E. Allen
IBM fellow emerita

Neil A. Armstrong
Chairman of the board emeritus, EDO Corporation

Norman R. Augustine
Retired chairman and chief executive officer, Lockheed Martin Corporation

William F. Ballhaus, Sr.
President, International Numatics, Inc.

David P. Billington
Gorden Y. S. Wu Professor of Engineering, Princeton University

Frederick H. Dill
Distinguished engineer, Hitachi Global Storage Technologies

Essex E. Finney, Jr.
Retired associate administrator, Agriculture Research Service, U.S. Department of Agriculture

George M. C. Fisher
Retired chairman and chief executive officer, Eastman Kodak Company

John H. Gibbons
Former assistant to President Clinton for science and technology; former director, Office of Science and Technology Policy

Daniel S. Goldin
Senior fellow, Council on Competitiveness; former administrator of NASA

Ronald E. Goldsberry
Chairman, board of directors, OnStation Corporation

Mary L. Good
Donaghey University professor and dean, Donaghey College of Information Science and Systems Engineering, University of Arkansas, Little Rock

Irwin M. Jacobs
Chairman and chief executive officer, QUALCOMM, Inc.

Jack E. Little
Retired president and chief executive officer, Shell Oil Company

Robert W. Lucky
Retired corporate vice president, Applied Research, Telcordia Technologies

Rear Admiral John B. Mooney, Jr.
U.S. Navy, retired; former chief, Office of Naval Research

William J. Perry
Michael and Barbara Berberian Professor, Stanford University; former Secretary of Defense

Henry Petroski
Aleksandar S. Vesic Professor of Civil Engineering and professor of history, Duke University

Alan Schriesheim
Director emeritus, Argonne National Laboratory

William E. Splinter
Interim dean, College of Engineering and Technology, University of Nebraska

Ivan E. Sutherland
Vice president and Sun fellow, Sun Microsystems, Inc.

Paul E. Torgersen
John W. Hancock, Jr., Chair and president emeritus, Virginia Polytechnic Institute and State University

Charles H. Townes
Professor of physics, University of California, Berkeley

Charles M. Vest
President, Massachusetts Institute of Technology

John T. Watson
Director, Clinical and Molecular Medicine, Heart, Lung, and Blood Institute, National Institutes of Health

James C. Williams
Dean of Engineering and Honda Chair, Ohio State University

Wm. A. Wulf
President, National Academy of Engineering

BOOK DEVELOPMENT COMMITTEE

Paul E. Torgersen (*Chair*)
John W. Hancock, Jr., Chair and president emeritus, Virginia Polytechnic Institute and State University

Neil A. Armstrong
Chairman of the board emeritus, EDO Corporation

Ross B. Corotis
Denver Business Challenge Professor, Department of Civil, Environmental, and Architectural Engineering, University of Colorado, Boulder

Samuel C. Florman
Chairman, Kreisler Borg Florman Construction Company

Mary L. Good
Donaghey University professor and dean, Donaghey College of Information Science and Systems Engineering, University of Arkansas, Little Rock

Robert W. Lucky
Retired corporate vice president, Applied Research, Telcordia Technologies

Henry Petroski
Aleksandar S. Vesic Professor of Civil Engineering and professor of history, Duke University

Morris Tanenbaum
Retired vice chairman and chief financial officer, AT&T Corporation

Wm. A. Wulf
President, National Academy of Engineering

REVIEWERS

John Ahearne
Michael Benzakein
Craig Brown
Charles Brownstein
Jeff Brunetti
Thomas Budinger
Robert Casey
Robert Colburn
Michael Corradini
Robert Costigan
William Cox
Thomas Crouch
Daniel Erickson
Jonathan Esslinger
Morris Fine
Millard Firebaugh
Elsa Garmire
Stephen Gehl
John Genova
Michael Geselowitz
Paul Gilbert
Tony Giometti
William Hale
Wesley Harris
Neil Hawks
Michael Higgins
William Hunt
Russell Lange
Roger Launius
Thomas Maddock
Byron Martin
Robert Maurer
John Mayo
Gene Meiran
Judith Mills
Bern Nagengast
Walter O'Brien
Lawrence Papay
Rustum Roy
George Sandusky
Robert Skinner
Lyle Stephens
Julia Weertman
George Wise
Kurt Yeager

INDEX

CREDITS

All illustrations, unless otherwise noted, copyright Uhl Studios, Inc., Dan Stuckenschneider, David Uhl.

Top (t), bottom (b), center (c), left (l), right (r).

Page **i:** PhotoDisc, **ii:** PhotoDisc, **iv:** PhotoDisc, **viii-1**: Craig Mayhew and Robert Simon, NASA GSFC, **2:** Ernst Haas/Getty Images, **3:** (t) Getty Images, (b) Schenectady Museum; Hall of Electrical History Foundation/Corbis, **4:** (t) Bettman/Corbis, (b) Bettman/Corbis, **5:** Stephen Simpson/Getty Images, **8:** (tl) Bob Rowan; Progressive Image/Corbis, (tr) Owaki-Kulla/Corbis, (b) Russell Illig/Getty Images, **9:** American Electric Power, **10:** (t) Corbis, (b) Swim Ink/Corbis, **11:** (t) PhotoDisc, (r) Department of Energy/NERC, **12-13:** Douglas Slone/Corbis, **14:** Bettman/Corbis, **15:** (t) General Motors, (b) Ford Motor Company, **16:** Getty Images, **17:** (t) MTU Friedrichshafen Archives, (b) The Mariners' Museum/Corbis, **18:** Ford Motor Company/ Courtesy of Ford Photographic Archives (4), **19:** (l) Spencer Rowell/Getty Images, (cr) Andy Sacks/Getty Images, (br) Michael Rosenfeld/Getty Images, **20:** (t) IFA/eStock Photo/PictureQuest, (b) Getty Images, **21:** Ford Photographic Department, **22:** (t) Bettman/Corbis, (bc) Getty Images, (br) General Motors, **23:** (tr) Bettman/Corbis, (c) Owen Franken/Corbis, (b) Nogues Alain/Corbis Sygma, **24-25:** Bruce Dale/National Geographic Image Collection/Getty Images, **26:** Bettman/Corbis, **27:** (t) Digital Stock Corporation, (b) Bettman/ Corbis, **28:** Bettman/Corbis, **29:** (t) General Electric Aircraft Engines, (b) Bettman/Corbis, **31:** (radar map) Air Traffic Control System Command Center, Federal Aviation Administration, **32:** (l) Corbis, (r) Stuart Westmoreland/ Corbis, **33:** Courtesy of Northrup Grumman, **34:** (b) Naval Historical Center, (r) Courtesy of Roberson Museum, **35:** (l) John Batchelor, (t) Air Force Flight Test Center History Office, (b) NASA, **36-37:** Darryl Torckler/Getty Images, **38:** Minnesota Historical Society/Corbis, **39:** (t) Lester Lefkowitz/Corbis, (b) Getty Images, **40:** (t) Corbis, (b) Getty Images, **41:** (t) Charles E. Rotkin/ Corbis, (b) Lloyd Cluff/Corbis, **44:** (tl) Bojan Brecelj/Corbis, (tr) Dean Conger/ Corbis, (b) Water Health International, **45:** Hans Stakelbeek, **46:** (c) Tom Ehrenreich/http://www.railroadextra.com, (bl) Security Pacific Collection/Los Angeles Public Library Special Collections, (br) Water Resources Center/ University of California, Berkeley, Library, **47:** (t) Otto Lang/Corbis, (b) Ed Kashi/Corbis, **48-49:** PhotoDisc, **50:** Lucent Technologies Systems, **51:** (tc, tr) Getty Images, (b) Glenn Mitsui/Getty Images, **52:** (tl) Jerry Cooke/Corbis, (tr) Lucent Technologies Systems, (b) Bettman/Corbis, **53:** (c) Alfred Pasicka/ Science Photo Library, (bl) Intel Corporation, (bc) Corbis/Sygma, (br) Intel Corporation, **54-55:** Photos courtesy of International Business Machines, **56:** (t) Intel Corporation, (b) Digital Vision/Getty Images, **58:** (l) Getty Images, (r) Lucent Technologies Systems, **59:** (l) Corbis/Sygma, (tr) PhotoDisc, **60-61:** Catherine Karnow/Corbis, **62:** Clive Mason/Allsport/Getty Images, **63:** (t) Getty Images, (b) Marconi Radio Club W1AA, **64:** Getty Images (3), **65:** (tl, tr) Bettman/Corbis, (b) Popular Mechanics/Hearst Communications, Inc., **68:** (l) Photodisc/Getty Images, (r) Digital Vision/Getty Images, **69:** Louis Fabian Bachrach, **70:** Getty Images (2), **71:** (l) Getty Images, (t) Science Photo Library, **72-73:** Andy Sacks/Getty Images, **74:** Michael Melford/Getty Images, **75:** (t) Wisconsin Historical Society, (b) Getty Images, **76:** Getty Images (2), **77:** (l) Getty Images, (r) Donovan Reese/Getty Images, **80:** Corbis (2), **82**: Getty Images (2), **83:** (t) Corbis, (b) Photodisc/Getty Images, **84-85:** Corbis, **86:** Popular Electronics Magazine, **87:** (t) Los Alamos National Laboratory/Science Photo Library, (b) Lucent Technologies Systems/Bell Labs, **88:** Getty Images, **89:** (t) Getty Images, (c) Intel Corporation, (bl) Microsoft, Inc, (br) Bettman/ Corbis, **92:** (tl) Thinkstock, (tr) Brian Pieters/Masterfile, (bl) Jon Feingersh/ Masterfile, **93:** Microsoft Corporation, **94:** Bettman/Corbis, **95:** (l) Courtesy of Doug Engelbart/Bootstrap Institute, (r) Getty Images, **96-97:** PhotoDisc, **98:** Bettman/Corbis, **99:** (t) Will & Deni McIntyre/Getty Images, (b) Minnesota Historical Society/Corbis, **100:** (t) Lucent Technologies Systems, (b) Sheila Terry/Science Photo Library, **101:** (l) Lucent Technologies Systems, (r) AFP/ Corbis, **102:** (tl) Getty Images, **103:** (br) Getty Images, **104:** (l) Getty Images, (tr, br) PictureNet/Corbis, **106:** (t) Corbis, (r) Getty Images, **107:** Lucent Technologies Systems (2), **108-109:** Photodisc, **110:** Getty Images, **111**: (t) Rob Melnychuk/Getty Images, (b) Photodisc, **112:** Getty Images, **113:** (l) Bettman/Corbis, (r) H. Armstrong Roberts/Corbis, **116:** (tl) Minnesota Historical Society, (tr) PhotoDisc, (b) Gina Minielli/Corbis, **118:** (t) Corbis, (b) Carrier Corporation, **119:** Carrier Corporation, **120-121:** Ron Chapple/ Getty Images, **122:** Photodisc/Getty Images, **123:** (t) Corbis, (b) Getty Images, **124:** Getty Images (3), **125:** (t) Owaki-Kulla/Corbis, (b) Transportation Research Board Photo Library, **127:** (tr) Wim Kooijman/DuoFoto, **128:** (tl) Larry Fisher/Masterfile, (tr) Ed Quinn/Corbis, (b) PhotoDisc, **130:** PictureArts/ Corbis, **131:** Chesapeake Bay Bridge-Tunnel, **132-133:** NASA, **134:** Getty Images, **135:** (t) NASA, (b) Getty Images, **136:** (tl, tr, br) NASA, (bl) Getty Images, **137:** NASA (2), **140:** NASA (2), **143:** NASA, Andrew Fruchter and the ERO Team [Sylvia Baggett (STScI), Richard Hook (ST-ECF), Zoltan Levay (STScI)], **144-145:** Boden/Ledingham/Masterfile, **146:** Mario Tama/Getty Images, **147:** (t) Cable Risdon of Cable Risdon Photography, **147:** (bl) Courtesy of Paul Baran, (bc) Hank Morgan/Science Photo Library, (br) Courtesy of Physical Library, UK Crown Copyright 1974, reproduced by permission of the Controller of HMSO, **148:** (t) California State University, Dominguez Hills, (b) Ryan McVay/PhotoDisc, **149:** Chung Sung-Jun/Getty Images, **150:** (bl) Corbis, **152:** (tl) PhotoDisc, (tr) Chris Hondros/Newsmakers/Getty Images, (b) Robert Burroughs/Liaison/Getty Images, **154-155:** PictureArts/Corbis, **156-157:** PhotoDisc, **158:** Bettman/Corbis, **159:** (t) PhotoDisc, (bl) Bettman/ Corbis, (br) Vittorio Luzzati/Science Photo Library, **160**: (l) Courtesy of Lawrence Berkeley Library, (c, r) PhotoDisc, **161:** Courtesy of the Carnegie Observatories, Carnegie Insitution of Washington, **162:** Photos courtesy of NASA, **163:** Photo courtesy of STSci and NASA, **164:** (tl) Roger Ressmeyer/ Corbis, (tr) NASA, (b) Digital Vision/Getty Images, **166:** George Eastman House, **167:** (tl) B & M Productions/Getty Images, (br) NASA, **168-169:** Jonnie Miles/Getty Images, **170:** Getty Images, **171:** (t) Richard Hutchings/ Corbis, (b) Getty Images, **172:** (tl) Getty Images, (tr, b) Schenectady Museum; Hall of Electrical History Foundation/Corbis, **173:** (t) The Hoover Company, N. Canton, OH, (bl, br) Getty Images, **176:** (tl) Courtesy of Rod Spencer, (tr) PhotoDisc, (b) Douglas McFadd/Getty Images, **178:** (t) Petrified Collection/Getty Images, (b) Courtesy of the Schenectady Museum, **179**: (t) Getty Images, (b) Courtesy of Sunbeam Products, Inc., **180-181:** Pete Saloutos/ Corbis, **182:** PhotoDisc, **183:** (t) Deep Light Productions/Science Photo Library, (bl) Courtesy of Jefferson University Photo Collection, (bc,br) The Bakken Library and Museum, Minneapolis, MN, **184:** (tl) Courtesy of the Cleveland Clinic, (tc) Bettman/Corbis, (tr) Digital Vision/Getty Images, **185:** (l) Robert J. Herko/Getty Images, (r) M. Freeman/PhotoLink/Getty Images, **186-187:** Photos courtesy of Medtronic, Inc. **188:** (tl) PhotoDisc, (tr) Spencer Platt/Getty Images, **190:** (l) Getty Images, (r) Texas Heart Institute, **191:** Simon Fraser/Science Photo Library, **192-193:** Greg A. Syverson/Getty Images, **194:** Getty Images, **195:** (tl, tr) PhotoDisc, (bl, br) Whiting Refinery, **196**: (t) Bettman/Corbis, (bl, br) Getty Images, **197:** (t) Corbis, (b) PhotoDisc, **200:** (t) Sunoco Corporation, (bl) PhotoDisc, (br) Digital Stock Corporation, **202:** (c) Getty Images, (br) Corbis, **203**: (l) Getty Images, (r) Oceaneering Inc., **204-205:** Roger Ressmeyer/Corbis, **206:** HRL Laboratories, Malibu, CA, **207:** (t) David Parker/Science Photo Library, (bl, br) Lucent Technologies Systems, **208**: (t) Michael Rosenfeld/Getty Images, (b) David Parker, Science Photo Library, **209:** Roger Ressmeyer/Corbis, **212**: (l) Corbis, (r) Rich LaSalle/ Getty Images, **214:** (t) PhotoDisc, (c) Bettman/Corbis, (b) Lucent Technologies Systems, **215:** Lucent Technologies Systems (2), **216-217:** Roger Ressmeyer/ Corbis, **218:** Getty Images, **219:** (l) Corbis, (c) Getty Images, (r) A. Barrington Brown/Science Photo Library, (b) Getty Images, **220:** (t) Getty Images, (bc, br) Roger Ressmeyer/Corbis, **221:** Sister Betty.org/Wolf Sterling, **224:** Pallet irradiator illustration courtesy of MDS Nordion, **226:** Getty Images (2), **227:** Getty Images, **228-229:** Alfred Pasieka/Science Photo Library, **230:** Getty Images, **231:** (t) Klaus Guldbransen/Science Photo Library, (bl) Bettman/ Corbis, (br) Underwood & Underwood/Corbis, **232:** (t) Getty Images, (bl) Dupont Industries, (br) Science Photo Library, **233:** (t) David Parker/Science Photo Library, (b) Getty Images, **234:** Photo courtesy of Orthopaedic Associates of Portland, **235:** Photo courtesy of General Electric Aircraft Engines, **236:** (tl, tr) PhotoDisc, (b) Hewlett-Packard Laboratories/Science Photo Library, **237:** Staff Photographer-RPI, **238-239:** Astrid & Hans Frieder Michler/Science Photo Library, **240:** Wood Ronsaville Harlin, Inc.

JOSEPH HENRY PRESS
500 Fifth Street, NW
Washington, DC 20001

The Joseph Henry Press, an imprint of the National Academies Press, was created with the goal of making books on science, technology, and health more widely available to professionals and the public. Joseph Henry was one of the founders of the National Academy of Sciences and a leader in early American science.

Any opinions, findings, conclusions, or recommendations expressed in this volume are those of the authors and do not necessarily reflect the views of the National Academies.

Printed in the United States of America.

Library of Congress Cataloging-in-Publication Data

Constable, George.
A century of innovation : twenty engineering achievements that transformed our lives / George Constable and Bob Somerville.
p. cm.
ISBN 0-309-08908-5 (hardcover)
1. Technological innovations. 2. Engineering—Technological innovations. I. Somerville, Bob. II. Title.
T173.8.C68 2003
609'.04--dc22

2003019019

Managing editor	Stephen Mautner
Project editor	Roberta Conlan
Designer	Francesca Moghari
Photo research	Tom DiGiovanni Jane Martin Angie Lemmer (Germany)
Illustration	Uhl Studios, Inc. Dan Stuckenschneider David Uhl
Illustration research	Stephanie Henke
Text research	Ruth Goldberg Mary Mayberry
Timeline research and writing	Janice Campion Jane Coughran Darlene Koenig Rosanne Scott Brooke Stoddard
Production manager	Dorothy Lewis
Production editor	Heather Schofield
Printer	Chroma Graphics, Inc. Largo, Maryland

Great Achievements project staff

NAE executive officer	Lance Davis
Project director	Robin Drummond Gibbin
Project associate	Maribeth Keitz